Crop Post-Harvest:

Science and Technology

Volume 2

Durables

Case studies in the handling and storage of durable commodities

Crop Post-Harvest:
Science and Technology

Volume 2

Durables

Case studies in the handling and storage of durable commodities

Edited by

Rick Hodges and Graham Farrell

Blackwell
Science

Natural
Resources
Institute

© 2004 Blackwell Science Ltd
a Blackwell Publishing company

Editorial offices:
Blackwell Science Ltd, 9600 Garsington Road, Oxford OX4
2DQ, UK
  Tel: +44 (0)1865 776868
Blackwell Publishing Professional, 2121 State Avenue, Ames, Iowa
50014-8300, USA
  Tel: +1 515 292 0140
Blackwell Publishing Asia Pty Ltd, 550 Swanston Street, Carlton,
Victoria 3053, Australia
  Tel: +61 (0)3 8359 1011

First published 2004

Library of Congress Cataloging-in-Publication Data is available

ISBN 0-632-05724-6

A catalogue record for this title is available from the British
Library

Set in 9.5/11.5 pt Times New Roman
by Sparks Computer Solutions Ltd, Oxford
http://www.sparks.co.uk
Printed and bound in India by
Gopsons Papers Ltd, Noida

The publisher's policy is to use permanent paper from mills
that operate a sustainable forestry policy, and which has been
manufactured from pulp processed using acid-free and elementary
chlorine-free practices. Furthermore, the publisher ensures that
the text paper and cover board used have met acceptable environmental accreditation standards.

For further information on Blackwell Publishing, visit our
website:
www.blackwellpublishing.com

This series of volumes is dedicated to the memory of
Dr Philip C. Spensley, Director of the Tropical Products Institute 1966–1982
for his leadership in a great period of expansion of post-harvest research and development

Part of a three-volume set from Blackwell Publishing and the Natural Resources Institute, University of Greenwich:

Crop Post-Harvest: Science and Technology Volume 1: Principles and Practice
Edited by P. Golob, G. Farrell and J.E. Orchard
0 632 05723 8

Crop Post-Harvest: Science and Technology Volume 2: Durables
Edited by R.J. Hodges and G. Farrell
0 632 05724 6

Crop Post-Harvest: Science and Technology Volume 3: Perishables
Edited by D. Rees, J.E. Orchard and G. Farrell
0.632 05725 4

Contents

Contributors

Armitage, D.M., Central Science Laboratory, Ministry of Agriculture, Fisheries and Food, Sand Hutton, York YO41 1LZ, UK.

Arthur, F.H., USDA Stored-Product Insect Research and Development Laboratory, Savannah, GA 31405, USA.

Barker, N.D., GrainCorp Operations Ltd, P.O. Box A268, Sydney South, NSW 1235, Australia.

Bartosik, R.E., Agricultural and Biological Engineering Department, Purdue University, IN, USA, and INTA Balcarce, Provincia de Buenos Aires, Argentina.

Bawalan, D.D., Philippine Coconut Authority, Diliman, Quezon City, Philippines.

Baxter, E.D., Brewing Research International, Nutfield, Surrey RH1 4HY, UK.

Bowman, R., Ricegrowers Co-operative Limited, P.O. Box 561, Leeton NSW 2705, Australia.

Boxall, R.A., Natural Resources Institute, University of Greenwich, Central Avenue, Chatham Maritime, Kent ME4 4TB, UK.

Bucheli, P., Nestlé R&D Centre Shanghai Ltd., No. 13 Qiao Nan, Cao An Road, Jia Ding District, Shanghai 201812, P.R. China.

Butts, C.L., USDA-ARS, National Peanut Research Laboratory, P.O. Box 509, 1011 Forrester Drive, SE Dawson, GA 39842–0509, USA.

Cardona, C., Centro Internacional de Agricultura Tropical, Apartado Aérea 6713, Cali, Colombia.

Darby, J., Stored Grain Research Laboratory, CSIRO Entomology, G.P.O. Box 1700, Canberra, ACT 2601, Australia.

Devereau, A., 6 Lindi Avenue, Grappenhall, Warrington WA4 2SL, UK.

Farrell, G., Plant Clinic Limited, Invicta Innovations, East Malling Research, Kent ME19 6BJ, UK.

Gebre-Tsadik, M., Ministry of Agriculture, P.O. Box 105, Harar, Ethiopia.

Golob, P., 128 Sea Road, Chapel St Leonard, Lincolnshire PE24 5RY, UK.

Guèye-NDiaye, A., Département de Biologie Animale, Faculté des Sciences, Université Cheikh Anta DIOP, BP 5005, Dakar, Senegal.

Hodges, R.J., Food Management and Marketing Group, Natural Resources Institute, University of Greenwich, Chatham Maritime, Kent ME4 4TB, UK.

Jayaraj, K., Technical Officer (Storage & Research), Indian Grain Management & Storage Institute, Hyderabad, Andhra Pradesh, India.

Jayas, D.S., Department of Agricultural Engineering, University of Manitoba, 15 Gillson Street, Winnipeg, Manitoba, R3T 5V6, Canada.

Johnson, J., San Joaquin Valley Agricultural Center, 9611 S. Riverbend, Parlier, CA 93648, USA.

Jonfia-Essien, W.A., Ghana Cocoa Board, Cocoa House, Post Office Box 933, Kwame Nkrumah Avenue, Accra, Ghana.

Knight, J., Imperial College of Science, Technology and Medicine, Silwood Park, Ascot, Berkshire SL3 7PY, UK.

Kutukwa, N., c/o Environment and Tourism, 2706 Hungwe Close, Ruwa, Zimbabwe.

Mbata, G.N., Department of Biology, Fort Valley State University, Fort Valley, GA 31030, USA.

Noyes, R.T., BioSystems & Agricultural Engineering, 224 Agriculture Hall, Oklahoma State University, Stillwater, OK 74078–6021, USA.

Patternaik, B.B., Department of Food & Public Distribution, Ministry of Consumer Affairs, Food & Public Distribution, Government of India, Krishi Bhavan, New Delhi, India.

Ramam, C.P., Agricultural Technologist & Officer-In-Charge, Indian Grain Storage Management & Research Institute, Hyderabad, Andhra Pradesh, India.

Rodríguez, J.C., INTA Balcarce, Provincia de Buenos Aires, Argentina.

Sembène, M., Département de Biologie Animale, Faculté des Sciences, Université Cheikh Anta DIOP, BP 5005, Dakar, Senegal.

Seo, Y., Dept. of Bioenvironmental and Agricultural Engineering, College of Bioresource Sciences, Nihon University, Kameino 1866, Fujisawa 252–8510, Japan.

Shamsher, H.K., Grain Associates, Unit # 06, USAID Plaza, Fazal-e-Haq Road, 29-Blue Area, Islamabad, Pakistan.

van S. Graver, J., Stored Grain Research Laboratory, CSIRO Entomology, G.P.O. Box 1700, Canberra ACT 2601, Australia.

Weifen, Q., Nanjing University of Economics, Nanjing 210003, Jaingsu Province, P.R. China.

White, N.D.G., Research Station, Agriculture and Agric. Food Canada, 195 Dafoe Road, Winnipeg, Manitoba R3T 5V6, Canada.

Wilkin, D.R., Imperial College of Science, Technology and Medicine, Silwood Park, Ascot, Berkshire SL3 7PY, UK.

Woods, J.L., Department of Agricultural and Environmental Science, University of Newcastle upon Tyne, Newcastle upon Tyne, Tyne & Wear NE1 7RU, UK.

Zuxun, J., Nanjing Economic College, 128 Teilubei Street, Nanjing 210003, Jaingsu Province, P.R. China.

About the Editors

Dr Rick Hodges has worked with durable commodities for over 25 years, as manager of the British Government's grain storage project in Mali after the severe Sahel drought of the early 1980s and then as a pest and quality control adviser to the Indonesian National Logistics Agency (BULOG). His research has focused on improvements in commodity management, especially on more cost-effective and environmentally acceptable approaches to insect control, and on helping African smallholder farmers tackle the larger grain borer. He is currently Reader in Post-harvest Entomology at the Natural Resources Institute, University of Greenwich, and Secretary of the Global Post-harvest Forum (PhAction).

Dr Graham Farrell trained in plant pathology and pest management. He worked in Africa for 10 years in natural resources project management and technical support, particularly in the design, implementation, monitoring and institutionalisation of demand-driven participatory programmes for the British Government, and in the logistics of food aid for the European Union. Dr Farrell's professional interests revolve around crop protection extension and translating outputs of research into materials used by dissemination agencies to strengthen the abilities of farmers and growers in articulating their technical and information needs. He is currently director of Plant Clinic Limited, a UK company providing advice on plant health, based on sustainable principles.

Preface

Durable commodities are the raw products from which food can be made and are the staples on which most humans rely; with but a few exceptions they are the seeds of plants. The storage and handling of durable commodities has been practised by man for millennia and has enabled mankind's populations to become sufficiently dense to support urban development and our current civilisation. In the modern world, appropriate storage and handling of durable commodities facilitate effective marketing chains and there underpin economic wellbeing and food security.

The development of grain storage has been a battle against biodeterioration; this is caused by adverse temperature and moisture conditions and mediated by micro-organisms (especially fungi), arthropods (particularly insects and mites) and to a lesser extent rodents and birds. While the problems are biological in nature, from the earliest times solutions have mostly been found in the realms of engineering. In the twentieth century, the use of pesticides of one sort or another blossomed. However, consumer demand in the late twentieth and the early twenty-first century has led to a strong move away from pesticides and a resurgence of engineering solutions to deliver products free of pests and pesticide residues, with the characteristics required by the end-user, and increasingly products are becoming traceable back to source.

Volume 1 in this series, on the principles and practice of post-harvest technology, provides details of how crops should be dried, handled, protected from pests and stored by smaller-holders or large-scale enterprises. In contrast, this second volume presents a series of case studies on how durable crops are actually stored and marketed. Authors were asked to contribute manuscripts covering a specific range of issues so that comparisons could be made between post-harvest systems used for the same

commodities in different countries. Thirty-eight authors contributed to this volume; all have practical knowledge of the systems they are describing but their backgrounds are quite varied, as would be expected of a subject that requires a multidisciplinary approach. Authors also differ in the amount of information available to them and the extent to which they are involved in managerial or developmental aspects. The result of this is a difference in emphasis between contributions, adding to the richness and interest of the text.

Many of the chapters present contrasting approaches to the storage of the same commodity. For example, the volume begins with rice storage in three different countries. In China, several storage methods are used across a range of climatic conditions with the objective of ensuring sufficient food supply to its large population. In Japan, rice storage is highly specialised and designed to provide a product quality specifically demanded by the Japanese consumer, while Australia provides high-quality products prepared according to the requirements of various domestic and export markets. Different approaches are taken to the same commodity according to a blend of economic, social and cultural requirements. The intention of this juxtaposition is to demonstrate differences in the degree of development and sophistication of these systems and the diversity of technology that can enable successful handling and storage of durable commodities. This approach is adopted for several commodities, but for the remainder it has only been possible to present single accounts. With the exception of cured fish, all are plant products; Chapter 14 on cured fish also differs in providing a detailed consideration of pest management issues alongside a profile of the cured fish industry of Senegal. Two sections, relating to peanuts and cured fish in Senegal, were translated from French by one of us (RH).

Many of the techniques and issues mentioned here are presented in detail in Volume 1 and, where appropriate, cross-references are made. For information on the common, scientific and family names of pests, pathogens and plants of importance to post-harvest technology, the reader should refer to Appendices 1, 2 and 3 of Volume 1.

The moisture content of commodities is a very important issue for correct handling and storage. In this volume where moisture contents are quoted these should be assumed to be on a wet weight basis (Vol. 1 p. 74); if they are on a dry weight basis this will be stated.

Every effort has been made to contact the copyright holders of photographs and diagrams. However, the editors would be pleased to make amendments in future editions.

Rick Hodges and Graham Farrell
Chatham, UK
November 2003

Acknowledgements

Many people have contributed to this Volume and the authors particularly wish to express their thanks as follows. Chapter 2 – Dr Dirk Maier and Ing. Roberto Hajnal; Chapter 4 – the late Mike Proudlove; Chapter 7 – Roy Kirsten, Fred Erlenbusch, Larry Teixeira, Sam England, Key Yasui, Gary Lucier and Edmond Bonjour; Chapter 9 – Sidy B. NDiaye, T. Diop, K. Coly, A. Dieme and A. Ba; Chapter 13 – Joseph Buatsie, G.A. Quayson, Beatrice Padi and Jack Botchway.

The editors would also like to thank the following for their support in reading completed chapters: Tony Swetman, Peter Golob, Martin Nagler and Andrew Devereau.

Chapter 1
Rice

J. Zuxun, Q. Weifen, Y. Seo, J. Darby and R. Bowman

Rice is believed to have originated in China and India. This crop has fed more people over a longer period than any other, making its domestication one of the most important developments in human history. There are two species of cultivated rice, *Oryzae sativa* and *Oryzae glaberrima*. Of the two, *O. sativa* is by far the more widely grown. *Oryzae sativa* is a complex of seven forms with a strong divergence between South Asian *indica* rices, which are usually long-grain types grown in the tropics or subtropics, and *japonica* (or *sinica*) rices of Chinese origin that are usually round-grain types grown in the subtropics or temperate zones. Both types may be either normal savoury rice or glutinous rice: the latter becomes sweet and sticky on cooking and is used in making desserts and confectionery.

The rice plant develops grain clusters called panicles; the plants are cut and threshed to release the grains. The intact grain has an abrasive, silaceous seed coat or husk and is called paddy or rough rice (Plate 1). Dehusking reveals the brown bran and germ that underlie the seed coat, and in this state the grain is called brown rice (Plate 1). Most people prefer to eat white rice (Plate 1) and this is prepared by milling or 'polishing' to remove the bran and germ, leaving just the endosperm of the grain (Vol. 1 p. 384, Wood 2002). Further specialised polishing produces a shinier, whiter-looking rice. Unlike most other cereals, rice is mostly eaten as whole grains and for this reason physical properties such as size, shape, uniformity and general appearance are very important. The quality of the product is affected by its physiochemical properties as well as factors such as the proportion of broken grains, which must be minimised for best eating quality and price premium.

This chapter deals with rice storage in China, Japan and Australia. The situations in these countries contrast strongly. Rice growing is an ancient tradition in both China and Japan but only began in Australia in the 1920s in New South Wales (NSW). In China, a very large and diverse storage system is engaged to meet the food security needs of a population of about 1.3 billion. China is the largest of about 100 paddy-producing countries worldwide with an annual production of about 166 million tonnes or approximately 37% of world output (FAO 2000). Both *indica* and *japonica* rice are consumed and, as paddy is a seasonal crop, storage is extremely important. The rice is generally stored in bins, silos or horizontal stores of Soviet design, with an increasing emphasis on low-temperature and controlled atmosphere storage.

Japan produces about 9 million tonnes of paddy annually and has a system that is highly specialised and developed to provide a product quality specifically demanded by the Japanese consumer. *Japonica* varieties are favoured, and freshness is very important as the Japanese eat boiled rice by itself. In order to preserve freshness, rice is stored using advanced methods to control conditions of temperature and humidity, in particular. Freshness is also achieved by storing and distributing mostly brown rice that is then milled and polished just before consumption.

Australia has an annual production of paddy of only about 1.6 million tonnes, of which 80% is *japonica* and the rest *indica*. In the early days the rice industry used a bag storage system where rice was harvested at low moistures (16.5%) prior to further drying, storage and milling. But today, only bulk systems are found, employing large-scale machinery, advanced aerated storage facilities, automatic control and purpose built *in situ* moisture sensing systems. The paddy crop is aeration dried in 6–7.5 m deep bins using either ambient or burner-heated inlet air and then stored for periods of up to one year before shipment to the millers and then the market. This capital-intensive industry provides high quality products prepared according to the requirements of various export and domestic markets and so delivers various quality requirements. These include minimised kernel breakage, a range of polished rice appearance and presentation characteristics, cooking textures and performances, and eating qualities.

China

Historical perspectives

Chinese archaeological excavation at the Yu Zanyan ruins (Hunan Province) has revealed 10 000-year-old carbonised paddy rice (Jin Zuxun 1984). It is believed that Chinese people began to store paddy from ancient times: at He Mu Du ruins, Zhejiang Province, 7000-year-old carbonised paddy was stored on frames made of branches (Li Lungshu & Jin Zuxun 1999).

Grain storage science was established in China at the beginning of the last century. In 1965, the State Science and Technology Committee formally approved a national grain storage research institute to meet the growing needs of grain production and storage (Tang Weimin & Hu Yushan 1997). Since then, paddy storage research has developed very quickly. China stores both *indica* and *japonica* rice that may be 'normal' or glutinous and may also be classified according to harvest time – early, medium or late season.

Physical facilities

The three main types of paddy warehouse in China are horizontal warehouses of Soviet origin (Fig. 1.1), silos (Fig. 1.2) and bins. In some places there are also under-

Fig. 1.1 Soviet-style warehouse for the storage of paddy and milled rice in China.

Fig. 1.2 Reinforced concrete grain silo used for the storage of paddy in China.

ground storage facilities of various designs (Lin Zaiyun & Wu Lina 1999). The general horizontal warehouse in China can house paddy in bag stacks 4 m high, 21 m wide with length chosen according to circumstances. In large horizontal warehouses bag stacks may be 6 m high, 24–27 m wide and more than 50 m long. Paddy storage bins are roughly 15–18 m high and 30 m in diameter and hold about 500 tonnes.

China has a large territory and paddy grows in wet to semi-arid tropical regions through to warm temperate areas, and in earth ranging from heavy clay to poor sandy soils. In these widely differing ecological conditions, appropriate storage facilities need to be chosen to ensure the safe keeping of paddy. Thus Chinese territory can be divided into the following seven types for the purpose of grain storage (Tang Zijun *et al.* 1999):

(1) The cold arid region of the Qingzang plateau
(2) The arid region of Monxin
(3) The cold humid region of northeast China
(4) The warm temperate, sub-humid and semi-arid area in the north
(5) The subtropical, humid monsoon region of central China
(6) The subtropical area of the southwest, and
(7) The hot humid region in the south

Horizontal warehousing is used for paddy storage in the south in areas 5, 6 and 7 while horizontal warehouses or round bins are chosen for paddy in areas 1, 2, 3 and 4. Silo complexes are used at ports and railways. The facilities used for paddy storage include cleaners and balances, dryers, transporters, aerators, cooling equipment, fumigation equipment, electronic grain inspection and control equipment, sampling equipment, computerised control systems and expert systems that automate grain management.

The status of milled rice storage varies in different parts of China. With the establishment and development of the grain market, milled rice began to be processed, handled and sold at the same time. Large and medium-sized cities have to store a certain amount of rice to guarantee supplies. Milled rice is usually stored in horizontal warehouses where effective storage techniques are easily applied. The equipment for rice storage is the same as that for paddy except that rice storage needs facilities for packaging rice for the consumer market (Jin Zuxun & Yu Yingwei 2000).

Grain cooling is an important strategy to achieve good quality preservation. Warehouses, bins, silos or underground stores may be defined as low temperature, when they are maintained below 15°C, or quasi-low-temperature which are at 15–20°C (Lu Qianyu 1998). Low-temperature storage requires that stores are reasonably airtight and insulated. The degree of insulation influences the effectiveness and cost of storage. Normal insulation materials are dilated pearlite, polystyrene foam, chaff, and so on, to give an overall coefficient of heat conduction below 0.5 kcal m^{-2} h^{-1} °C^{-1}. The cooling is usually achieved by using air conditioners or special refrigeration units (cereal coolers). The units are installed close to the ceiling of the store so that the cold air does not play directly onto the stock, which otherwise might result in moisture condensation. Where possible grain is pre-cooled by exposure to cool weather conditions before placing it in store and/or opening the ventilators and doors during cold weather.

Objectives of storage

Rice milled from paddy is primarily for domestic consumption and supplies often exceed demand. As a consequence, paddy and rice storage is a long-term strategy within the country. With the reform of domestic grain circulation and China's entry into the World Trade Organization, international and local trade for paddy and milled rice are expected to increase.

Major sources of quality decline

The outer husk of paddy makes it particularly resistant to both insect damage and changes in humidity and temperature. Under normal storage conditions, newly harvested paddy respires strongly, although after one year the rate of respiration decreases and keeping quality improves. Paddy respiration gives out heat and moisture that can result in condensation. This leads to mould growth and grain sprouting in the upper portions of stacks (20–30 cm below the surface). Even if paddy is received at an acceptable moisture content, the temperature of the upper part of the stack often rises abruptly, usually about 10–15°C above ambient. Paddy may crack if dried in the sun or at high temperature, or if subjected to rapid absorption or desorption of moisture, leading to grain breakage. If maintained in store at high temperatures then paddy may become rancid as a result of the accumulation of free fatty acids (Table 1.1).

Paddy that is not dried or cooled in a timely manner after harvest will yield yellowed grains at polishing. Extensive research has shown that the higher the temperature and moisture, the more severe the yellowing (Table 1.2). Besides a change in colour, the quality of yellowed rice differs significantly from unyellowed rice in a number of other respects (Table 1.3), resulting in an overall lower quality rating for yellowed rice (Table 1.4). However, no mycotoxins such as aflatoxin (Vol. 1 p. 128, Wareing 2002) have yet been detected in yellow rice.

China's storage loss of paddy is about 0.2% in national reserves while rural household storage losses are around 7–13%. In general, the main causes of grain storage loss in China are insects in the south and moisture in the north. Paddy is the principal crop in the southern part of China where the temperature and moisture are very suit-

Table 1.1 Effect of temperature on the fatty acid value of paddy after 3 months of storage.

Moisture content (%)	Original FFA*	Final FFA*		
		15°C	25°C	35°C
13.2	13.8	21.1	21.7	25.7
15.2	14.6	22.1	23.3	23.3
17.2	16.9	24.4	23.5	44.5
19.6	18.9	24.6	46.8	43.3

*Free fatty acid (mg KOH/100 g dry basis).

Table 1.2 Yellowing of rice grain endosperm related to paddy storage period, temperature and moisture of *indica* rice.

Temperature (°C)	Moisture (%)	Period		
		1 month	2 month	3 months
30	13.1	No yellowing	No yellowing	No yellowing
30	13.6	No yellowing	No yellowing	No yellowing
30	14.7	No yellowing	No yellowing	No yellowing
30	15.7	No yellowing	No yellowing	No yellowing
35	13.1	No yellowing	No yellowing	No yellowing
35	13.6	No yellowing	No yellowing	No yellowing
35	14.7	No yellowing	No yellowing	Yellowing starts
35	15.7	No yellowing	No yellowing	Yellowing starts
40	13.1	No yellowing	No yellowing	Yellowing starts
40	13.6	No yellowing	No yellowing	Yellowing starts
40	14.7	No yellowing	No yellowing	Yellow discoloration
40	15.7	No yellowing	No yellowing	Yellow discoloration
45	13.1	No yellowing	Yellow discoloration	Yellowing darkens
45	13.6	No yellowing	Yellow discoloration	Yellowing darkens
45	14.7	Yellowing starts	Yellow discoloration	Yellowing darkens
45	15.7	Yellowing starts	Yellow discoloration	Yellowing darkens
Control (15°C)	13.1–15.7	None	None	None

Data from Zaoqing Grain Bureau, Guangdong Provincial Grain Bureau.

Table 1.3 Quality changes when *indica* rice has yellowed during paddy storage.

	Sprout rate (%)	Moisture (%)	Brown rice rate (%)	Acid value (%)	Fatty acid value*	Reducing sugar (%)	Viscosity (η)
Control	86	10.00	78.75	1.63	11.08	0.233	1.30
Yellowed (22%)	39	10.40	77.30	1.66	111.30	0.414	1.20
Yellowed (45%)	35	10.45	77.00	120.50	120.50	0.520	1.11

*mg KOH/100 g dry basis.

	Moisture content (%)			
Temperature (°C)	13.1	13.6	14.7	15.7
Original	84.6	84.6	85.0	84.2
30	75.8	75.6	72.9	68.1
35	66.1	65.0	63.4	63.6
40	64.4	63.5	61.4	59.7
45	60.9	60.9	59.5	59.5

Table 1.4 Evaluation scores* of *indica* rice that has become yellowed during paddy storage.

*Rice with evaluation score of 70 or below is considered unsuitable for storage. See text for the quality control equation used to calculate the evaluation score.

able for storage insects. The principal pests of paddy and milled rice are the beetles *Sitophilus oryzae, Sitophilus zeamais, Rhyzopertha dominica* and *Oryzaephilus surinamensis* and moths *Sitotroga cerealella* and *Plodia interpunctella* (Vol. 1 p. 96, Hodges 2002). Moulds are the most significant micro-organisms attacking paddy (Table 1.5) and their growth and propagation is closely related to moisture content and temperature. Paddy harvested around the lake area of south China will usually have high moisture content due to frequent rainfall.

China's paddy storage loss is low in terms of quantity since grain storage experts have accumulated extensive experience on the storage of this form of rice. Nevertheless, quality loss of paddy and milled rice is substantial. Long storage periods due to bumper harvests in consecutive years lead to quality deterioration. Paddy shows a sharp fall in quality by the end of the second year in storage and further deterioration occurs during the third year for both *indica* and *japonica* rice (Table 1.6). There is also some decline in nutritional values over exceptionally long periods, particularly in protein and vitamin B3 (niacin) (Table 1.7).

Milled rice keeps less well than paddy for two reasons. Its nutritious components are exposed to biodeteriora-

Table 1.5 Fungi associated with paddy in China.

Species	Detection rate (%)	Species	Detection rate (%)
Aspergillus flavus	36.6	*Alternaria*	1.61
A. niger	4.78	*Penicillium*	19.86
A. ochraceus	3.19	*Fusarium*	13.75
A. versicolor	19.40	*Rhizopus*	2.91
A. candidus	16.45	*Mucor*	3.90
A. oryzae	14.30	Other moulds	2.63
A. glaucus	5.34		

Table 1.6 Influence of storage period on the quality of *indica* and *japonica* paddy. Source: *Seed Storage and Inspection*, Zhejing Agriculture University (unpublished).

	Indica rice				Japonica rice			
	Year 0	Year 1	Year 2	Year 3	Year 0	Year 1	Year 2	Year 3
Viability (%)	97.5	93.0	47.0	0	97.5	89.0	4.0	0
Lipid (%)	2.9	2.5	2.0	1.9	3.8	3.0	2.6	2.4
FFA*	15.0	16.4	18.2	45.5	182.0	182.0	195.0	255.0
Salt-soluble N (%)	0.34	0.21	0.22	0.17	0.22	0.22	0.19	0.14

*Free fatty acids (mg KOH/100 g dry basis).

Table 1.7 Effect of storage time of paddy on nutrient value.

						Vitamin		
Years	Moisture (%)	Carbohydrate (%)	Lipid (%)	Protein (%)	FFA*	B1	B2	B3
1	13.3	73.9	2.36	8.6	26.8	0.27	0.08	3.28
2	12.5	72.1	2.48	9.03	52.1	0.26	0.08	—
3	13.3	73.9	2.45	9.20	42.1	0.26	—	—
4	12.8	71.2	2.43	8.46	44.4	0.25	0.08	—
5	11.7	72.6	2.40	9.18	60.0	0.25	—	3.16
6	13.2	68.9	2.55	9.39	54.0	0.26	0.08	3.16
7	13.2	67.8	2.43	7.44	43.0	0.24	—	2.08
8	13.6	68.2	2.41	7.10	63.9	0.25	0.06	2.08
9	13.5	—	—	8.75	54.0	—	—	—
10	13.5	—	—	5.1	54.0	—	—	—

Data from Sichuan Provincial Grain Bureau. *Free fatty acid (mg KOH/100 g).

tion and it absorbs moisture quickly, since it is rich in hydrophilic colloids such as starch and protein. Moist rice is easily attacked by fungi and insects. Milled rice is also susceptible to ageing, leading to discoloration, increase in acidity value, natural flavour loss, viscosity reduction and deterioration in cooking qualities. If the rice has not been well polished then some bran layer may remain. This is rich in lipids, and oxidation resulting from unfavourable storage conditions will lead to rancidity through the formation of free fatty acids. The flavour of cooked rice is connected mostly with carbonyl compounds. After long periods of storage, grain will have an altered carbonyl profile and hence altered flavours on cooking (Table 1.8).

Commodity and pest management regimes

Management of paddy

Quality control of paddy entering a store is extremely important. Before entry into store, paddy must conform to safe moisture content standards. These standards vary according to the type of paddy, harvest season and temperature of storage (Table 1.9). In China, early and

middle season harvested paddy are easily within the safe storage moisture limit for they are harvested when the air temperature is high. However, late harvested paddy always has a relatively high moisture content as it is harvested when air temperatures are relatively low. Therefore, when the late harvested paddy is stored it is necessary to reduce the moisture in winter and spring. Generally speaking, moisture limits for *japonica* are higher than for *indica* rice. Late harvested paddy limits are higher than those for early harvested paddy, limits for winter harvested paddy are higher than for summer harvested and limits for paddy grown in the north are higher than for that grown in the south.

If paddy is above the safe limit then there are four ways of reducing the moisture content: drying at ambient air temperatures, mechanical ventilation, hot air drying and in-store drying (Vol. 1 p. 157, Boxall 2002). Grain depots are dependent on sun drying for about 50% of their drying capacity (Ren Yonglin & van S. Graver 1996); the rest is provided by other means. When using hot air drying, for example by rotary grain dryers, if the paddy is at below 18% m.c. then it can be reduced to 14.5–15% m.c. in a single step. The air temperature should be roughly 200°C and the temperature of the paddy at output from the dryer below 55°C. To avoid cracking, grain moisture should not be reduced by more than 3% in any one step of a drying process; for paddy at above 18% m.c., drying will require at least two steps with tempering and convection cooling after each step. Since the end of the 1980s, in-store drying has been undertaken in China. In-store drying at low temperature by slow aeration can maintain paddy quality cheaply as heating costs are avoided and, because it is done in store, there are no transportation, loading or unloading costs (Zhao Simeng 1996; Dai Tianhong & Cao Chongwen 1996).

Provided paddy is at an acceptable moisture content then its suitability for storage can be assessed using the following quality control equations that have been

Table 1.8 Change in proportions of volatile carbonyl compounds of cooked rice stored at different temperatures.

	Carbonyl compounds (%)	
	5°C	40°C
Acetaldehyde	50.8	25.1
Acetone	31.0	42.1
Butanone	11.0	8.9
Pentyl aldehyde	—	4.9
Hexyl aldehyde	7.2	19.1

Table 1.9 Safe storage moisture limit of paddy according to rice type, harvest season and prevailing storage temperature.

	Limits of moisture content (%)			
	Indica type		*Japonica* type	
Temperature (°C)	Early season	Middle and late season	Early season	Middle and late season
30	< 13	< 13.5	< 14	< 15
20	~ 14	~ 14.5	~ 15	~ 16
10	~ 15	~ 15.5	~ 16	~ 17
5	< 16	~ 16.5	< 17	< 18

devised to give an 'evaluation value' for *indica* and *japonica* rice:

$$Y_{\text{indica paddy}} = 65.7 + (0.07 \times \text{germination rate}) - (0.25 \times \text{free fatty acid value}) + (1.7 \times \text{viscosity})$$

$$Y_{\text{japonica paddy}} = 86.3 + (0.005 \times \text{germination rate}) - (0.25 \times \text{free fatty acid value}) + (1.7 \times \text{viscosity})$$

Grain is considered unsuitable for storage if the evaluation value falls below 70 or if the FFA value is above 25 mg KOH/100 g or the viscosity less than 4.5 mm² s⁻¹ (Table 1.10).

Conventional storage in warehouse, silos and bins

Most paddy is stored in conventional warehouses in jute or polypropylene bags. These bags are thoroughly cleaned before they are filled. In bulk or bag storage, foreign matter should be removed before the paddy is stored. Foreign matter can accumulate in 'belts' when conveyors are used to load paddy into a horizontal warehouse, or in 'hoops' if loading into bins or silos that do not have a cloth funnel. Organic foreign material is a hidden danger because of its high moisture absorption and high fungal load. Fine bran reduces the gap between the grain kernels and so any heat generated will disperse more slowly. The impurity limit is strictly enforced although there are different limits for horizontal warehouses (1%) and bins (0.5%).

Particular attention should be paid to early aeration and, once aeration is complete, making the store airtight. Timely aeration can prevent the paddy stack from forming dew, giving out heat and going mouldy. The warehouse should be made hermetic once the paddy has been aerated in winter so that it can be kept at low temperature during the summer. Natural aeration in winter is widely employed in China's horizontal warehouses, and sometimes use is also made of mechanical aeration. Axial and mixed flow fans are usually employed in horizontal warehouses while centrifugal fans are used in bins.

In conventional storage, both fumigants and residual insecticides are used for insect control. The most widely used fumigant is the gas, phosphine, which must be applied at concentrations of at least 70–100 ppm over extended periods to be effective. The dosage can be raised if *Rhyzopertha dominica* or pests that have developed some resistance to phosphine are present. Conventional fumigation under gas-tight sheet is applied to the bag stacks in horizontal warehouses. Circulated fumigation is applied in large horizontal warehouses or bins where these can be made sufficiently airtight. The pressure half-lives of the stores (500 Pa to 250 Pa) should be not less than 40 s in the horizontal warehouses and no less than 60 s in silos. Residual insecticides such as malathion, fenitrothion, pirimiphos-methyl, decamethrin or a mixture of decamethrin and malathion are sprayed onto paddy bag stacks.

Low-temperature storage

In smaller horizontal warehouses, natural-air low temperature and mechanical low temperature ventilation is used. Paddy in other stores may be chilled using air conditioners, refrigeration units or cereal coolers. Refrigeration units are the most expensive due to rigid requirements for insulation and the high cost of maintenance (Ju Jinfeng 1997).

Table 1.10 Quality control classification for storability of *indica* and *japonica* type paddy.

	Indica type			*Japonica* type		
	Suitable for storage	Unsuitable for storage	Aged	Suitable for storage	Unsuitable for storage	Aged
Evaluation value	≥ 70	< 70	—	≥ 70	< 70	—
Fatty acid value*	≤ 25	25–32	> 32	≤ 25	25–32	> 32
Viscosity (mm² s⁻¹)	≥ 4.5	4.5–2.5	< 2.5	≥ 10	10–4	< 4
Taste score	≥ 70	< 70 ≥ 60	< 60	≥ 70	70–60	< 60
Colour	Normal	Normal	Abnormal	Normal	Normal	Abnormal
Flavour	Normal	Normal	Odour	Normal	Normal	Odour

Data from National Grain Reserve Bureau and National Quality and Technology Monitoring Bureau, July 1999 (unpublished).
*Free fatty acids (mg KOH/100 g dry basis).

Cereal coolers for low-temperature paddy storage have been adopted in large horizontal warehouses and bins. They are movable cooling ventilators that can also reduce humidity (Fig. 1.3). The coolers consist of a chilling system and temperature and humidity regulating systems and come in three cooling capacities: 100 kW, 50–100 kW and less than 50 kW. They work by drawing air into an evaporator through a filter net and cooling it by heat exchange. When the humidity of the cooled air is excessive, the back heating installation will warm it and reduce the relative humidity to a specified level. The cooled air is then pumped into the paddy pile through ventilation ducts and air distributors to lower the temperature and to control the humidity of the pile. Effective results can be achieved by cooling the grain in winter using natural cold air and timely mechanical aeration, and in the second year, when it becomes warm, giving supplementary cooling using the cereal cooler.

Air conditioners are used to cool paddy in many large and medium-sized cities. The temperature of this kind of quasi-low-temperature air-conditioned warehouse is controlled at 15–20°C. It can inhibit respiration, limit damage by insects and moulds and maintain rice quality.

Controlled atmosphere storage

The gas changes occurring in paddy stored hermetically, at different moisture contents and temperatures, have been researched in China. The moisture content and temperature of freshly harvested paddy is usually high and, as there are also usually insects in the paddy, oxygen becomes depleted rapidly to levels that prevent insect population growth (Table 1.11). However, paddy that has already been well dried reduces oxygen concentrations much more slowly. Thus, in practice, paddy should be contained in an airtight enclosure and oxygen needs to be displaced by filling the enclosure with carbon dioxide or nitrogen. Generally, carbon dioxide is preferable to nitrogen as it is more readily available and, unlike nitrogen, is toxic to insects. Experiments have shown that even when paddy, at less than 13% m.c., is stored under carbon dioxide for more than four years, loss in grain viability is only slight.

Open-air storage

Open-air storage is needed when there is insufficient warehouse capacity. It is a temporary storage method to protect paddy from sunlight, moisture ingress, insects, mice and squirrels. The paddy may be stored in bags or in bulk and placed on a plinth made of bricks, earth or concrete. The size and shape of the plinth depends on the amount of grain to be stored; there are several common sizes to support various quantities of grain (Table 1.12). The materials used to cover the paddy vary. Bags stacks may be covered by reed mats, straw, bamboo, fibreboard or polypropylene, while the circular cone of bulk paddy may be covered with straw, tarpaulin, PVC polyvinyl alcohol fibre, double-coated plastics or reed matting. Open-air storage has improved due to the development

Fig. 1.3 Cereal cooler used for cooling paddy or milled rice stocks, shown here attached to a sealed warehouse in China.

Table 1.11 Factors affecting oxygen concentration of paddy under hermetic conditions.

Temperature (°C)	Moisture (%)	Length of hermetic sealing (days)	O_2 concentration (%)
31	17.0	10	0.2
21	17.0	15	0.2
21–22	17.6	10	0.2
21–22	16.7	20	0.2
21–22	15.6	30	0.2
13	17.0	30	8.4

Table 1.12 Size of open-air storage plinths and the amount of paddy that can be stored on them.

Shape of plinth	Size of plinth (m)	Piling type	Storage height	Amount (tonnes)
Round	7 (diam.)	Bulk	6 m	100
Round	8 (diam.)	Bulk	4.5 m	150
Square	6 × 10	Bulk	4 m	—
Square	10 × 10	Bag	15 layers	200

of suitable materials and, provided reasonably gas-tight covers are used, can be combined with fumigation and cooling by ventilation as and when necessary.

Management of milled rice

The quality of the incoming rice must be ensured, and milled rice of different grades and variety should be distinguished. Rice with moisture content and impurity levels above the national limits is not generally accepted into store. The safe moisture content varies according to temperature of storage (Table 1.13), and this becomes particularly important if rice is to be stored through the summer as in southern parts the maximum temperatures may reach at least 35°C. The standards enforced vary according to the period of harvest and the grain type (Table 1.14).

Table 1.13 Safe moisture contents of milled *indica* rice under different storage temperatures.

Temperature (°C)	Moisture content (%)
0	18
5–10	≤ 16
20	≤ 14
25	≤ 13.5
30	≤ 13
35	≤ 12
40	≤ 11

Milled rice should be monitored during storage. Moisture content, temperature, insect damage and quality decline indexes should be checked regularly. When controlled atmosphere storage is being used then gas composition concentration should be checked regularly.

Conventional storage

Conventionally, milled rice is stored at ambient temperature and humidity but for long-term storage timely aeration and sealing with a cover is necessary to maintain quality. Heat and moisture can be reduced by ventilation in the cold season when the rice is received and the cooled rice then insulated by sealing under a cover. Newly processed milled rice should be cooled to ambient temperature before it is stacked in a store. If the moisture of the milled rice is too high then it should be ventilated before it is sealed under covers.

Low-temperature storage

Low temperature can ensure the freshness of the milled rice, prevent insect damage and maintain its natural colour, flavour and taste. Stored milled rice can be chilled cheaply by allowing cold winter air to enter stores through ventilators and open doors. This can reduce temperature quickly but is slow to reduce moisture content. The same effect can be achieved using mechanical ventilation. This method, although quick in reducing temperature and less expensive than air conditioning, may

Table 1.14 Moisture content standard for milled rice stored through the summer according to rice type and season of harvest in China.

Type of milled rice	Moisture standard		
	Safe grain	Semi-safe grain	Unsafe grain
Late season *japonica* 'normal' and glutinous	< 14.5	14.6–15.0	> 15.1
Early season *japonica*	< 13.5	13.6–14.0	> 14.2
Late season *indica* 'normal' and glutinous	< 13.0	13.1–13.5	> 13.6
Early season *indica*	< 12.5	12.6–13.0	> 13.1

cause excessive moisture loss and is entirely dependent on the prevailing climatic conditions. Sometimes, window air conditioners are installed in warehouses to keep the temperature around 20°C. This technique is less expensive than refrigeration but the rate of temperature drop is slow, especially in the middle and bottom layers of bag stacks. Cereal coolers, mentioned earlier, have been adopted for low-temperature milled rice storage. These machines are very effective in maintaining milled rice at high quality and have a bright future.

Controlled atmosphere storage

There are four types of controlled atmosphere approaches applied in milled rice storage: storage under naturally low oxygen conditions created in an hermetic environment, filling with nitrogen, filling with carbon dioxide or storage under vacuum. These may be used for bag stacks where hundreds of tonnes of rice are sealed into plastic envelopes – a similar technology has been developed for milled rice storage in Indonesia (Nataredja & Hodges 1989) – or for small quantities sealed into packets made of special films such as polyester/polyethylene composite. Safe hermetic storage of milled rice depends on appropriate m.c., temperature and oxygen content. Rice at 13.5% m.c. at 30°C can be stored safely in air with a normal oxygen concentration. Rice at 14.5% m.c. should be stored in an atmosphere containing only 0.5% oxygen. If the moisture content of the rice is above 15.5%, it cannot be stored safely even if the oxygen concentration is kept as low as 0.5%, unless the temperature is reduced to 20°C.

If nitrogen or carbon dioxide is to be used then bag stacks of milled rice are sealed into a plastic envelope and air is extracted to create a partial vacuum. The envelope is then gassed with either nitrogen or carbon dioxide. Generally speaking, the nitrogen concentra-

tion should be more than 95%, the moisture content of the milled rice should be less than 16.5% and the room temperature should be under 20°C. This method is effective in killing insects and inhibiting micro-organism growth. Using the same technique, but instead applying carbon dioxide at the rate of 1 kg per tonne of rice, insect and mould attack can be restrained and grain quality maintained. For the domestic market, 5 kg and 10 kg packets of milled rice filled with carbon dioxide are available. The carbon dioxide concentration in the pack at the time of sealing is about 70% but falls to about 40%. Since carbon dioxide is strongly adsorbed onto milled rice, the rice packed this way appears to be under vacuum. This method is very effective in preserving freshness.

For storage under vacuum, polyester/polyethylene composite of 0.13–0.14 mm thickness is used as the material for vacuum packs. Milled rice with a moisture content less than 15.5%, packaged in a vacuum and held this way for two years, maintained better quality than rice stored in conventional sacks (Table 1.15). However, the choice of packaging material is important: polyester/polythene mix works well whereas polyvinyl chloride appears a little less effective. Materials vary considerably in their gas permeability and this needs to be taken into account when planning modified atmosphere or vacuum storage (Table 1.16).

Economics of storage

Evaluations of the cost of grain storage are very difficult to obtain. However, the estimated annual cost of paddy storage in horizontal warehouses, including fumigation and handling, is relatively low, up to about RMB40 ($US5) per tonne. If ordinary mechanical ventilation for lowered temperature is included this rises to about RMB80 ($10) per tonne (Table 1.17); milled rice is

Table 1.15   Quality comparison of vacuum-packed and conventionally packed *japonica* rice.

Variety	Packaging material	Weight (kg)	Sensory evaluation
Special 2	Polyester/polyethylene	20	Normal flavour, no insects or moulds. A little yellow when opened
Standard 1	Polyester/polyethylene	20	Normal flavour, no insects or moulds. A little yellow when opened
Special 2	Polyvinyl chloride	20	Both quality and colour aging
Standard 1	Polyvinyl chloride	20	A little rancid odour and discoloration
Special 2	Open weave sack	50	Strong rancid odour and insects present
Standard 1	Open weave sack	50	Strong rancid odour and insects present

Table 1.16 Gas and moisture permeability of various films at 20°C and 65% relative humidity (g m^{-2} 24 h^{-1}).

Polymer	N$_2$	O$_2$	CO$_2$	Moisture
Polyvinyl chloride	0.16	0.37	3.7	5–6
LD polyethylene	1.90	5.50	25.2	24–48
HD polyethylene	0.27	0.83	3.70	10–25
Polypropylene	0.70	1.30	3.70	8–12
Polyester	0.0052	0.020		20–24
Polyvinylidene chloride	0.00094	0.0053	0.23	1–2

Thickness of the materials tested was 0.2–0.3 mm. Polyvinylidene chloride cannot be used for food packaging but can be used to cover the cereal pile.

Table 1.17 Costs of using different rice storage methods (RMB per tonne*).

Storage method	Horizontal warehouse		Bin	Silo
	Paddy	Milled rice	Paddy	Paddy
Conventional	20–40	40–60	60–100	60–100
Mechanical ventilation for low temperature	20–80	40–60	80–100	100–200
Cereal cooler for low temperature	100–300	150–250	20–40	100–250

*RMB8 = $US1.

somewhat more expensive to store. If cereal coolers are used then costs rise substantially, although it is not clear to what extent these increased costs are offset by improved quality preservation. For paddy, the use of cereal coolers in bins and silos is more efficient than in horizontal stores, which helps offset the higher costs associated with bin and silo storage.

Future developments

Paddy and milled rice storage will continue to be essential activities in national food security and trade. The ultimate goal is environmentally sound storage. In China, the policy of three lows (low loss, low pollution and low cost) and three highs (high quality, high nutrition and high efficiency) is widespread. This can only be guaranteed by use of scientific management practice, modern equipment, intelligent inspection, large-scale business methods, good information and high quality personnel.

For the future, China's paddy and milled rice storage will focus on low-temperature methods and controlled atmosphere storage that conform to environmentally sensitive storage strategies. Of the two approaches, low temperature is the more important and includes natural low-temperature ventilation, mechanical ventilation

and cereal coolers. Carbon dioxide and nitrogen filling and vacuum techniques can be applied to small packs of milled rice. Controlled atmosphere storage can be adopted in horizontal warehouses that are sufficiently airtight. The implementation of the policy of 'high quality, good price', China's more open grain market and emphasis on environmental protection will ensure that China becomes a market leader in the supply of high quality milled rice (Anon. 1999; Jin Zuxun & Qiu Weifen 1999).

Japan

Historical perspectives

Rice is the most important agricultural product in Japan and has been an important factor in the development of Japanese society and culture. Even though the Japanese have eaten rice since ancient times, it was only after the nineteenth century that it became the national food staple. The crop predominates because of its suitability and profitability for many of the country's part-time, small-scale farms. In addition, given that 70% of the water used in the irrigation of rice comes from rivers and lakes, there is a general belief that paddy fields play an important role in preventing floods and water erosion. This factor alone is thought to provide at least partial justification for the high domestic support that rice production receives (AAF Canada 2001).

The Japanese first learned to grow rice about 2300 BP in the Yayoi Period. The practice was probably adopted from Korea and China and, as would be expected, *japonica* varieties commonly grown at this time in southern China and on the Korean peninsula have been found in Japanese archaeological remains. Rice growing seems to have come first to the northern part of the island of Kyushu in western Japan, but since rice was originally a tropical plant it did not grow well if the summer was unusually cool. In the past, rice crops have failed on many occasions in the northern part of Honshu, Japan's main island, causing famine. However, varieties of rice were developed that could tolerate cool summers better, and northern Honshu (the Tohoku region) has become one of Japan's leading rice growing areas. Thus rice became the Japanese food staple, supplemented with meat, fish, vegetables and nuts.

Since rice can be stored, village leaders and other powerful people amassed large quantities of this cereal, which led to a gap between rich and poor. Pit stores 2 m × 2 m found in archaeological sites at Fukuoka possibly reveal the influence of contacts with northern China, where such pits were popular for storing sorghum. Other old store types include tall granaries and wooden stores known from the Yayoi Period (Fig. 1.4). In every prefecture of eighth-century Japan, rice was stored in one of two types of tall, square granaries depending on whether the rice was kept as grain or as unthreshed panicles.

Fig. 1.4 Toro ruin's restored warehouse in Japan built between 57 and 285 in the Yayoi Period.

However, poor management led to high levels of loss, and so by the tenth century storage at the prefecture level had declined, to be replaced by village-level storage. In ancient Japan, widows and poor people had the right to gather any paddy remaining in the fields after harvest. The rice crop harvested from leased paddy fields was also used to help poor people or during lean years. Until a little over a hundred years ago, rice was also used to pay taxes in Japan. Sun drying was the only drying method, either at the homestead or in the fields. The panicles were hung on frames or spread on the ground in the backyard. However, large losses often occurred because of rain during drying, and so an Imperial decree in the mid-ninth century urged agricultural officers to prepare drying frames. Rice was transported in bags made from straw (Tongming 1998).

Rice is the only food grain in Japan where 100% of the domestic needs are covered by local production. Japan's potential production capacity of rice is approximately 14 million tonnes. Since this amount exceeds domestic needs there have been reductions in the area planted, such that in 2001 the government set aside 1.01 million ha (40% of the potential rice paddy production area). Total rice production in 2000 was forecast at 9.2 million tonnes, while domestic consumption was forecast to be 10 million tonnes. Per capita annual consumption in 1998 amounted to 65 kg. This represents a decrease to almost half of the consumption of 1962, when consumption reached its peak at 118 kg, and is due to diversification of eating patterns by Japanese consumers (AAF Canada 2001).

Rice imports were banned after 1945 to help revive Japan's farming industry following World War II. However, in the 1990s overseas pressure demanded that Japan replace all of its import bans and quotas on rice with tariffs, a policy known as tariffication, through the GATT Uruguay Round of multilateral negotiations. On 30 September 1993, the Japanese Government decided to go into the international rice market on a major scale for the first time in three decades. This coincided with Japan's worst rice harvest in the post-war period. Due to a cool, wet growing season and an outbreak of blight the harvest was reduced by 36%. Foreign rice was purchased on an emergency basis to make up the shortfall. Consequently, Japan became the largest rice importer in the world. On 14 December 1993, the Japanese Government accepted a limited opening of the rice market under the GATT plan (Anon. undated).

The country thus converted its rice import regime to a tariff-only basis on 1 April 1999, consistent with its obligation under the WTO Agreement on Agriculture. In 2000, Japan agreed to open its rice markets to the purchase of 7.2% of total domestic consumption which led to the import of 693 000 tonnes of rice, mainly from the USA (330 000 t), Thailand (149 000 t), Australia (108 000 t) and China (88 000 t) (Kagatsume *et al.* 2002).

Physical facilities

Warehousing

In Japan, rice is stored as rough rice in bins and silos or as bagged brown rice in cold storage (Arthur *et al.* 2003). Most recently constructed grain warehouses in Japan are steel reinforced structures employing insulating materials to reduce the effects of heating from outside; this retains fresh grain quality and prevents insect invasion (Fig. 1.5). These warehouses are equipped with air conditioning systems to control temperature and humidity and this avoids the need for fumigation. The Japanese Food Authority classifies warehouses equipped with temperature-control systems into three categories.

Fig. 1.5 Fukagawa Government warehouse built in Japan in 1996.

Low-temperature warehouses

This type of warehouse has a heat-insulated structure, and is equipped with a standard cooling system that keeps the temperature of stored rice at 15°C or lower, and at a stable moisture content of 15%. The low-temperature warehouses account for more than 50% of total capacity (Table 1.18) (Nagaya 2000).

Although standard air conditioning systems are used, these can be expensive to run, and so there have been experiments to test snow for cooling and maintaining the humidity of stored rice. At a site in Hokkaido, a warehouse was divided into snow and rice storage areas, with an air cycling arrangement whereby warm air from the rice storage area was cooled in the snow storage area. Air from the rice area had a relative humidity of 65%. This was mixed with air in the snow area, r.h. of 100%, lowering the r.h. to 70%. Air at this humidity, when returned to the rice area, maintained rice at 15% m.c. In another warehouse, stocks of paddy were kept in the presence of snow at 0°C for up to ten years with very little decline in quality (Kobiyama undated).

Semi-low temperature storage

The structure and facilities of this type of store are the same as those of the low-temperature warehouse. However, in this scenario, the temperature of stored rice is maintained at 20°C or lower.

Normal temperature

This type of warehouse minimises the effects of exterior temperatures. In particular, it is capable of keeping the temperature inside the warehouse at 23°C or lower during the summer and periods of hot weather.

Table 1.18 Numbers and capacity of rice warehouses in Japan. Source: Nagaya (2000).

Operator	No. of operators	No. of buildings	Warehouse capacity ('000 tonnes)			
			Low temp.	Semi-low temp.	Normal temp.	Total
Agricultural co-operative	1 213	9 465	2 422	784	3 822	7 028
Trader	1 029	1 427	143	10	352	505
Commercial	654	2 116	3 893	10	849	4 752
Government	10	67	151	1	44	196
Total	2 906	13 075	6 609	805	5 067	12 481

Range of warehouse designs

A number of different warehouse designs are used. With all types it is recommended that the structure is built with appropriate insulation and made airtight to maintain stable temperatures inside the warehouse all year round.

Steel-reinforced concrete

Steel-reinforced concrete warehouses have excellent earthquake-proof, fireproof and airtight properties. However, they are relatively good conductors of heat and where there are west facing roofs and walls it is recommended that these are insulated.

Concrete blocks

Lately, many warehouses are being built using a method by which the foundations and pillars or beams are built with steel-reinforced concrete, with concrete blocks placed in between. If the external walls of these constructions are not plastered, external humidity will penetrate the building. Concrete block warehouses are prone to cracking, so there is a risk of gas leaks when fumigating. It is necessary to take measures to maintain airtightness, including testing the building's airtightness every five years. The earthquake resistance of these warehouses is poor.

Stone

Stone warehouses have good heat and fire resistance, similar to that of steel-reinforced concrete constructions. Currently, these structures can be seen in stone-producing areas, where stone procurement is convenient. As with concrete block structures, stone warehouses have poor earthquake resistance, and airtightness tests are required at all times.

Clay

In Japan, clay-walled warehouses are traditional structures. They have excellent fireproof and insulating properties, with a double roof and thick walls (20 cm or more) that reduce solar heating sufficiently to maintain temperatures inside at 20–25°C during the summer. However, due to the lack of construction materials, shortage of qualified workers, and difficult maintenance, construction of these warehouses has practically ceased and their number is gradually decreasing.

Wood

Wooden warehouses have a simple structure consisting of an external wood frame or straight wood beams, but they are inadequate as grain warehouses. They are susceptible to solar heating and humidity and are affected by insects and rats. However, in dry and cold highlands the interior temperature and humidity of wood warehouses remain low so grain quality is well maintained.

Recommendations for warehouse construction

Roof

The roof is the single most important element in warehouse structure affecting the quality of stored rice. Since 37% of the heat penetrating the warehouse comes through the roof, it is necessary to build double structures or employ materials with superior insulating properties to limit heat transfer.

Floor

To prevent ground moisture penetrating the warehouse interior the floor must be raised. The floor must be built carefully, making sure it can withstand heavy weights such as stored commodities and cargo-handling machinery. The foundations should be solid and the building earthquake-proof.

Doors

It is recommended that the warehouse has at least two entrances. However, door size is not standardised, and must be determined according to the size of the machinery to be used. The door must be insulated and fireproof, and must be equipped with a rat barrier and an additional screen to prevent the entry of insects.

Silo storage

Japan imports large amounts of cereals, such as wheat and maize, for both food and animal feed. Large silos have been built, especially at ports, to allow efficient loading and unloading as well as transportation of the bulk cargo. Away from the ports, silos with elevators handle domestic rough rice and wheat. Silos are made of reinforced concrete or steel plates.

Objectives of storage

In Japan, rice storage forms an integral part of a marketing system designed to ensure a stable and sufficient rice supply at an acceptable quality. Japanese eat boiled rice by itself, unlike people elsewhere who eat it mixed with other foods, which are often strongly spiced. It is therefore important in Japan that rice is tasty, fragrant, shiny and sticky. Freshness is key to ensuring these characteristics, consequently rice is stored using advanced methods to control storage conditions, particularly temperature and humidity. Three bodies play significant roles in rice storage and marketing: government, independent traders (the voluntary sector) and rice producers. The government purchases rice to maintain a stockpile that, in principle, will be sold after one year of storage. The purchase and selling prices of government rice are determined every year upon hearing the opinions and suggestions of the members of the Rice Price Council. The Government of Japan procured rice harvested in 2000 for $US2.04/kg, a reduction of 2.7% from the previous year, and sold rice at $2.34/kg, a reduction in selling price of 1.6%. These were the fourth consecutive price reductions.

There has been considerable recent liberalisation of the Japanese rice market. In December 1994, Japan introduced the Law for Stabilisation of Demand–Supply and Price of Staple Food, otherwise know as the Staple Food Law, in order to reduce the gap between rice supply and consumer demand. This law deregulated the rice market, giving independence to rice producers and improving the efficiency of rice distribution. To facilitate the rice market the government prepares basic guidelines in November of each year so that producers have the necessary, accurate information on the demand for rice. This promotes a good balance between supply and demand. Under the Staple Food Law, the government authorises private traders to distribute rice. The price of this rice is determined by means of tenders so that the price is quoted properly and justly, and fairly reflects market value. For this purpose, the government has authorised the Voluntarily Marketed Rice Price Formation Centre to formulate an appropriate price level to serve as a reference for the trading of voluntarily marketed rice and to contribute toward the stable buying and selling of the commodity.

Management of the national rice stockpile

Since 1993, the last time Japan experienced a seriously poor rice harvest, the government has stockpiled rice to ensure supplies in lean years in line with the provisions of the Staple Food Law. The minimum national rice stockpile is 1.5 million tonnes, including rice kept by the private sector. This figure was determined from research that identified the actual need during poor harvests in the past. This enables the government to make good shortfalls in supply in the event of serious shortages. In implementing the stockpiling scheme, the government takes due consideration of the volume of rice handled by both the government and by the public sector. Under the Staple Food Law, the government is mandated to purchase a certain portion of rice from producers, i.e. those who comply with their target area under the production adjustment program, and to ensure the smooth management of rice stockpiles.

In November 1997, the government formulated a New Rice Policy to counter the excessive accumulation of stocks and a drastic decline in the price of privately marketed rice. The intention was to match supply and demand to give a stable market price. Then in 1998 under new policies, the government introduced regulations, the Operational Rules for Reserve Stocks, to achieve a desirable stock level. Based on this rule, the government reduced the amount it buys by taking into account unsold stocks. According to the March 2000 stock volume basic plan, carry-over stocks of rice stockpiled by the government were forecast at 2.29 million tonnes at the end of October 2000. The continued downward trend in consumption of rice by Japanese consumers is likely to further complicate the efforts of the government to reduce the large rice carry-over stocks (AAF Canada 2001).

Rice marketing channels

Distributors of both private sector and government-marketed rice are required to register with the competent body in each prefecture, to ensure that rice is shipped and supplied from producers to consumers in an efficient and systematic manner.

In the former distribution system, under the Food Control Law, rice was collected and distributed at predetermined locations in a linear manner. In November 1995, the Staple Food Law was enforced, which diversified the marketing channel, bringing new routes into play (Fig. 1.6) (Nishioka 2000). In this system, any wholesaler or retailer could market rice and became a registered entity provided basic requirements were met. This has encouraged rice trading so that even convenience and liquor stores are included. With regard to rice other than orderly marketed rice, as shown in the lower

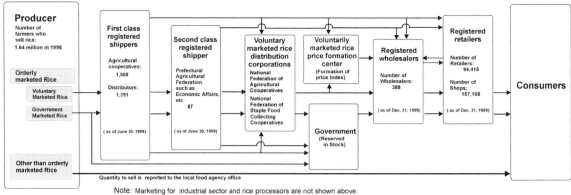

Fig. 1.6 Rice marketing channels in Japan.

part of Fig. 1.6, rice producers can sell rice freely as long as they report the quantity they sell. Consumers can buy rice directly from producers. In 2000, the distribution of rice by the various marketing channels was as follows – voluntary 4.48m t, government 0.50m t and producers 5.34m t.

Commodity management

Rice is harvested at moisture content of 23–24%. The opportunity for harvesting is short but this approach is important in maintaining quality as rice rapidly fissures at moisture contents less than 23%, but at higher moisture contents protein values may be above permitted levels. So farmers must time their planting and field size such that the harvest can be gathered in promptly. Once harvested, the grain mass of high-moisture rice heats up rapidly. Thus guidelines stipulate a maximum period of 8 hours for rice to move from the field to the dryer, because longer periods lead to the development of unacceptable off odours (Mutters 2000).

Initially, paddy rice is stored in country elevators and silos. It is then husked to give brown rice. Japanese rice mills tend to be the most sophisticated in the world, but rarely handle more than 100 t per day; this is relatively small when compared to US facilities that can mill 1000 t per day. The mills are usually operated by co-operatives that pack rice into woven polypropylene bags; locally produced rice is distributed entirely in the form of brown rice.

The climate from June to August in Japan is rainy and hot so that quality deterioration in rice stored under natural conditions could occur due to respiratory activity, fat oxidation, fermentation, insect pests, rats, bacteria, mould or a combination of these. Physical conditions in storage are thus controlled to maintain low temperature and low humidity to prevent such changes and so avoid the use of insecticides or fumigants which pose environmental and health risks. The rate of quality decline in rice is reduced at lower moisture content. Insects such as the weevils, *Sitophilus* spp, propagate only very slowly when the m.c. is 11.5% or less; likewise mould and other micro-organisms when moisture is 13.7% or less. However, for rice to have the characteristics preferred by the Japanese consumer it needs to be marketed at about 15% m.c. The significance of this moisture is that rice at 15% does not crack when put into water before cooking, whereas rice, at say 12%, will crack and become mushy. Thus in Japan, rice is stored long-term at a relatively high moisture content. Currently, 16% is considered acceptable whereas in most other countries 14% is the maximum for safe storage. Good preservation at this high moisture is achieved by holding the rice at a relatively low temperature.

Almost all insects found in warehouses, including weevils (*Sitophilus* spp) and Indian meal moth (*Plodia interpunctella*), invade and propagate after harvest. However, insects and micro-organisms such as mould are almost inactive at temperatures below 15°C. In view of this, the recommended storage temperature for rice in warehouses is at 15°C or below. The respiration rate of rice at 15% m.c. and 30°C is 3.5 times greater than at 10°C. Rice stored at close to 0°C can be maintained over long periods with little or no quality decline (Anon. 1995). However, such storage is not cost effective due to high energy consumption. The Japanese Food Agency (JFA), in order to preserve freshness of rice for long periods, holds milled rice at a moisture content of approximately 15% (70–75% r.h.) in warehouses maintained at 15°C or below. In established storage facilities no attempt is made to exclude air, but experimental trials have successfully demonstrated the value of storage under CO_2 at ambient temperature (Mitsuda 1999).

The Japanese are very particular about rice quality and believe that rice is best when served fresh after milling. Thus there are many stores in Japan where small machines polish brown rice as it is purchased, though this obsession with quality has a price and Japanese rice is about five times the price of rice in other countries.

During the summer, bagged brown rice may be moved from cold environments to storage under ambient conditions, where it is at risk of attack from storage pests. Paddy is also at risk from insect pests. For both types, aeration with ambient air might be useful as an insect management technique but has not been employed by the majority of pest management programmes because of lack of research. A modelling exercise by Arthur *et al.* (2003), using historical weather data, demonstrated that weather conditions in much of Japan would allow the use of aeration to cool rough and milled rice and that the data could contribute to the development of risk management models for stored paddy and bagged brown rice.

Besides controlling physical conditions in stores to maintain rice quality, store hygiene is an important measure. Since insects hide in grain residues left on warehouse floors and in accumulations of dust, efforts are made to keep facilities clean and neat to avoid creating insect breeding grounds.

Inspection and grading of rice is a key element of commodity management. Rice is inspected at least once in the marketing chain (Fig. 1.7). Rice quality is assessed on a combination of subjective and objective factors. Its acceptability is determined by quality testing against a set of fixed criteria, but the ranking of these is largely dependent on the consumer. Since 1993, rice for the Japanese market has been subject to official quality inspection against standards set by the Ministry of Agriculture Food and Forestry. In these standards, 'perfect rice' describes grains that are completely filled and have a satisfactory shape and appearance. For brown rice, appearance alone is the major quality criterion. Physical characteristics considered important include the number of unbroken grains, the number of damaged grains, moisture content and percentage of whole grains. Most inspection of brown rice is done visually. However, there are cases in which it is difficult to judge the rice grade by visual inspection alone so that various analyses and measurements are needed. At the actual inspection site, the grade assigned to a rice consignment is first made

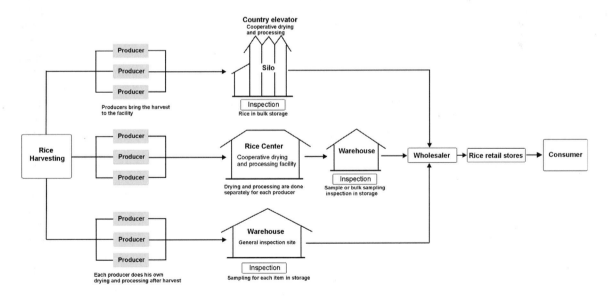

Fig. 1.7 An inspection flowchart showing that rice is inspected at least once in each of the Japanese marketing chains.

by visual inspection; this requires that the inspector is a skilled judge.

The inspector takes a sample in the palm of the hand and judges the appearance, proportion of largely undamaged grain (head rice), moisture content and the feeling of weight. Judgements of moisture content are based on the feel by inserting a sampling spear into the bag of grains, on the sound when the sample is poured into a carton or inspection box, on the feeling when a sample is taken into the palm of the hand and on the feel when a sample is clenched in a fist. Since minimum test weights are prescribed for each type of rice and grade, it is necessary for the inspector to develop his skills by performing many measurements and comparing the feel with measured weights.

A grade is assigned first of all by the general appearance of a sample. The sample may then be subject to various analyses; if these agree with the original grade assignment then this is adopted. If not, then the lowest grade according to the assessment criteria is applied. As an example, the inspection standard of the JFA for lowland non-glutinous brown rice is shown in Table 1.19 (Mutters 1998, 2000; Anai *et al.* 1989).

Freshness is rated so highly that tests have been devised for its assessment (Matsukura *et al.* 2000). Good eating quality is related to high stickiness, sweet flavour, gloss and palatability. To determine these parameters, taste panels are employed to evaluate different rice lots. Taste panel scores are then used to calibrate taste analysers which relate the scores to physicochemical properties. In white rice, amylose, protein content and moisture content are the main determinants of acceptability. In addition, scores for brown rice are also influenced by the

fatty acid content, fatty acids being contained in the bran layer, which is removed during milling. The palatability criteria used for further grading of table rice involve six sensory tests: for appearance, aroma, hardness, stickiness, taste and overall evaluation (Mutters 1998).

Economics of storage

The annual cost of brown rice storage in low-temperature warehousing is about $100/tonne.

Australia

Historical perspectives

The Australian rice industry began in the 1920s, following a search for crops capable of cultivation in the newly developed Murrumbidgee River Irrigation Area (MIA) of New South Wales (NSW) (Anon. 1994). One crop is grown per year over the spring and summer months. Initially, paddy was handled and stored in bags, the industry lacked machinery, had little farming expertise or experience, and production was low. It was recognised though, that the grain at high moisture was not satisfactory for storage. So a maximum moisture limit was recommended for all stored rice, and as a result grain was field dried to16.5% moisture content (Anon. 1994; Bramall 1985).

The crop size grew steadily in the 1930s and 1940s although the grain was still harvested and bagged in the field, then stored in large sheds owned by the Rice Marketing Board (RMB), a government statutory authority of the state of NSW. As the crop size continued to grow,

Table 1.19 Japanese standard for brown rice. Source: Mutters (1998, 2000), Anai *et al.* (1998).

	Minimum values			Maximum values				Foreign grains		
Grade	Weight (g/L)	Perfect grains (%)	Characteristics*	Moisture (%)†	Damage (%)‡	Dead (%)	Coloured (%)	Rough rice	Other	Foreign matter
1	810	70	See footnote	15	15	7	0.1	0.3	0.3	0.2
2	790	60		15	20	10	0.3	0.5	0.5	0.4
3	770	45		15	30	20	0.7	1.0	1.0	0.6
Off grade§	< 770	—		15	100	100	5.0	5.0	5.0	1.0

*Thickness of skin, grain fullness, hardness, size uniformity, shape, lustre, skin rub, white core, white belly and chalkiness.
†At present 1% is added to the moisture content, so the upper limit is 16%.
‡Includes germinated, diseased, insect damaged, cracked and malformed grains.
§Off grade indicates brown rice that failed to qualify for a particular grade.

mechanisation was becoming widespread and farm yields were increasing. As a result, growers experienced substantial problems achieving the required 16.5% moisture content for bagging. Furthermore, paddy rice dried in the field to 16.5% or lower was shown to crack and this reduced the milling outturn.

In 1951, the RMB undertook an investigation to ascertain if bulk handling methods would suit the rice industry. These were subsequently introduced and bulk handling has raised efficiency, reduced costs and improved harvest management (Bramall 1985). Bulk handling methods allowed the paddy crop to be harvested at higher moisture content, a useful feature as weather conditions are typically wetter during harvest, and facilitated controlled in-store drying to reduce milling losses.

The introduction and development of bulk handling methods has proceeded over an extended period. In 1956, the RMB converted a bag shed to bulk storage and installed a suction 'top to bottom aeration system consisting of simple semicircular ducts and axial fans. This system proved that drying with unheated air was effective in the climatic conditions of the MIA. Industrial aeration drying systems were further developed based on this trial, making use of technical support from Australia's Commonwealth Scientific and Industrial Research Organisation (CSIRO) and an awareness of the research work at the University of California USA (Anon. 1994; Bramall 1985). These aeration systems consisted of 6 m deep bulk storage sheds containing a series of bins constructed from timber frames with cor-

rugated iron cladding. Below-floor and above-floor duct layouts were used with three to six fans per bin. Different sized centrifugal fans were used on different bins to provide a range of aeration drying rates of 8–15 litres per second per tonne ($L\ s^{-1}\ t^{-1}$). The bins were loaded using grain slingers to create a reasonably level bin surface. From this early work, it was determined that paddy rice with initial moisture contents as high as 20% could be satisfactorily dried in such sheds using unheated aeration with a specific aeration rate of approximately 11 $L\ s^{-1}\ t^{-1}$. Considerable supervision of this operation was necessary as the rate of drying depended on the prevailing weather conditions.

From its inception, the rice industry of NSW has generally shown a steady increase in production and yields (Fig. 1.8); only 64 hectares were harvested in 1925, rising to just over 180 000 hectares in 2001. Storage capacity has increased accordingly. Today, the industry is characterised by large steel or concrete storage sheds containing bins up to 7.5 m deep with capacities from 600 to 4000 tonnes. Automatic overhead loading systems ensure level bin filling at any depth. Aeration systems have aeration drying rates up to 18 $L\ s^{-1}\ t^{-1}$. Modern facilities have integrated automatic drying control and, where additional drying is required, gas burner-assisted heating is used.

Since the 1960s until 2001, control measures for stored product insect have improved. A key part of this improvement involves the reliance on keeping the grain cool over winter, so limiting insect growth. Initially,

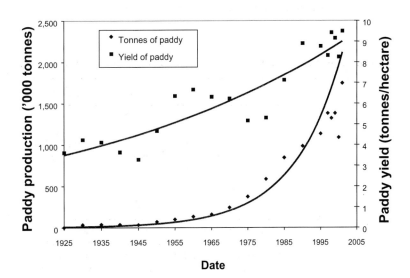

Fig. 1.8 Increase in paddy production and yields of the Australian rice industry (Anon. 1994; Annual General Reports of the RMB).

hygiene practices were simple and inconsistently applied. The use of insecticides was commonplace. Today, hygiene programmes are a key part of the design and operation of the storage and handling facilities but there is no treatment with residual insecticides. However, space treatment with dichlorvos, a non-residual insecticide, is used to disinfest structures and to treat localised infestation outbreaks. Some dedicated sealed stores have been acquired for disinfestation of paddy rice with the fumigant phosphine.

During the late 1990s, deregulation of the government-controlled water and irrigation supply resulted in a rapid increase in rice production. This has placed immense pressure on the industry's storage and drying infrastructure. Despite an intense expansion of storage capacity, facilities have not kept pace with this growth and unaerated wheat industry stores have been increasingly used for rice. Using unaerated storage has increased the quality problems with mould and insects. The industry is extremely conscious of the capital costs and risks involved in investing in stores at a rate that matches the expansion in production, and is actively seeking more economical store types.

Overall, today's rice industry in Australia is capital intensive, highly automated and completely reliant on bulk handling. The RMB owns the storage facilities, and by state law all rice grown in NSW is required to be delivered to the RMB. The rest of the production chain milling, logistics, marketing, and so on is owned by the Ricegrowers Co-operative Limited (RCL). Growers are directors of both the RMB and RCL. The rice growers of NSW own and operate the total post-harvest rice production chain, from receival through to marketing.

Physical facilities

As a result of the historical development of the storage facilities at RCL, a range of performance features exists across its stores today. Stores from the 1960s and 1970s were characterised by loading capacities of 100–150 th^{-1}, 6 m high walls, minimal automation, manual bin levelling systems, 6–9 L s^{-1} t^{-1} aeration rates, no burner-assisted drying, and no aeration control (Figs 1.9 and 1.10). Stores from the 1980s and 1990s operate on a large scale and with greater sophistication; they have loading capacities of 250 th^{-1}, plus 7.5 m high walls, central control rooms for fully automated operation, automatic bin levelling systems, 11–16 L s^{-1} t^{-1} aeration rates, burner-assisted drying and advanced aeration control (Figs 1.11, 1.12 and 1.13).

Fig. 1.9 Paddy rice shed built in 1960s, Australia (Murrami, NSW), with 6 m corrugated iron walls, above-floor ducts and aeration fans for a 9 L s^{-1} t^{-1} aeration rate.

Fig. 1.10 Paddy rice shed built in 1970s, Australia (Gogeldrie, NSW), with 6 m plate iron walls, below-floor ducts and aeration fans for a 7 L s^{-1} t^{-1} aeration rate.

Fig. 1.11 Paddy rice shed built in 1980s, Australia (Murrami, NSW), with 7.5 m plate iron walls, above-floor ducts and aeration fans for a 12 L s^{-1} t^{-1} aeration rate.

Fig. 1.12 A strategic paddy drying shed built in 1990s, Australia (Gogeldrie, NSW), with 7.5 m concrete walls, above-floor ducts and a 16 L s⁻¹ t⁻¹ aeration rate.

Fig. 1.13 Paddy rice shed built in 1990s, Australia (Gogeldrie, NSW), showing the uniform bin filling achieved with an automatic bin levelling system (same facility as Fig. 1.12).

From the 1970s until 2001, the design and features of the storage facilities have steadily improved. Strategically positioned stores with faster overall drying capability have been built to ensure a sufficiently rapid supply of paddy to keep the mills running at optimum capacity. To keep pace with demand there has been a steady increase in capital investment. Expressed in present-day terms, first-year investment capital costs ranged from \$US170 to \$300 per paddy tonne, depending on the particular design. Overall, the stores were built to last at least 40 years and 1960s designs are still functioning as well today as when they first opened.

Objectives of storage

Aeration drying and safe storage of paddy rice for up to 15 months are key components of the rice processing chain. The aerated storage facilities are operated with the understanding that they feed into the milling operation and so great care is taken to ensure appropriate drying. Overall, the objective is the timely and cost-effective production of dry paddy rice at consistent moisture content with breakage susceptibility reduced to a minimum. Specific storage objectives are as follows:

- Paddy rice should be available throughout the year, at 13.7–14.0% m.c., for milling.
- The paddy supplied to the mill should be a consistent product, there should be no moisture content differential within storage bins, and in particular the rice at the base should not become over-dried.
- All paddy rice should be dried and stored such that breakage during this and any subsequent processing stage is minimised.
- Paddy should be dried and stored to prevent contamination by fungi or their derivatives.
- The operating costs of the storage system should be minimised.
- Paddy rice should be dried and stored to maximise presentation, cooking characteristics and eating functionalities sought by market such as mouth feel, stickiness, etc.
- Paddy rice should be dried and stored without infestation by stored product insects and the use of insecticidal chemicals minimised.
- Paddy is stored in safe and secure structures with minimal risk of external contamination by birds, rodents or any other non-rice foreign materials.
- Operation of the storage system should be straightforward and safe.

Major sources of quality decline

The Australian industry supplies rice to a variety of international and domestic markets, and a range of quality characteristics are in demand. Cereal quality assurance with respect to foreign matter, weed seeds, moisture content, mould, chemical residues and insects is well addressed as a result of the advanced facilities used. Specific quality factors demanded by certain markets can be more challenging and RCL focuses on meeting such demands. These factors include kernel breakage levels, a range of appearance and presentation characteristics of the polished rice, cooking texture and performance, eating qualities, and a range of downstream processing characteristics needed for the production of crackers, flours, cakes, and so on.

Rice kernel breakage is a major quality issue that can be adversely affected by certain aeration drying practices. Increased breakage can result when freshly harvested paddy rice is dried rapidly in half-filled burner-assisted aeration bins. Poorly controlled aeration facilities can result in a significant range of moisture from the base to the top of bins, providing grain at varying moisture to the mills and so increasing kernel breakage. Occasionally, inadequate aeration drying may necessitate the use of column dryers and an increased grain breakage can result. For the most part, however, well-controlled aeration provides the low breakage levels sought by the industry.

The appearance of polished rice, its cooking characteristics and its eating qualities change during storage. Commonly measured factors include water absorption, viscosity, cooking stability, mouth feel, stickiness, colour, lustre, whiteness and aroma. Although such properties are primarily related to the type and variety of the rice, changes to these in storage can be significant. Furthermore, one market will regard the same change as a quality loss, another as an improvement. In general, reducing the moisture and/or the temperature of the paddy rice slows most rates of change. When controlled properly, aerated storage facilities can manipulate the temperature or moisture of paddy rice to influence such changes. Advanced control features are not installed throughout the entire industry at present so the control of these quality traits is not yet widespread.

Paddy rice that is stored through the year until the following summer can experience sporadic insect infestation, especially at the periphery of grain bulks. This is a result of the passive heating of the grain near the walls and surfaces, raising the grain temperature and providing favourable conditions for insect growth. Also, the RMB storage facilities are not sealed, so allowing insect ingress.

Commodity and pest management regimes

A major quality control component of the Australian rice industry is receival sampling that involves the inspection of all truckloads of paddy rice delivered to the RMB storage facilities. This is a key trading point in the rice production process between the RMB and growers.

For acceptance into the storage system, samples are inspected for a range of quality factors including foreign seeds, foreign matter, insects, field stains/moulds and other signs of damage. To determine the value of the rice, samples are also assessed for variety, moisture content and breakage levels and grower payments are adjusted accordingly. The more market-specific cooking and appearance characteristics are not used in the routine assessments. However, extra-ordinary assessments may be used on occasion. The size, quality and logistics of the yearly rice crop are defined during the receival programme. All information is entered into a centralised database for detailed operation's management.

Combined in situ moisture and temperature sensors are used to monitor stored rice, in particular the progress of aeration drying. These sensors are robust units that handle the harsh conditions characteristic of deep-bed grain storage and provide an adequate measurement of moisture content. The sensors are positioned at three depths in the grain bulk and the progression of drying fronts through the grain bed is determined by comparing sensor readings. The sensors are used as a management tool to monitor the drying and storage processes. However, a good understanding of the aeration processes is needed by operational staff before they can make informed management decisions using the results of these sensors.

To complement the in situ sensor results, a storage sampling programme is implemented across the industry throughout the drying and storage process. Samples are taken from a depth of up to 1 m from the surface for quality and moisture analyses. Typically, paddy is sampled during the drying stage, at the end of drying, about every two months during long-term storage and, for certain markets, immediately before unloading the bin. Extra sampling may be undertaken to diagnose any storage concerns.

The insect control measures implemented are appropriate for the climatic conditions experienced during the NSW rice harvest and storage period. Rice in southern Australia is harvested during autumn when grain temperatures typically range from 12° to 25°C. It is then stored over winter and spring where colder temperatures are common. Such temperatures slow the growth and development of stored product insects. The main insect species encountered within the grain bulks are the beetles *Sitophilus* spp, *Rhyzopertha dominica*, *Tribolium* spp, *Oryzaephilus* spp and *Cryptolestes* spp. *Trogoderma variabile* occurs in the shed structures and surrounds while unused and uncleaned handling equipment can support infestations by the moths *Ephestia* spp and *Sitotroga cerealella*. Apart from at harvest, a regular hygiene programme is practised at all storage facilities, including a thorough pre-filling bin clean and preparation within the few months before harvest. Dryacide (Vol.1 p. 271, Stathers 2002), a product based on diatomaceous earth, is applied as slurry to the structural surfaces to reduce infestation pressure in some facilities. More conventional residual insecticides and space treatment with dichlorvos can be applied to external handling equipment.

Bird problems are reduced by exclusion using wire mesh across all air and light entries, and by laying poison baits. Rodent pests are controlled by the use of poison baits, minimising refuges and maintaining a hygiene programme.

Economics of operation of the system

In the NSW rice industry, paddy rice is not left in the field to dry completely for quality reasons, but instead is aeration dried in storage by four to ten percentage points. The slow deep-bin aeration drying system that is used consumes substantial quantities of energy. Although this is relatively expensive, it lowers milling costs and raises the market return on finished products.

The operating cost of the storage facilities involves electrical power, labour for operating, maintenance and hygiene programmes, overall management of the stores and, at approximately 15% of facilities, fuel gas (LPG). The largest single expense is the electricity required to run the aeration fans, which amounts to $A6–10 per paddy tonne at electricity tariffs of 13–15 cents per kilowatt-hour. The actual cost of aeration drying in a particular season and in a particular design of store is dependent on several factors, including:

- weather experienced during aeration drying;
- weather experienced during harvest;
- moisture content of harvested paddy rice;
- type and variety of paddy rice;
- speed at which the paddy rice is dried (size of aeration fans);
- depth of paddy rice through which aeration air is pumped; and
- use of burner-assisted heating within the drying programme.

In summary, the electrical energy consumed per tonne is a result of a design trade-off between how fast the rice is dried and the accepted risk to quality due to rice remaining at high moisture. The higher the moisture and the longer it remains undried, the greater the risk, but the faster it is dried, the higher the costs.

Future developments

As the size of the harvested crop is increasing year on year, alternative store types are being adopted by RCL for particular scenarios. Rice that has been dried and stored through a complete season in RCL facilities is being moved to bunkers (see Chapter 3) for subsequent storage in order to free up drying facilities. Bunkers are lower performance stores but provide acceptable facilities for dry, cool grain, and aerating bunkers of rice that are not completely dry and cool is being considered for the future. As an alternative, the storage of uncooled rice in facilities sufficiently airtight to allow phosphine fumigation is currently under investigation.

Developments in aeration control are being pursued at RCL to improve performance in several areas. A new control method, adaptive discounting (ADC), is being trialled on natural air and burner-assisted aeration drying facilities. The aims are to improve the milling quality of the dried rice, increase aeration efficiency, enable chemical-free insect control and provide a user-friendly operating interface.

Alternative approaches to insect control in bulk paddy through to the finished product, and for structural treatments, are being pursued. This is in response to increased infestation associated with larger harvests, requiring the use of contract storage, and with increased storage over summer months. Preparations are being made for the phase-out of methyl bromide by 2005; consequently, the use of alternative fumigants such as carbonyl sulphide, phosphine and ethyl formate are being investigated.

References

AAF Canada (2001) Japan. *Agriculture and Agri-Food Canada, Bi-weekly Bulletin,* **14**, 3. http://www.agr.gc.ca/policy/winn/biweekly/English/biweekly/volume14/v14n16e.pdf.

Anai, S., Hosokawa, A. *et al.* (eds) (1998) *Rice Inspection Technology.* The Food Agency, Ministry of Agriculture, Forestry and Fisheries, Tokyo, Japan.

Anon. (1995) *Rice Post-harvest Technology.* The Food Agency, Ministry of Agriculture, Forestry and Fisheries, Tokyo, Japan.

Anon. (1999) *New Technologies for Grain Storage*. Department of Warehouses and Storage, China National Grain Reserve Bureau, Beijing, P.R. China.

Anon. (undated) Japan rice trade. Trade and Environment Database, American University, Washington DC, USA. http://www.american.edu/ted/japrice.htm.

Anon. (1994) Ricegrowers battled greedy millers, then a protected world. *Southern Edition*, 24 March, 9–11.

Arthur, F.H., Takahashi, K., Hoernemann, C.K. & Soto, N. (2003) Potential for autumn aeration of stored rough rice and the potential number of generations of Sitophilus zeamais Motschulsky in milled rice in Japan. *Journal of Stored Products Research*, **39**, 471–487.

Boxall, R.A. (2002) Storage structures. In: P. Golob, G. Farrell & J.E. Orchard (eds) *Crop Post-Harvest: Science and Technology. Volume 1 Principles and Practice.* Blackwell Publishing, Oxford, UK.

Bramall, L.D. (1985) Paddy drying in Australia. In: B.R. Champ & E. Highley (eds) Preserving Grain Quality by Aeration and In-store Drying: Proceedings of an International Seminar, 9 11 October 1985, Kuala Lumpur, Malaysia. 118 129. ACIAR Proceedings 15. Australian Centre for International Agricultural Research, Canberra, Australia.

Dai Tianhong & Cao Chongwen (1996) Feasibility analyses on low temperature grain drying in the main grain productive areas of China. *Grain Storage*, **25**, 23–26.

FAO (2000) *Food Outlook. November, 2000*. United Nations Food and Agriculture Organization, Rome, Italy.

Hodges, R.J. (2002) Pests of durable crops – insects and arachnids. In: P. Golob, G. Farrell & J.E. Orchard (eds) *Crop Post-Harvest: Science and Technology. Volume 1 Principles and Practice.* Blackwell Publishing, Oxford, UK.

Jin Zuxun (1984) *Facilities and Techniques of China's Grain Storage in Ancient Times*. Agriculture Press, Beijing, P.R. China.

Jin Zuxun & Qiu Weifen (1999) Strategies of China's grain storage research for the 21st century. *Journal of the Chinese Cereals and Oils Association*. Special Edition.

Jin Zuxun & Yu Yingwei (2000) Studies on the modernized types of China's warehouses. *Journal of the Chinese Cereals and Oils Association*, **15**.

Ju Jinfeng (1997) The current situation and prospect of grain storage under low temperature and grain drying under normal temperature. *Grain Storage*, **26**, 15–28.

Kagatsume, M., Chang Ching-Cheng & Wu Chia-Hsun (2002) A general equilibrium analysis of Japan's rice tariffication. In: *Proceedings of the 5th Annual Conference on Global Economic Analysis*, 5–7 June 2002, Taipei, Taiwan.

Kobiyama, M. (undated) The utilization of snow/ice cooling energy. EIT No. 31. Energy and Information Technology Inc. http://www.eit.or.jp/pdf/eit31e.pdf.

Li Lungshu & Jin Zuxun (1999) The great achievement of grain storage scientific research in China. In: J. Zuxun, L. Quan, L. Yongsheng, T. Xianchang & G. Lianghua (eds) *Proceedings of the 7th International Working Conference on Stored Product Protection*, Beijing, China. **1**, 40–49. Sichuan Publishing House of Science and Technology, Chengdu, Sichuan Province, P.R. China.

Lin Zaiyun & Wu Lina (1999) Underground grain storage engineering. In: J. Zuxun, L. Quan, L. Yongsheng, T. Xianchang & G. Lianghua (eds) *Proceedings of the 7th International Working Conference on Stored Product Protection*, Beijing, China. **2**. Sichuan Publishing House of Science and Technology, Chengdu, Sichuan Province, P.R. China.

Lu Qianyu (1998) Storage techniques of cereal grains. In: J. Zuxun, L. Quan, L. Yongsheng, T. Xianchang & G. Lianghua (Eds) *Proceedings of the 7th International Working Conference on Stored Product Protection*, Beijing, China. **1**, Sichuan Publishing House of Science and Technology, Chengdu, Sichuan Province, P.R. China.

Matsukura, U., Kaneko, S. & Momma, M. (2000) Method for measuring the freshness of individual rice grains by means of a colour reaction of catalase activity. *Journal of the Japanese Society of Food Science and Technology*, **47**, 523–528.

Mitsuda, H. (1999) Grains storage technology friendly to the environment. Japan Food Technology Association, Kyoto, Japan. http://sbpark.com/mitsuda/mitsu8e.html.

Mutters, R.G. (1998) Concepts of rice quality. Rice Quality Workshop, University of California Cooperative Extension Rice Project, University of California, Davis CA, USA. http://agronomy.ucdavis.edu/uccerice

Mutters, R.G. (2000) Producing high quality rice for the export market. University of California Cooperative Extension Rice Project, University of California, Davis CA, USA. http://agronomy.ucdavis.edu/uccerice/QUALITY/wasa0100.pdf.

Nagaya, S. (2000) *Storage Techniques and Storage Facilities for Unmilled Rice*. Textbook No. 5 of Post-harvest Rice Processing, Tsukuba International Centre, Japan International Cooperation Agency, Tokyo, Japan.

Nataredja, Y. & Hodges, R.J. (1989) Commercial experience of sealed storage of stacks in Indonesia. In: *Fumigation and Controlled Atmosphere Storage of Grain*. ACIAR Proceedings No 25, Canberra ACT, Australia.

Nishioka, M. (2000) *The World Supply and Demand for Rice and Rice in Japan*. Textbook No. 1 of Post-harvest Rice Processing, Tsukuba International Centre, Japan International Cooperation Agency, Tokyo, Japan.

Ren Yonglin & van S. Graver, J. (1996) Grain drying in China: problems and priorities. In: B.R. Champ, E. Highley & G.I. Johnson (eds) *Grain Drying in Asia*. Proceedings of an International Conference held at the FAO regional Office for Asia and Pacific, Bangkok, Thailand. ACIAR Proceedings **71**. Australian Centre for International Agricultural Research, Canberra, Australia

Stathers, T.E. (2002) Pest management – inert dusts. In: P. Golob, G. Farrell & J.E. Orchard (Eds) *Crop Post-Harvest: Science and Technology. Volume 1 Principles and Practice.* Blackwell Publishing, Oxford, UK.

Tang Weimin & Hu Yushan (1997) Introduction to development and extension of technical research for grain storage in China in recent years. *Grain Storage*, **26**, 15–21.

Tang Zijun, Wang Minjie & Wu Suqiu (1999) Preliminary study on China's grain storage region according to its climate. In: J. Zuxun, L. Quan, L. Yongsheng, T. Xianchang & G. Lianghua (eds) *Proceedings of the 7th International*

Working Conference on Stored Product Protection, Beijing, China. **2**. Sichuan Publishing House of Science and Technology, Chengdu, Sichuan Province, P.R. China.

Tongming, H. (1998) Rice cultures south of Yangtze River and Japan: variations and background in harvest, drying and storage. *Agricultural Archaeology*, **1**, 335–343.

Wareing, P.W. (2002) Pest of durable crops – moulds. In: P. Golob, G. Farrell & J.E. Orchard (eds) *Crop Post-Harvest: Science and Technology. Volume 1 Principles and Practice.* Blackwell Publishing, Oxford, UK.

Wood J.F. (2002) Food processing and preservation – flour. In: P. Golob, G. Farrell & J.E. Orchard (eds) *Crop Post-Harvest: Science and Technology. Volume 1 Principles and Practice.* Blackwell Publishing, Oxford, UK.

Zhao Simeng (1996) *Grain Drying Technology.* Henan Provincial Science and Technology Press, Henan, P.R. China.

Chapter 2
Maize

P. Golob, N. Kutukwa, A. Devereau, R. E. Bartosik and J. C. Rodríguez

Maize (*Zea mais*) is the third most important grain worldwide (Plate 2), after wheat and rice. It is generally believed that the earliest form of maize arose from natural hybridisation of two or three grasses native to Mexico and Guatemala. Maize development is solely due to human cultivation as no wild maize plants have yet been found. Examples of the earliest forms of maize are primitive and small-grained, and it has been suggested that this important staple has been cultivated for 6000 years, maybe even longer. Portuguese traders brought maize to Africa; the earliest written reference to the plant in West Africa dates from 1502.

This chapter considers maize storage and marketing in three contrasting situations – Tanzania (which exemplifies East and Central Africa), Zimbabwe and Argentina. In Tanzania, subsistence agriculture is predominant with over 60–80% of grain remaining on-farm for home consumption and small-scale local sales. In Zimbabwe, a commercially orientated state marketing board offers price support and food security by purchasing from large-scale farmers as well as from smallholders who use storage systems similar to those in Tanzania. In Argentina, a commercial maize marketing system competes on the world market without the subsidies commonly granted to other producers in North America and western Europe.

The maize post-harvest system of Tanzania is similar in many respects to that of several other Anglophone countries in sub-Saharan Africa including Malawi, Zambia, Uganda and Kenya. There are some differences in post-harvest handling between countries and also between areas in these countries; such differences reflect mostly ethnic tradition and are not necessarily a result of economic or technical influences. The economic development of these countries has converged during the last ten years towards a system of free enterprise, with gov-

ernment controls being reduced continually. The current situation in Tanzania is that farmers sell their crop to traders and millers. In addition, a parastatal National Milling Corporation purchases grain from traders for milling and operates in competition with private millers. National food security is assured by a Strategic Grain Reserve of typically about 50 000 tonnes of maize in bag stores spread around the seven regions of Tanzania and operated by government.

Most maize producers in Tanzania and other countries in the region cultivate local, unimproved maize varieties; however, in regions producing surplus, farmers cultivate and sell some hybrid and composite maize. Liberalisation of agricultural markets has resulted in higher production and greater opportunities for farmers to participate in markets. The maize may be stored on-farm, as cobs (Plate 2) or as shelled grain, in a variety of local granaries. In the past, insect infestation of stored maize has not been a serious cause of losses but the potential for such losses has increased in recent years for two reasons. First, due to liberalisation larger stocks are now stored for longer periods. Second, a very damaging pest of stored maize, the larger grain borer *Prostephanus truncatus* (Vol. 1 p. 97, Hodges 2002), was introduced into Africa from central America, which resulted in much greater losses when maize was stored by farmers for longer than three months. As a result of this farmers have been receptive to the adoption of new storage methods such as the shelling of cob maize and admixture of insecticides. This has increased the use of storage structures suitable for shelled grain, particularly sack storage.

In Zimbabwe, most maize marketing is controlled by the Grain Marketing Board (GMB). This is a parastatal organisation under the Ministry of Agriculture. It runs a network of 68 depots countrywide, of which 14 store bulk maize in silos and the remaining 54 depots handle only bagged grain. GMB also occasionally operates collection points for maize and other crops. The Board sets a support price for maize purchases, on which private grain traders base their prices. It also stabilises the selling price of maize and its by-products by releasing competitively priced maize to the market during deficit periods.

At the bag depots, including collection points, storage facilities are open-air cover and plinth (CAP). Bulk and bag depots together store about 5.14m t maize, of this 0.9m t is bulk and 4.2m t in bag. Of the grain in bag, about 0.3m t is stored in sheds, the remainder is CAP. The depot network is served by either laterite or tarred roads and many of those on tar roads are also served by rail. The depots and collection points receive maize directly from producers registered with the GMB. Each depot, depending on availability of stocks, can supply grain for export. GMB, like other maize marketing agencies in other countries, suffered considerable losses due to maize discolouration called stackburn. The prevalence of this problem increased significantly with the introduction of woven polypropylene (WPP) bags. The incidence of stackburn has been much reduced by the adoption of a number of simple technical strategies.

The GMB holds Zimbabwe's strategic grain reserve (SGR). In 2001, this amounted to 390 000 t but declined to 150 000 t in 2002 because of political unrest. Part of this reserve is in physical stocks while the remainder is in the form of cash. If required, the latter will facilitate the financing of maize imports. The Board also has the sole mandate to import and export maize and to this end it issues permits to private importers and exporters. Although GMB is responsible for the SGR it also trades grain commercially and provides storage facilities for third parties, particularly millers, brewers, stockfeeders and commodity brokers. It thus generates funds and remits dividends to the Government of Zimbabwe, its sole shareholder.

Maize, soyabeans and wheat are the main crops of Argentina and storage facilities, handling and transportation equipment are common to these commodities. About 17m t of maize are produced annually of which about 7m t are exported. The storage capacity of Argentina has been increasing in line with increases in grain production. Private and co-operative elevators are the backbone of the modern Argentine grain handling and storage system with a wide range of storage capacities from 15 000 to 100 000 t. Relatively few farms have their own storage facilities; those that do usually have a number of corrugated steel bins of 100–400 t capacity. Temporary storage capacity is an important component in the system and a successful recent development has been the adoption of silobags that offer hermetic storage.

The biggest challenge for the future is for Argentina to take advantage of its grain production capacity by modernising its grain processing industry further and by improving its post-harvest grain handling systems with respect to storage capacity and capability of segregating grain of different quality standards. This will increase the efficiency and use of resources, and help Argentine farmers and grain handlers to become more competitive on the world market.

Tanzania (East and Central Africa)

Historical perspectives

Maize is the main food staple throughout Tanzania except in the dry central and southeast zones of Dodoma, Singida, Lindi and Mtwara Regions where sorghum, millet and cassava are the main food crops, and along the coast where the humid climate is suitable only for rice. Even in relatively dry localities where the maize crop fails regularly, such as Igunga district in Tabora Region, it remains the preferred staple. In the southern highland zone of Iringa, Mbeya, Rukwa and Sumbawanga and in the northeast, in Arusha Region, maize is produced in surplus and is also a cash crop. In other regions, including Kigoma, Tabora and Shinyanga in the west and Morogoro and Kilimanjaro in the east, maize is grown primarily for home consumption but may also be used as a cash crop in years of surplus.

Most maize producers in Tanzania and other countries in the region (with the particular exception of Zimbabwe) cultivate local, unimproved maize varieties. These varieties produce characteristically small, low-yielding cobs that bear irregular rows of white, hard, flinty maize kernels. Cobs have a thick, tight husk that provides an excellent physical barrier against penetration by insect pests. The kernels themselves are also resistant to insect attack and are easy to mill by hand. In the maize surplus regions, producers cultivate hybrid maize and some composites for sale. These are much higher yielding than local varieties but they are very much more susceptible to damage by pests both in the field and in store. Furthermore, the soft floury nature of the kernels makes them awkward to mill. Thus these improved varieties are grown solely for sale and even farmers growing hybrids as a cash crop still cultivate local maize to store on their farms for family consumption.

Much of East, Central and Southern Africa experiences a unimodal rainfall pattern that results in a single annual cropping season. One crop of maize is planted in each year, except in some highland areas where bimodal rainfall allows two crops. Most families in the region rely on their own production to supply their food needs, i.e. the majority of farmers are at or near subsistence, growing only small quantities for sale, although production has risen as a result of the liberalisation of agricultural markets. Maize must therefore be stored throughout the year at the homestead. It has been estimated that at least 60–80% of all maize produced remains in the village, primarily with the farmer (Temu 1977; Compton *et al.* 1993; Golob *et al.* 1999). The remainder passes through marketing chains and ends up with consumers in Dar es Salaam, a major maize-deficit area, in regional and district towns, or even with local consumers who require additional maize to meet shortfalls.

The control of maize marketing goes back to the World War II era when the British colonial government, as part of the war effort, introduced bulk purchase schemes to secure cheap supplies of food. A statutory board was established with monopoly powers of buying and selling. After the war, control of maize was maintained in order to guarantee a price for large-scale commercial farmers, but which would also ensure sufficient food for the local population. In Tanzania, this system was perpetuated by the formation of the Grain Storage Department (GSD) in 1949. The GSD was the sole buyer of maize, rice, millet, sorghum and beans, with responsibilities for storage, distribution, imports and exports (Temu 1977).

The GSD had a working stock of 15 000 t of maize (three months' requirements) distributed over the country, with a maximum carry over reserve of 10 000 t. As will be described later, the primary function of the GSD was little different to that of the current Strategic Grain Reserve. Each year before planting, the GSD fixed the maize price. However, following two successive bumper crops, probably resulting from the incentive of higher producer prices, the GSD was closed and in the seven years before independence there was free trade in maize and other cereals.

After independence, agricultural marketing in Tanzania, like most other facets of social and political life, came under government control. In 1963, the National Agricultural Products Board (NAPB) was formed which had the following functions (Temu 1977):

- to guarantee an incentive price to farmers;
- to ensure an efficient marketing and distribution system;
- to prevent exploitation by private traders;
- to maintain a famine reserve; and
- to contribute to agricultural or general economic development.

There was, at this stage, continued emphasis on large-scale estate farming to produce export crops and so earn foreign exchange. NAPB's customers were not individual producers but rather co-operative unions that in turn were made up of a number of primary societies whose members were producers. The marketing chain (Fig. 2.1) remained essentially the same for the following three decades, although the functions of the different players changed.

The Arusha Declaration in 1967 marked the beginning of the period during which all government policies were orientated towards the national objectives of self-reliance, which was primarily achieved first by '*ujamaa*' (compulsory collectivisation) and then by 'villagisation'. Government intentions were directed toward assisting the small-scale subsistence farmer who comprised the majority of the population. This period emphasised the overarching role of the government, which assigned to itself the roles of agricultural manager, entrepreneur and investor (World Bank 2000).

During the next 20 years most of the large estates producing for the export market were nationalised and many new government-owned farming enterprises were started. Most direct taxation on small farmers was abolished though a produce levy was still paid to marketing boards. Inputs, especially fertiliser, were heavily subsidised and pan-territorial and pan-seasonal pricing was introduced for maize and other commodities, through a single market channel. After 1967, co-operative unions were seen as a vehicle for socialist policies and every individual had to become a member of a primary co-operative society. This did not conform to the principles of international co-operativism whose essence is voluntary membership. Farmers did not feel they had a stake in the organisations and had no sense of ownership.

The unions were dissolved by the government in 1976 and replaced with crop marketing parastatals. The National Milling Corporation (NMC) was given the legal monopoly to procure, process and distribute eight major food staples, including maize. NMC also regulated all movement of maize within the country, including that destined for export. The uniformity of the pricing structure, being the same throughout the season and throughout the country, led at times to failure to meet consumer demand, failure to purchase maize and other crops offered for sale, and failure to pay farmers. NMC,

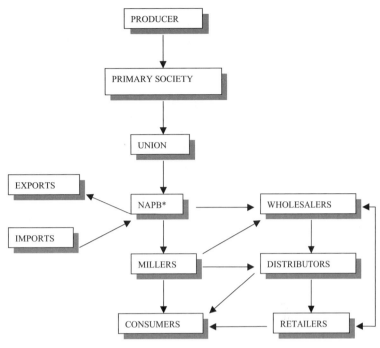

Fig. 2.1 Marketing chain in Tanzania after independence.

* National Agricultural Products Board

which had poor operating and management practices, ran at a substantial loss and was heavily subsidised by the government. The country was unable to supply its food needs and had to rely on imports and food aid to meet its shortfalls. Shortages were principally caused by the excessive cost of transportation from areas of plenty to those facing deficit and were not due to an absolute deficit in the country as a whole.

Co-operatives were reintroduced in 1982 and became agents of NMC for input supply and maize marketing. However, the situation did not improve, because of mismanagement and misappropriation, combined with a lack of democracy throughout the co-operative system.

In response to the problems with NMC and the unions, the government gradually deregulated markets and prices. To begin with, control on grain movement was lifted on quantities from 30 kg in 1984 to 500 kg in 1996, and in 1997 all restrictions on movement were lifted (World Bank 2000). In 1989, the single marketing channel through the NMC was abolished and replaced by free enterprise. Finally, in 1997 consumer subsidies were abolished. NMC now only purchases grain from the private sector and only for milling; it operates in competition with private millers.

Initially, agricultural market liberalisation caused confusion among producers and traders and benefited only a small proportion of the population (Tyler & Bennett 1993; Coulter 1994). Only farmers close to road and rail arteries feeding the main urban centres, especially Dar es Salaam, were able to sell their grain quickly and at a reasonable price (Coulter & Golob 1992). Traders were quite willing to transport maize long distances to the coast, but only from those producers to whom they had easy access. Farmers in major maize-producing regions of Ruvuma and Rukwa, which are at least 1000 km from the coast, were at a serious disadvantage due to high transportation costs. Without ready access to markets, maize production in some areas declined, exacerbated by the lack of inputs or their high cost (World Bank 2000). In recent years, maize has been transported from surplus districts to neighbouring deficit districts and central government has encouraged farmers in the remoter areas of Rukwa and Mbeya to export maize to Malawi, Zambia and the Democratic Republic of the Congo (SADC 2002).

Liberalisation of grain markets has increased the importance of on-farm storage. For this reason the arrival in western Tanzania in the late 1970s of a maize storage pest from central America, the larger grain borer

(*Prostephanus truncatus*), is of special importance. This pest is particularly damaging to stored maize cobs and losses in store over a season can be twice those expected from the usual pests such as weevils (*Sitophilus* spp) (Dick 1988). Not only did the larger grain borer threaten farmers directly but its presence also threatened regional maize trade. To restore trading confidence in the region an FAO-managed regional project harmonised phyto-sanitary procedures within Southern Africa, facilitated by the development of a manual (*The Phytoguide*) set-ting out essential phytosanitary procedures to prevent the spread of the pest (Semple & Kirenga 1994).

Physical facilities

Harvesting, drying and shelling

Traditionally, maize was stored at home on the cob and in most places with unimodal rainfall systems almost com-plete drying of cobs occurred in the field. In some areas, such as central and southern Malawi, maize plants that reached maturity were cut and the entire plants heaped into stooks (Fig. 2.2) or bundles in the field to dry. Stook-ing, which is done just as the rains cease, also allows the moist soil to be turned early, in preparation for the next growing season, before it becomes too dry and hard.

Many farmers simply leave the plants standing in the field until sufficiently dry for the cobs to be removed and carried home to the store, although the cobs may be bent over when the plants have reached maturity in the field to prevent further water migration into the kernels. Alter-natively, maize plants may be cut and the cobs placed on raised platforms, either in the field or at the homestead to assist drying. This aeration may be supplemented by heat from the kitchen fire. In the case of the *dari* the drying platform is located beneath the roof eaves of a house and above the kitchen; in the *chanja* system the raised dry-ing platform is located outside the house (Fig. 2.3). Both these structures are found in Tabora Region, Tanzania. It is also common in western Tanzania to tie pairs of cobs together by the outer husk leaves and then to place the pair across a vertical drying frame, the *nkuli*, which may be 2–6 m high (Fig. 2.4).

Fig. 2.3 Drying platform, *chanja* (Malawi).

Fig. 2.2 Building a maize stook in the field (Tanzania).

Fig. 2.4 Vertical drying frame, *nkuli* (western Tanzania).

Harvesting and transportation is undertaken using manual labour; only on a very few large commercial farms and estates is harvesting and drying done mechanically. Labour is drawn from family members and may be supplemented by hire. However, as harvesting operations occur at the same time on all farms in a locality, casual labour may be in short supply and it may be necessary for families to help each other.

This diversity in dealing with the mature maize crop is generally illustrative of the post-harvest system in this part of Africa. There was a tendency in the past to ascribe specific practices to the population of a region. For example, in Malawi all farmers were said to store maize on the cob with husk intact. This may have been more or less correct in the past but now the storage system and farming as a whole have become much more diverse so that it is dangerous to make this type of generalisation. Nevertheless, traditional practices are still commonplace and many farmers continue to use methods that have changed little over many generations.

After drying, maize is usually stored on the cob with the husk intact, although cobs may be dehusked to aid further drying. Increasingly, farmers shell maize before storage. Husks are stripped from the cob manually, usually by women and children but sometimes also by men (Marsland & Golob 1999). Heaped cobs may be sorted and those that are small or showing signs of damage are set aside for immediate consumption. The remainder is shelled usually by beating the cobs with clubs, sometimes with the cobs in a sack. This may be done on the bare earth or on a drying platform. Small hand-held maize shellers may be used but shelling by this method is tedious and relatively uncommon (Fig. 2.5). Many

women use their hands alone to shell the grain, rolling the cobs under their thumbs. With family members involved, one tonne of grain can be shelled in a day. In areas where maize is produced in surplus, mechanical shellers are used, sometimes tractor mounted, or the tractor may simply be driven over the cobs to shell the grain. Most shelling methods result in significant damage to the grain. Although this does not usually affect market prices, as there are no price differentials for quality, a high proportion of broken grain makes the crop more difficult to sell. Storage of damaged grain also encourages insect and fungal attack.

On-farm stores

Traditional methods of storage still predominate in the region (Vol. 1 p. 192, Brice 2002). They have been developed to suit the needs of a simple, subsistence farming system. However, as production systems become modernised these storage methods may not be able to cope with increases in production, especially if governments continue to encourage farmers to store on-farm.

The principal store used in East and Central Africa is a cylindrical basket (Fig. 2.6). This may be constructed from a variety of different materials. These include

Fig. 2.5 Maize cobs being shelled using one of several different designs of hand sheller.

Fig. 2.6 Cylindrical storage basket, *kihenge*, made from twigs (Tanzania).

woven, split bamboo (e.g. the *nkokwe* in Malawi), interwoven twigs from a number of different tree species (e.g. the *kihenge* in Tanzania), reeds, sorghum or elephant grass and rope made from twisted grass (the *ntanta* in Rukwa Region, Tanzania, or the *chigwa* in the Lower Shire Valley, Malawi). Baskets may be plastered on their internal or external walls or both with a mixture of cow dung, termite mound soil, ash or earth. They are often raised at least 30 cm above the ground on a wooden platform by supporting poles or rocks. In general, if the basket is located outside a house then it is covered by a thatched roof. Increasingly, for security reasons and where farmers have large properties, the basket is put inside the house but without its roof. This structure is used for the storage of both shelled and unshelled maize as well as other cereals and pulses. When well designed, built and maintained, it makes a good store capable of preserving grain quality, and should not require replacement for at least five years. However, in some areas building materials, especially twigs and bamboo, are becoming increasingly scarce so farmers are resorting to alternative store types.

One such alternative is the *kilindo*, a small cylindrical structure having the appearance of a drum without a top (Fig. 2.7). It is made during the rainy season from bark cut from the trees of the *miombo* forests. After removing the bark, its innermost layer is peeled away and then folded and stitched to form a wide tube with a diameter of 1 m or more. Another piece of bark is attached to one end of the tube to seal it and create a storage container.

After this operation, the structure is left to dry and then supported in the same way as the typical cylindrical basket, either within or outside the house. The *kilindo* is found in forested areas, especially in Tabora Region and parts of Sukumaland where it is used for storage of a range of food grains including maize and other grain crops.

Dual-purpose drying and storage structures are also found in some areas. These cribs (*nchete* in Malawi) are rectangular in shape, being constructed with open or open-weave sides to allow extensive air flow. They may be raised up to 2 m off the ground to allow cooking beneath or, if very large, they may be located directly on the ground (Fig. 2.8). Large bamboo or sisal poles are often used in their construction.

Other large storage bins may be made of mud blocks. They are square or rectangular in cross-section; in southern Zambia they are often cylindrical and may be raised above ground on a wooden platform or stones. In Tanzania, farmers who do not possess the knowledge, skills or tradition to make a *kihenge* often use these mud bins. A smaller, pot-like structure, the *konti*, made of a mixture of moulded cow dung and ash, which can accommodate four bags of maize, is commonly used in Babati and Mbulu districts of Arusha Region in Tanzania.

Smaller containers, gourds, earthenware pots and baskets made of interwoven palm fronds or other soft grass leaves are widely used throughout the region for storage of small quantities of grain, particularly seeds, flour or any semi-processed food kept for immediate consumption. Metal containers, from 200-litre capacity oil drums to large silos, are found in households located near the towns or shopping centres.

Fig. 2.7 Bark storage container, *kilindo* (western Tanzania).

Fig. 2.8 Drying and storage crib, *nchete*, with legs fitted with rat-guards (Malawi).

Table 2.1 Frequency of use of storage structures in 1994 and, in parentheses, 1984. The data are percentage of farmers interviewed (n = 390: 47–79 per region). Source: Golob *et al.* (1999).

| Region | Current use | | | |
	Sacks	*Kihenge*	Platforms outside	Platforms/Lofts inside
Arusha	18 (25)	20 (26)	40 (25)	0 (0)
Iringa	47 (8)	12 (12)	4 (0)	0 (0)
Kilimanjaro	25 (4)	0 (0)	32 (20)	11 (46)
Morogoro	26 (6)	29 (41)	8 (1)	0 (14)
Rukwa	35 (3)	4 (23)	2 (0)	5 (20)
Tabora	13 (4)	7 (14)	5 (12)	41 (31)

In this example there has been a substantial change to sack storage to protect grain against *Prostephanus truncatus.*

Increasingly, as farmers sell grain to markets and traders, they are turning to jute, hessian and woven polypropylene sacks to store maize (Table 2.1). Sacks are becoming more popular for long-term storage inside the house as producers become more conscious of the need to prevent theft of their food stocks. Storage in sacks requires the harvest to be shelled as soon as possible and that the grain be sufficiently dry. Secondary drying can be achieved by exposing the grain in a thin layer to the sun on a baked earth or concrete drying floor. Although grain stored in sacks is particularly susceptible to rodent damage, the convenience of handling outweighs the drawbacks. Sacks made from man-made fibres are cheaper than those of jute or hessian, which have to be imported, and these are becoming increasingly popular especially among urban and peri-urban populations.

Traditional storage provides adequate protection to grain against the destructive effects of rain, ground moisture and pest attack. However, as new maize varieties have been introduced and larger quantities cultivated, traditional methods are proving less adequate. Improvements to store design have been attempted either by modifying existing designs or by introducing new types based on materials such as cement and galvanised iron.

Storage objectives in relation to income generation and marketing

Agricultural market liberalisation, which began in Tanzania in 1989, provided incentives for farmers to produce more for sale. Prices normally rise some months after harvest and farmers get the best prices if they can store their maize for nine months or more, provided there are no significant losses due to pest damage.

Storage duration

In the western plateau and the southern highlands where rainfall is unimodal, farmers must retain grain in store for ten months or more to meet the needs of their own subsistence. In the bimodal regions of Arusha and Kilimanjaro, storage can be for much shorter periods but stocks are often carried over between seasons so prolonging the storage period. In a normal year, two-thirds of farmers store for longer than six months, whereas in drought years the same proportion have exhausted their stock within six months (Fig. 2.9).

In most regions, production levels are the main determinants of storage duration. Families do not tend to eke out small harvests; rates of food consumption remain similar in normal and in drought years. A relatively large proportion of the crop is sold immediately after harvest so that debts and school fees can be paid. The only other factor having a significant effect on storage duration is the presence of *P. truncatus*. In Tabora and Kilimanjaro

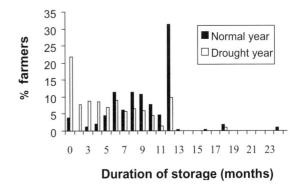

Fig. 2.9 Average duration of storage of maize in normal and drought years during the 1990s (Golob *et al.* 1999).

Regions, in particular, farmers attempt to avoid the effects of this pest by reducing the storage period and thus curtailing exposure to insects.

Production of hybrids

Raising production most easily increases income and this can be achieved by cultivating hybrids. However, hybrid maize tends to be offered for sale as soon as possible after harvest owing to its high susceptibility to insect damage during storage. There is evidence that some farmers store hybrids through the year and this practice is increasing as hybrid flint varieties, with hard grain kernels, increase in popularity and availability. In drought years, when prices are relatively high at harvest and infestation pressure is likely to become severe as storage progresses, sales of hybrid maize immediately after harvest increase, most being sold within the first few months.

Hybrid production depends on having good access to agricultural inputs and to markets. Deregulation has made inputs more costly and more difficult to access so that many farmers, particularly the poorest, are unable to grow hybrids and their production has declined (Morris *et al.* 2001). Furthermore, many farmers live in remote communities and cannot readily sell their produce. These farmers still grow only local varieties which they retain on-farm for food, occasionally selling small quantities when they have an excess, to raise cash or for gifts or to make beer. They derive most of their income by selling their own labour.

Improved store capacity and store design

If there are to be increases in production and more prolonged storage then increases in storage capacity will be required. Current storage methods could absorb a doubling in production; for example, in a study in Mbeya Region of Tanzania, only one person of a group of seven in Nuduli village thought he would be unable to store double the quantity he was currently harvesting, and everyone in a group of 25 in Tawi Kilolo village thought they would be able to cope (Coulter & Golob 1992). Production increases beyond this may require additional facilities if the grain is to remain on the farm. Extra storage could be obtained either by constructing more traditional stores, a practice common in several regions, or by increasing the use of sacks. However, building more traditional storage structures may not be the most efficient way to increase capacity on the farm.

With this in mind, many organisations have attempted to introduce new designs or improve the performance of traditional structures by modifying them; however, this has been largely unsuccessful.

In Zambia, much effort has been put into developing a concrete *kihenge*-type store, the *ferrumbu* (Fig. 2.10). Designed in the mid-1970s using cement and chicken wire, this store has undergone modification to reduce costs: chicken-wire support was replaced with a wooden framework and then the quantity of cement was reduced by replacing some of it with soil. Despite intensive efforts by the local extension services over the years there has been very limited uptake of the *ferrumbu*. Similar structures together with drying floors have been introduced into southern Tanzania but with equally limited uptake.

In East and Central Africa, very little success has been achieved in improving the main farm storage structure. However, new practices adopted by producers living in peri-urban areas, where space is at a premium, may eventually prevail. They use smaller, more convenient stores such as sacks, old oil drums and, increasingly, high-density polyethylene containers or tanks, which are manufactured for water storage. These tanks are extremely robust and durable and capable of holding up to 4 t of maize, though smaller tanks are the norm. There is little doubt they will find similar use in rural areas in due course.

Perhaps the most commonly recommended improvement is to install metal rat-guards onto the support poles of the structure (Fig. 2.8). This innovation requires other changes to be made, including the raising of the store support platform, and therefore using longer support

Fig. 2.10 Ferrocement storage bin, *ferrumbu* (Zambia).

poles; fewer but much stouter poles that can support a greater weight are required. The need for these changes, together with the fact that metal to make the guards is often expensive or unobtainable, has resulted in very limited adoption of this recommendation. Furthermore, it is crucial that the guards are fitted correctly, otherwise rodents are able to circumvent them; guards made of grass and mud mixtures have proven ineffective.

Other modifications, such as the fitting of a hatch at the base of the basket (Fig. 2.11) for easy unloading rather than having to access the maize by removing the thatched roof, strengthening the base with cement, or plastering the walls with a mud mix, have all been adopted by farmers to a limited extent. Success has been achieved when the particular recommendation has been well targeted at communities where there is prior knowledge of the practice.

Major sources of quality decline

Deterioration after harvest

Once the maize crop has matured in the field it is subject to deterioration by a variety of different factors. The crop remaining in the field can be lost through damage by mammals, including goats, cattle, baboons and even elephants if the farms are in remote locations. A study in Malawi showed that, at this stage, rodents are thought to be responsible for the greatest losses, although insect pests and theft are causes of concern (Marsland & Golob 1999). Theft is particularly prevalent in marginal maize growing areas and is one of the main factors for the switch

from storing in baskets located outside the house to using sacks kept in the home (baskets may be moved inside the home when there is sufficient room). If the rainy season is prolonged, as occurs regularly in the southern Malawi highlands, then attempts to dry in the field may result in mould development and mycotoxin formation. Mould and mycotoxin problems have become acute in Zambia in several years in the past and have caused the parastatal buying agents at the time, Namboard, to introduce additional quality grades in order to avoid wholesale rejection of the national crop; the low-graded material was bought for animal feed.

Damage that occurs once the crop is mature may be manifest when the crop is placed in store. Growth of storage fungi and insect damage may well start before the maize is removed from the field. However, as storage begins when the climate is dry, fungal development stops and may not be a problem. Insect populations may increase during storage albeit only slowly during the five months or so of dry weather that occurs when storage begins.

The most important storage insect pest of maize on farms in Tanzania is the larger grain borer (*P. truncatus*) (see Box 2.1) or in its absence the maize weevil (*Sitophilus zeamais*) and angoumois grain moth (*Sitotroga cerealella*) (e.g. Giles & Ashman 1971; Hindmarsh & MacDonald 1980; Golob 1984). These insects attack cobs before harvest. Other insects, such as the beetles *Rhyzopertha dominica*, *Tribolium castaneum*, *Carpophilus* spp and *Gnatocerus* spp, and the moth *Plodia interpunctella* may also be found on farms but are of relatively minor importance.

Fig. 2.11 Improved maize store with a bottom hatch for easy maize discharge (Tanzania).

Box 2.1 The larger grain borer in Africa

The larger grain borer (*Prostephanus truncatus*) is a very serious pest of farm stored maize and dried cassava (see Vol. 1 p. 97, Hodges 2002). It was a neotropical species that first arrived in East Africa in the late 1970s and has since spread widely in other parts of the continent (Farrell 2000). Its preference for developing and breeding on cob maize makes it well suited to exploit the traditional system of storage, and after its arrival storage losses were estimated to have double from about 5% to 10% in those places where the pest is not well controlled. Unusually for a storage pest, *P. truncatus* appears to have the bulk of its population widely but thinly distributed in alternative hosts, including dry twigs and branches of many tree species. It is also capable of attacking and boring through the timber of stores and houses, wooden utensils, leather, soap and other food commodities. *P. truncatus* forms massive populations on maize and cassava rapidly, presumably because these offer a concentrated source of starch, unlike the beetle's other hosts where populations density remains relatively low.

Mature maize cobs in the field may be infested by adult beetles flying in search of new hosts, although such attack often starts after the maize or cassava has been placed in store. It is known that many typical storage pests are strongly attracted from considerable distances to the odours of stored products; however, this is not the case in *P. truncatus*, which is not attracted to stored food at long range and possibly not even at very short range. Primary host selection is a chance event and made, mostly or entirely, by males boring test burrows into anything sufficiently soft. When a male has found a suitable food source it releases a chemical attractant (pheromone), which leads to secondary host selection by the females and other males so attracted. Since primary host selection is a matter of chance, it is strongly affected by the number of beetles flying around looking for new hosts. The higher the number of such beetles, the greater the risk to farmers. If the number of flying beetles can be determined, either directly by trapping or indirectly using models that predict the abundance of *P. truncatus* based on climatic factors (Hodges *et al.* 2003), then the risk to farmers can be predicted (Birkinshaw *et al.* 2002). Farmers can also be encouraged to sample their own stores to detect the pest and take suitable action against it (Meikle *et al.* 2000).

P. truncatus is the only storage pest for which classical biological control has been advocated. A predatory beetle *Teretrius nigrescens*, from the pest's geographical range in Central America, has been released in both East and West Africa for this purpose. Details of this are presented in a case study in Vol. 1 p. 308 Box 6.2.

Off the farm, in central storage facilities, neither *P. truncatus* nor *S. cerealella* are serious pests, the former because it is relatively easily controlled by fumigation and the latter because it does not do well in large grain stacks as the adults are unable to penetrate beyond a depth of a few centimetres. *Sitophilus zeamais* and the moths *P. interpunctella*, *Corcyra cephalonica* and *Cadra cautella* predominate in bag stacks.

Post-harvest losses

Losses may occur at all stages in the post-harvest system – during transport to the home, in storage and at processing – but it is very difficult to assess losses which occur at each of these stages (Vol. 1 p. 141, Boxall 2002). Most estimates have focused on measuring weight loss during storage. In Malawi, losses in stored local maize were found to be 3% or less for maize stored for up to ten months in the Lilongwe plain and Lower Shire Highlands (Golob 1981a, b) and similar losses have been recorded in Zambia (Adams & Harman 1977) and Kenya (De Lima 1979).

Although low levels of loss were reported in these earlier studies, it was widely felt that the spread of high-yielding varieties could increase storage losses significantly. Evidence on this point is mixed. The results of a survey in Karonga and Chitipa Districts in northern Malawi indicate 'very low weight loss figures … [which] … reconfirm the findings from former surveys in the country' (GTZ 1994). The mean weight loss after nine months was 5.1%; insects were the main cause of loss, followed by moulds. Losses caused by rodents were insignificant. A key factor behind these figures was that the hybrids were either sold or consumed first, before major losses could occur. In contrast, a recent nationwide survey estimated an average maize loss over a six-month period of 17.7% (CODA 1996). The main reason for the discrepancy between the GTZ and CODA figures was that in the latter study the average loss total was raised significantly by a very high loss figure for hybrid maize (62.2%) while losses in local varieties were estimated to be 5.2%. Within Malawi, the CODA results have been questioned, as the methodology used was based on a single spot estimate from farmers and it does not appear to take account of the fact that, as noted by the GTZ study, many farmers sell or consume hybrids quickly.

Estimates of on-farm weight loss of unprotected maize are surprisingly scarce for Tanzania. Generally, those estimates that have been made overstate the problem because they do not take into account the effect of withdrawals from granaries during the season (e.g. Keil 1988; Henckes 1994). The real losses to unprotected maize during the course of a storage season lie somewhere between 10% and 30% by weight, primarily the result of *P. truncatus* infestation, significantly more than the 2–3% which is normally lost as a result of attack by indigenous insect pests (Tyler & Boxall 1984). These estimates were derived from storing open-pollinated, flint varieties. For many years, farmers have refused to store hybrid dent-varieties because these are extremely susceptible to insect damage. Such varieties have been shown to lose in excess of 25% by weight in less than six months in store when unprotected against insect attack. Even when treated with insecticide, damage can still be considerable, especially if cobs are stored (Golob 1984). The advent of hybrid, flint varieties may enable farmers to store high-yielding varieties for longer without significant losses.

Losses during processing have received some attention. It is reported that around 40% of the weight of the original maize grain is lost during local processing into refined white flour. When corrected for crude extraction rate, the processed grain had lost 45% of crude protein and 55% of energy.

Commodity and pest management

In general, farmers take little or no management action during the time they store their maize. Housekeeping around the store is minimal, though some farmers may clean the interior of the basket between storage seasons. The fabric of the basket may be repaired before the storage season begins but this will depend on the inclination of the farmer. As a rule, local maize remains in sufficiently good condition during storage that there is no incentive to introduce store improvements. Neither rodents nor fungi present a significant risk. Only *P. truncatus* presents a threat that the farmer is prepared to take action against. However, in many places the severity of attack by this pest varies greatly from year to year (Hodges *et al.* 2003) so there may be little incentive for farmers to invest in a permanent change in practice. Very few farmers employ prophylactic treatments so action is only taken once the pest is seen on the maize. As a result, sacks are left untreated, allowing carryover pest populations to be transferred from one locality to another through trade. This has been the major reason for the spread of the pest both within Tanzania and throughout neighbouring countries.

Pest management for farm storage

Farmers storing flint varieties where *P. truncatus* is absent do not need to apply pest control measures unless storage lasts for longer than five months or at least until just before the rainy season begins. Even then, with only small quantities of grain remaining after six or more months of storage, producers often find it is not worthwhile to protect their grain (Golob 1984). Furthermore, insecticide treatment may not always offer worthwhile protection; in Malawi, although weight losses were lower for maize dusted with Actellic 2% than for untreated maize, the difference was not statistically significant (GTZ 1994). This might be explained by inappropriate application technique, poor quality of insecticide, or that insect damage to non-hybrids was too low to be influenced by chemical application.

Nevertheless, tradition dictates that farmers use certain measures to protect their produce whether storage is extended or not (Golob & Webley 1980). In this region of Africa, it is very common to find wood ash, produced in the kitchen fire, being admixed as a storage protectant against insect infestation. Cobs may be subjected to smoke and heat from the kitchen fire or, when outside the house, from a fire lit underneath the maize platform to facilitate cob drying. The fire raises the temperature of the cobs, often scorching the outer husk leaves, and helps to disinfest the maize of storage pests. This method was particularly prevalent in areas where cobs were stored under the roof eaves before *P. truncatus* became established. *Sitophilus* species are said to be particularly sensitive to this treatment but it does not have any noticeable effect against *P. truncatus* under normal circumstances (Hodges *et al.* 1983).

The use of plants as grain protectants, whether dried or aqueous extracts, so prevalent in the Indian subcontinent or in West Africa, is much less popular in East and Central Africa (Vol. 1 p. 280, Belmain 2002). Some farmers in Kenya and Uganda use whole plants of the Mexican marigold (*Tagetes minuta*) to protect maize in stores. In Malawi, crushed dried tobacco leaves are mixed with grain for storage but mostly for protecting beans (Taylor 1995). In southern Tanzania farmers claim to use many different plants, particularly Leguminosae, as grain protectants (Table 2.2) (Mkoga & Shetto 1999).

Effective control of *P. truncatus* can be achieved by the application of synthetic pyrethroid insecticides to maize (Vol. 1 p. 247, Birkinshaw 2002). In Tanzania, currently a single compound, permethrin, is approved for use for this purpose. As permethrin does not control

Plant name.	Family
Neuratanica mitis	Leguminosae
Dolichos kilimandscharicum	Leguminosae
Vernona advensis	Leguminosae
Vernona amygdalina	Leguminosae
Tephrosia vogelii (rotenone)	Leguminosae
Pilliostigima thoningii	Leguminosae
Swartzia madagasxariensis	Leguminosae
Azadirachta indica (neem)	Meliacea
Melia azadarach	Meliacea
Tagetes minuta	Compositae
Artemesia afra	Compositae
Chrysanthemum cinneraerifolium (pannumethrum)	Compositae
Clemaptosis scabiosifolia	Ranunculaceae
Zanha africana	Sapindaceae
Gnidia kraussiana	Thymelaceae

Table 2.2 Plants with insecticidal properties used by farmers in the southern highlands in Tanzania as store protectants. Source: Mkoga & Shetto (1999).

other insect pests associated with stored maize, especially *Sitophilus* spp, it is mixed with Actellic (pirimiphos-methyl) and applied as a cocktail. It is sold under the trade name Actellic Super in a dust formulation (ASD). Dusts are convenient for farmers to use as they are of low concentration and require no further dilution and no specialised equipment for application. The active ingredients in ASD are 1.6% pirimiphos-methyl and 0.3% permethrin, and the recommended application rate is 100 g per kg of grain. This gives a dose of 16.6 ppm pirimiphos-methyl and 3.3 ppm permethrin. Although this rate of application exceeds the Codex maximum residue limits (Golob 2002), in practice 50% of the dust is lost during application. In Kenya, Uganda, Malawi and Zambia the recommended application rate for ASD is 50 g per kg of grain (half the nominal dosage used in Tanzania).

ASD has been very effective in controlling *P. truncatus* and all other important storage insect pests. During an extension and control programme in western Tanzania from 1984 and 1987 to help farmers cope with *P. truncatus*, losses sustained by 105 farmers in three villages were assessed during a storage season. When food removals for home consumption were taken into account the real food loss over a period of 7–9 months was less than 2% (Golob 1991). These low levels were the result of farmers taking action to control the pest by applying insecticide immediately they saw the beetle in their maize, preventing the buildup of pest populations. ASD is also sold in areas that are free of *P. truncatus* as a general storage protectant for maize and other grains.

ASD has superseded the insecticides that rely on only a single active ingredient for the control of storage in-

sects. In Malawi, it is the only currently recommended grain protectant, having replaced pirimiphos-methyl (Actellic) which itself was a replacement for malathion (Chibwe & Kalinde 1999). However, in other countries several alternative chemicals can be purchased and used for this purpose. For example, in Kenya, several dusts containing either malathion or pirimiphos-methyl are on sale and another, containing the pyrethroid bifenthrin, is produced specifically for *P. truncatus* control; these insecticides are also available in Tanzania.

Although synthetic insecticides are available for protecting stored maize against insect infestation, not all farmers use them. There are various reasons for this. First, many farmers are afraid that they will contaminate and poison their food. Secondly, they may be too expensive for many ordinary farmers to buy; in 1988 the cost of treating a bag of maize grain was 2% of its value, ten years later it had risen to 6% (Golob *et al.* 1999). In rural areas more than 20 km from a town, storage insecticide has been difficult to obtain and when available has often been of dubious quality. Since market liberalisation, supply and distribution of inputs in Tanzania have been the responsibility of the private sector. To maximise returns on investment, insecticide importers and distributors have restricted their distribution networks to wholesalers and retailers in large towns, putting the onus on farming communities to obtain the insecticide themselves. Thus farmers may have to travel 50 km or more to purchase them. Some small-scale traders may visit local markets to sell ASD but in Tanzania unscrupulous traders have, in the recent past, been adulterating the dust and selling it in unmarked packets (Urono 1999). In a survey of the quality of ASD, five out of thirteen samples were below

Table 2.3 Analysis of samples of ASD (Actellic Super Dust) collected from retail outlets in Kilimanjaro Region and Babati district, Arusha Region in Tanzania. Source: Urono (1999).

Retailer identification	Active ingredient (%)		Comment
	Pirimiphos-methyl	Permethrin	
A (Ndep)	0.6	0.37	Below specification
B (TF)	1.44	0.35	Within spec.
C (Kib T.S.)	1.36	0.34	Within spec.
D (Kilimanjaro)	1.49	0.37	Within spec.
E (Kilimanjaro)	1.61	0.40	Within spec.
F (Kilimanjaro)	1.71	0.37	Within spec.
Village	0.2	0.03	Below spec.
S (Village)	1.4	0.10	Below spec.
W (Village)	Not detected	Not detected	Below spec.
T (From FT)	1.9	0.28	Within spec.
Unknown (Babati a)	1.6	0.42	Within spec
Unknown (Babati b)	1.54	0.21	Below spec.
Unknown (Babati c)	2.4	0.26	Within spec.

specification for active ingredient, which in one case was not even detectable (Table 2.3). Hence, on occasion farmers have been dissatisfied with the control they have achieved with ASD. It is planned to introduce tamper-proof packaging to overcome this problem.

However, some of the control failures are the responsibility of the farmers themselves. The relatively high price of ASD persuades some farmers to under-dose. During field visits, farmers often admitted applying one packet of 200 g of ASD to three or more 100 kg sacks of maize grain instead of to just two as is recommended (Golob *et al.* 1999). Farmers also apply the dust to maize cobs, which is likely to be much less effective than application to shelled grain. Poor application procedures will result in insect survival and might also lead to the development of insect resistance.

The value of pest control for the Tanzanian farmer

There are potentially considerable gains to be made by farmers treating their maize stock with insecticide. An example of such benefits is presented in Table 2.4, where a farmer who harvests twenty bags of maize wishes to sell ten bags and keep the remaining ten bags for family consumption. The benefits of treatment depend on whether local or hybrid varieties are stored and on the extent of the rise in maize price after harvest. The calculations in Table 2.4 clearly illustrate why farmers do not store hybrids for long periods and why it is so important that ASD should be of a high quality and applied correctly to maximise its effectiveness. By keeping the grain in good

condition for six months the farmer will have earned an extra Tsh23 700 (Tsh 117 700–94 000) or 25% with local grain and Tsh40 500 (Tsh 110 500–70 000) or 57% with hybrid maize at sale in December (Table 2.4).

The example does not consider the costs of losses from the grain stored for home consumption. If these bags are left untreated they are likely to be so severely damaged that the farmer's stock will be exhausted after nine months, that is three months before the arrival of the new harvest. It is likely that the farmers would have to then buy back three bags of grain at Tsh24 000 (probably at a higher price as buying prices are always greater than selling prices), a cost of Tsh72 000 unbudgeted for at the start of the storage season. If the maize kept for food had been treated it would have lost a total of 4% by weight, which the farmer could accommodate in his/her original plans and so would not have to buy.

Large-scale storage

Traders

After agricultural marketing was first liberalised in the mid-1980s, the only long-term storage along the marketing chain was at the farm. Once the crop was sold it moved through the system very rapidly and reached the consumer in the large urban centres in a matter of days. More recently, traders and millers have been taking advantage of seasonal variation in prices by storing grain. There are large numbers of stores, which were used by NMC, the Regional Trading Company (formerly government suppliers

Table 2.4 Case study on the financial benefits of treating maize with pesticide in Morogoro (based on 1998/99 prices in Tanzanian shillings).

	July Tsh/bag	December Tsh/bag
A farmer harvested 20 bags of maize in July. Prices of maize on the market double in the 6 months after harvest	11 000	24 000
Soon after harvest the farmer decides to keep 10 bags for own consumption and sell 10 bags to raise cash. What will be the best strategy to maximise income from the maize for sale?		
Strategy 1: No insecticide treatment and sell maize at time of harvest	Tsh raised	Tsh raised
Sell 10 bags with no losses (all varieties fetch same price at harvest)	110 000	
Strategy 2: No insecticide treatment, store for 6 months then sell		
Local flint variety losses in 6 months = 15%		
Hybrid dent variety losses in 6 months = 25%		
Sell 10 bags of local grain in Dec., with 15% loss ($\equiv$ 8.5 bags)		204 000
Sell 10 bags of hybrid grain in Dec., with 25% loss ($\equiv$ 7.5 bags)		180 000
Strategy 3: Treat grain with ASD*, store for 6 months then sell		
Local flint variety losses = 2% + treatment costs of Tsh 7500		
Hybrid dent variety losses = 5% + treatment costs of Tsh 7500		
Sell 10 bags local grain in Dec. with 2% loss ($\equiv$ 9.8 bags) less ASD cost		227 700
Sell 10 bags hybrid grain in Dec. with 5% loss ($\equiv$ 9.5 bags) less ASD cost		220 500
Price advantage of selling in December		
Untreated local grain	= 204 000 – 110 000	
	= Tsh 94 000	
ASD-treated local grain	= 227 700 – 110 000	
	= Tsh 117 700	
Untreated hybrid grain	= 180 000 – 110 000	
	= Tsh 70 000	
ASD-treated hybrid grain	= 220 500 – 110 000	
	= Tsh 110 500	

*ASD = Actellic Super Dust.

of sugar and other household items) and other parastatal organisations that have been either sold or leased to the private sector (Parathasarathy 2000). A large amount of spare storage capacity exists in the urban areas of regional and district towns and none of the large- or small-scale traders or millers interviewed believed that storage capacity was in short supply. However, the total capacity available is not known so that it is not possible at present to predict whether there will be additional need in the future.

Small-scale traders have difficulty in maintaining grain quality during storage. They have neither the knowledge nor expertise to manage the grain and are forced to employ pest control companies to prevent damage. However, the control treatments undertaken are often ineffective because fake, substandard or adulterated insecticide is used, dosage rates are incorrect and the application techniques are poor. Essentially, traders are being cheated for want of knowledge and lack of regulation of pest control companies. Although registration of companies is mandatory, shortage of manpower in the inspectorate to monitor or prosecute offenders makes regulation unworkable.

Even when very large-scale storage is undertaken and the company employs in-house pest control teams, pest control practices are often incorrect and wasteful of resources. For example, the Export Trading Company has approximately 110 000 t storage capacity located in the main regional centres and at present has more than 25 000 t of maize in store. The company has its own pest control teams who spray with dichlorvos (Nuvan) every two weeks and fumigate with phosphine every month. In an open store environment, dichlorvos acts as a space treatment, killing flying insects. It does not kill the major part of the population that is in the stored food and thus its effects are largely cosmetic (Vol. 1 p. 242, Birkinshaw 2002); it is best used to disinfest an empty store, before it is filled, as in this situation it can be applied directly onto insects on surfaces and in cracks and crevices. Phosphine should not have to be applied more frequently than once every three months, if the correct exposure period of five days or more is used (Vol. 1 p. 321, Taylor 2002). However, the company only uses exposures of three days and so insects survive treatment and the population soon rises again, resulting in the need for further fumigation.

Strategic Grain Reserve

The Strategic Grain Reserve (SGR) was established to maintain a reserve of maize for release in times of food shortage and to provide food for public institutions, particularly the prison service. It has a capacity of about 180 000 t spread around seven regions. Three centres, Makambako, Songea and Sumbawanga, are located in the main supply areas and the remainder, at Dodoma, Shinyanga, Arusha and Dar es Salaam, are in deficit areas. The targeted reserve is 150 000 t, with a maximum stock achieved at just over 107 000 t when the SGR was first commissioned in 1991. Stocks are rotated to prevent deterioration, although grain has remained in store for up to 2–3 years. Very small quantities of maize have been issued as relief grain, so the SGR has generally been acting as a trading company and not as a buffer stock for last resort emergencies (Table 2.5).

Quality maintenance is undertaken by pest control teams that operate to a high standard. However, SGR is wasteful of materials, fumigating every three months even though its stores are sprayed twice a week with dichlorvos.

Future developments

Village stores

Quite clearly, on-farm capacity is sufficient to fulfil storage requirements for the immediate future. If production should rise significantly then this capacity may need to be increased. Given the variability in the size of stores and in levels of production, it is impossible to predict accurately the current capacity for on-farm storage. However, it is safe to assume that the average family can store at least 15 sacks of cereal or pulse grains. For 3.5 million farming families this gives a total working capacity of 4.75m t, while production is about 1.5m t.

Generally there is no shortage of storage capacity at present but in certain areas where land for the homestead is constrained by high populations, such as in Kyela in Mbeya Region, there may be problems. In such cases, farmers may construct additional storage structures or larger stores. This has already happened in many areas. Alternatively, families may choose to store grain in sacks, inherently a more efficient form of storage in terms of space, and construct a small platform in a room within the house on which to store and manage the produce. Farmers have started to take these initiatives and will continue to do so because the methods are convenient and the family retains control of its crop.

When space becomes a major constraint producers will have to store away from the farm. This would require the hire of a building, a cost that an individual may be unable or unwilling to bear. Farmers will, therefore, have to form groups in order to afford and make efficient use of off-farm stores. During the last 15 years, 960 village stores each of about 300 t capacity have been constructed around the country by government and donor projects to meet the needs of surplus production of maize, cotton, tobacco and other crops. Stores were built with projects contributing the materials and villages the labour. Larger stores of 2 700 t capacity were constructed in the southern highlands with Japanese assistance.

Table 2.5 Maize movements through the Tanzanian Strategic Grain Reserve, 1990–2000. (The balances of grain between years are commercial sales.)

Year	Opening stocks (tonnes)	Maize purchases (tonnes)	Maize issued as relief grain (tonnes)
1990/91	107 500	26 777	7 561
1991/92	14 300	83 805	12 158
1992/93	60 787	69 482	12 274
1993/94	94 710	27 998	1 172
1994/95	41 247	24 245	—
1995/96	15 661	73 190	—
1996/97	23 017	59 154	16 083
1997/98	20 482	43 882	10 121
1998/99	40 832	56 173	18 775
1999/2000	24 417	86 449	—

The village stores were underutilised from the moment construction was completed and many have fallen into disrepair. They were designed to be used by the co-operative system, and while the co-operatives took responsibility for managing and maintaining the stores they were used mainly for cash crops. When the co-operative system collapsed the stores were mostly abandoned. There were several reasons for this, including conflicts of ownership, single-purpose use and decline in agricultural production. Many stores are now used as village offices, reception areas for agricultural inputs, places of worship, nurseries, and so on. There is no reason why farmer groups should not use these stores when production increases, providing that they are repaired and generally made good. Each person might contribute a fixed amount of grain. The group would agree on the duration of storage and would either manage the stock themselves or employ a third party to do so. The group would need to be trained and equipped to keep grain in good condition. Funds to meet the costs of running the enterprise would be raised from profits accrued on sale of the crop. Banks may well be willing to advance credit to the group using the grain as collateral.

Such a scheme is being introduced in Dodoma Region by SNV (a Dutch non-governmental organization). Groups of 15 farmers are establishing grain banks with assistance of local government extension staff who provide technical support; each person contributes Tsh1000 as an entrance fee and shares are available at Tsh2000. Membership fees and shares may be bought with cash or in kind. The grain bank will buy staples like cereals and oilseeds from members as the first priority. The SNV Dodoma Micro Projects Programme canvassed views and opinions in the target villages before proposing the system. In particular, the grain bank will provide women with the opportunity to obtain cash, which they would not normally be able to do as men control conventional sales of agricultural produce.

Trader storage

With good training and regular refresher courses the standard of pest control in the country would improve. Tanzania is introducing a new Plant Protection Act (PPA). This will affect how pest control companies are able to operate. The PPA specifies that pest control companies that wish to fumigate must have staff who can demonstrate proof of being trained. Only certified staff should be able to lead fumigation teams. Certification will occur only when a trained person satisfies the course director

that they are competent. Staff who have been trained previously must complete a refresher course during which they will demonstrate their practical capabilities. Staff of established pest control companies will have a two-year grace period from the date of the enforcement of the PPA within which to obtain certification.

Strategic Grain Reserve

There has been much debate on the future of the SGR (Parathasarathy 2000; Hindmarsh 2002). However, as a supplier of last resort the government should maintain a reserve to meet emergency food needs because there is a time lag of two to three weeks between initiating procurement from South Africa, the nearest reliable source of surplus white maize, and delivery at Dar es Salaam. The quantities needed to meet this requirement should not be above 20 000 t, 20% above the maximum released in any one year since the SGR became operational (Table 2.5). This will result in the need for fewer stores. Furthermore, privatisation of the management of the stock could bring improvements in efficiency and should be considered.

Zimbabwe

Historical perspectives

In 1931, the Maize Board of Rhodesia, the oldest predecessor of the current Grain Marketing Board (GMB), was formed to buy and store certain agricultural products, particularly maize, in order to:

- guarantee a market for large-scale commercial farmers;
- guarantee reserve stocks of food staples;
- guarantee a source of agricultural raw materials for local industry; and
- ensure a livelihood for farmers dependent on a single crop.

The Maize Board was subsequently renamed the Grain Marketing Board of Rhodesia. This operated bag depots along the line of rail serving commercial farmland. Bulk storage started in 1954 when the Aspindale silo complex in Harare became operational. The depot network remained unchanged until political independence in 1980, when it expanded into communal and small-scale commercial farming areas that are mostly served by laterite roads. Subsequently, depots were opened in resettlement areas.

In the early 1980s, government policies resulted in increased production of grains and oilseeds by a range of farmers, including large-scale commercial, small-scale commercial, communal and resettlement farming communities. The increased production required an expansion of storage facilities and to date GMB operates a total of 68 permanent depots of which 12 have bulk handling facilities. In addition to the permanent depots, GMB also operates collection points in order to:

- procure agricultural commodities in short supply;
- shorten the distances producers have to travel to market their produce; and
- provide emergency storage during bumper harvests.

To date the GMB has the capacity to store a total of 5.1m t of which 0.9m t is stored in bulk and 4.2m t in bag. Of the stock maintained in bag, 0.3m t is stored in sheds and the remainder outdoors in cover and plinth storage (CAP).

In recognition of these rapid developments in production and storage systems, Zimbabwe has been given the portfolio of food security in the Southern African Development Community (SADC) with a secretariat in Harare and a mission to strengthen food security in the region.

Physical structures

Collection and distribution points

Collection and distribution points are normally established at rural service centres, usually in consultation with the traditional leadership and the local business community. Such places have little equipment, comprising gum pole dunnage, moisture meters and tarpaulins. The objective is to procure the maize quickly and transport it to the nearest permanent depot within three days at most. The maize is not graded at a collection point. If there is a maize deficit in the local area, these sites assume the role of distribution points.

Typical bag depot

A typical bag storage depot (Fig. 2.12) has a 3 m high perimeter security fence with security lighting in cases where electricity is available. The main gate is the only access for the entry of staff, clients and vehicles. Parking facilities are provided outside the main gate for trucks that stay overnight to deliver the following morning. A 2 m wide firebreak is maintained along the perimeter security fence, with 1 m inside and another 1 m outside. About 10 m into the depot, there is usually a grading room and office. The grading room has a weighing scale,

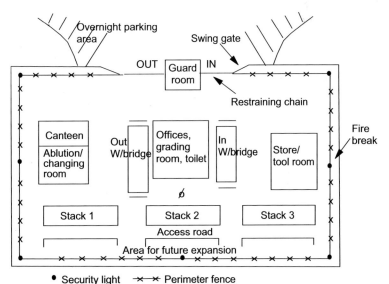

Fig. 2.12 Plan view of a typical bag depot layout (Zimbabwe).

a moisture meter, sieves and test density equipment. In cases where electricity is available, there will be 'in' and 'out' weigh bridges, of up to 30-t capacity, on either side of the grading room. Where there is no electricity, a ten-bag scale with a capacity of up to 1 t is available. The grading room and office are usually on a raised platform to facilitate easy inspection and sampling of incoming and outgoing grain.

At 50 m or more inside the depot, there are hardstands of compacted earth or concrete on which bag stacks are built. Commonly, they measure 54 m by 18 m with a slightly convex surface and with adequate drainage and vehicular access around them. Bag stacks of 5000 t used to be constructed (Fig. 2.13) and equipped with cotton

Fig. 2.13 Typical bag stack with covers drawn back to allow airing (Zimbabwe).

tarpaulins; more recently stacks larger than 1000 t are not constructed for reasons that will be mentioned later. Between the hardstands and the road, there is usually a clear area where electrical or diesel-powered bag stacker machines are positioned during offloading and loading. Fire points are strategically located around and within the depot. The infrastructure and equipment at a typical bag depot is completed by an ablution block-cum-changing room, a storeroom-cum-workshop/tool room, a tarpaulin rack and a number of trolleys.

Typical bulk storage depot

Bulk storage depots (Fig. 2.14) are equipped with silos that are typically of concrete construction, 30 m high and 15 m wide. These are free standing and are in a line with the elevator tower at the centre. At the top is the gantry that houses the conveyor belts responsible for directing grain to individual bins via a tripper. The gantry has a large glass surface forming the walls. In the event of a dust explosion this will easily blow out without damaging the concrete structures.

Each free-standing bin has a top and side manhole and the floor of the bin slopes towards the centrally placed grain outlet port. A sliding metal cover closes the bin outlet. All outlets are located in a tunnel that runs below the bins and houses the bottom conveyor belts that offload the grain from the bins for despatch. Elevators and a flight of steps are also available for use by people or to take other materials to various floors of the silo complex.

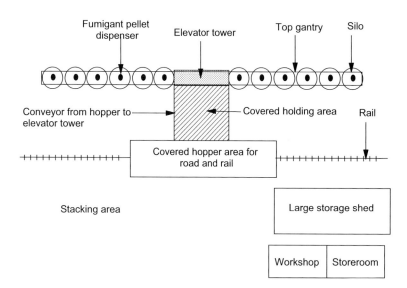

Fig. 2.14 Plan view of a typical silo complex (Zimbabwe).

Fire-fighting equipment, water hose reels and a cage to enter and leave the bins are also available. Dispensers for aluminium phosphide pellets are located directly above the top conveyor belts on either side of the grain elevator tower and dust extractors are found at every place where grain falls, at each end of the gantry and in the bottom tunnel. Ancillary buildings include storage sheds, a workshop and other multipurpose buildings.

The bulk storage complex at Aspindale is different from the typical facility described above in that the silo bins are nested (Fig. 2.15), with four distinct lines of circular bins that are joined together. In between these lines are star bins that make three distinct lines. The Aspindale gantry is therefore much wider and houses several lines of conveyor belt. The other basic facilities are the same as already described.

At Chiweshe and Mukwichi there are bolted steel silos. These are free standing with off-centre inspection hatches in addition to centrally placed grain inlet ports. Instead of rubber belt conveyors, the facilities use a system of chain conveyors for moving the maize. Maize is removed from silos by gravity or using grain augers affixed to the sides of individual bins. Another unique feature of the bolted silos is that they have open eaves to allow for ventilation and are therefore not fumigable. Disinfestation is achieved by spraying the grain with a contact insecticide before it is elevated to the top. Mixing of the maize and pesticide is achieved by the bucket elevators and chain conveyors. The gantry has a steel roof and walls of steel mesh. Ancillary buildings are as described.

Objectives of storage

GMB is the major grain handling authority in Zimbabwe. It is a parastatal marketing board with responsibility for maintaining a strategic grain reserve. It also procures its own commercial stocks for trading. The GMB is the sole importer and exporter of maize on behalf of the Government of Zimbabwe. At any given time, the GMB is expected to hold an equivalent of 0.5m t of maize in stocks and cash. The cash is held in anticipation of the need to import maize to replenish the strategic grain reserve.

Quality standards

All local and imported maize should conform to quality acceptance standards (Table 2.6). The parameters that are measured in order to arrive at a particular grade of maize include test weight (hectolitre mass), moisture content, extraneous matter, trash, broken and chipped maize, brown pigmentation, aflatoxin contamination and defectives. The variable showing the lowest quality is used to determine the overall grade of the maize consignment in question.

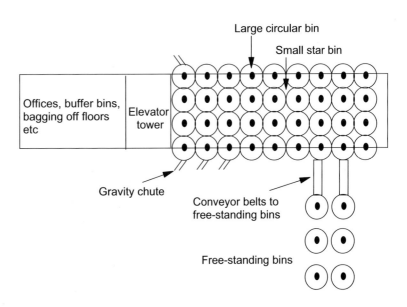

Fig. 2.15 Plan view of the Aspindale bulk storage complex (Zimbabwe).

Table 2.6 Maize quality standards used by GBM in Zimbabwe.

Attribute/grade	A	B	C	D	UG
Moisture content (%, max)	12.5	12.5	12.5	12.5	12.5
Test weight (kg/hL, min)	70.0	68.0	66.0	—	—
Extraneous matter (%, max)	0.5	0.75	1.0	4.0	—
Trash (%, max)	0.1	0.1	0.25	0.3	—
Broken and chipped (%, max)	8.0	—	—	—	—
Brown pigmented (%, max)	6.0	8.0	—	—	—
Defectives (%, max) (diseased, insect damaged, germinated, discoloured, undeveloped, stained, toxic diseased, other coloured and shrivelled)	6.0	12.0	17.0	22.0	—
Aflatoxin (ppb, max)	20.0	20.0	20.0	20.0	—

Note: In addition, maize must be free from live infestation and toxic weed seeds such as *Datura, Crotalaria*, Corn cockle, Cow cockle, Darmal, etc.

Criteria to judge successful operation

The success of the operation of the system is judged by the following criteria:

- Adequate buffer stocks must be available to the nation at all times.
- All maize produced in the country should have a ready market at the GMB. The GMB is a buyer of last resort.
- GMB is the custodian of national maize standards and strives to ensure that the quality and quantity that is procured is the same as the quality and quantity that is marketed.
- The ability to meet internal and external customer requirements in terms of both quality and quantity. A record of customer complaints and satisfaction is maintained.
- The ability to remit dividends to its sole shareholder, the Government of Zimbabwe.

Effects of seasonal factors

The weather conditions in any one growing season will determine whether harvests are surplus or deficit. During bumper harvests GMB, in its capacity as buyer of last resort, may make large procurements. These may necessitate construction of emergency CAP storage. Exports are encouraged and the quality standards remain in force. Pest control operations are delayed during surplus years, particularly because of the limited supply of fumigation sheets, resulting in increased cross-infestation between stocks.

In deficit years, maize imports to meet local demand are inevitable and quality standards are more difficult to enforce. More often than not, the maize arrives in the country at a moisture content higher than 12.5%. The potential for grain deterioration is high since the GMB does not have drying facilities.

Marketing of maize

Individuals, institutions and millers visit storage depots to collect quantities of a specific quality of maize. The minimum quantity to be purchased is one 50 kg sack and the maximum depends on stocks available for purchase; these are usually controlled in deficit years. In most cases, sales are made from the depot although maize may be transferred from outlying depots to zone centres, that is, urban centres for millers and other customers.

Major sources of quality decline

Quality problems at commodity intake

Insects pose the greatest threat to quality at intake. Certain storage pests, such as weevils (*Sitophilus* spp) and the moth *Sitotroga cerealella*, start their infestation while the crop is still in the field (Vol. 1 p. 94, Hodges 2002). Besides field infestation, producers may also deliver the previous season's crop, before the new harvest, and this can be heavily infested. It is common for some producers to mix old and new crops in the hope of attaining a superior grade for the old crop. Grain intake is usually a slow process that delays the initial fumigation while bag stacks are completed and silos filled, allowing insect infestation to build up. A potentially serious infestation problem facing Zimbabwe is posed by *P. truncatus*. At the time of writing it is generally believed that the destructive pest is already in the country but as yet there is no official acknowledgement. The GMB, the Commercial Farmers Union, the Plant Protection Research Institute and other stakeholders have come together to monitor for this pest along Zimbabwe's borders and in particular at all official border entry points.

Some harvesting or post-harvest operations predispose the maize to quality decline. Sometimes maize may be insufficiently dried, for instance when large-scale farmers harvest early to make way for winter cereals, or drying is so rapid or done at such a high temperature that the grain suffers from stress cracks. These cracks will cause breakage of the maize in subsequent handling operations. If artificially dried, so that warm grain is accepted into storage, condensation and associated high moisture problems may result.

Maize that is not thoroughly cleaned will have a high percentage of foreign matter and the grain will therefore not store well. Foreign matter may also damage machinery. Improperly set shelling blades will result in a high percentage of broken grain which is subsequently prone to infestation by secondary insect pests (Vol. 1 p. 94, Hodges 2002) and more rapid moisture uptake leading to mould growth. Mycotoxin contamination is not a serious problem during intake because Zimbabwe maize is usually well dried at harvest (12% m.c.) and so fungal growth is limited.

Any producers delivering wet maize are requested to dry it and those presenting grain with a high foreign matter content are asked to clean it. Quarantine fumigations are often required when imported grain is infested. Donated maize often presents quality problems, particularly wet grain heating. Imported maize is prone to mycotoxin contamination because it lands in a tropical environment with a high moisture content. Maize caking and bridging are common phenomena in silos and the writer remembers a time when picks, shovels and wheelbarrows had to be used to remove imported maize and soyabeans that had caked inside silos at Concession in 1993 soon after the serious drought of 1992.

Quality problems during storage

Insect pest infestation, either brought in with the maize or residual infestations in stores, packaging material, grain debris and handling equipment, continue to pose a threat to quality during storage. The lack of rodent and bird proofing results in quantity and quality losses, increased spillage, stack collapse and increased labour costs associated with retrieving spillage.

Maize in bag stacks may become discoloured by a process called stackburn. This is caused by a particular type of grain heating first noted within outdoor bag stacks of white maize in Zimbabwe. Stackburn was previously found mostly in the top layers of grain in bags lying directly under the tarpaulins covering stacks. Depot managers found that stackburn on tops of stacks occurred mostly in the period September–March when temperatures and humidity are high. If stackburn occurred inside stacks then it tended to start earlier when temperatures were invariably high inside the stack. In recent years, the central areas of stacks have become more affected. The term *internal stackburn* was used to differentiate this from the more common *top stackburn*. The increased incidence of internal stackburn coincided with a switch from storing in jute bags to using woven polypropylene (WPP). Although initial reports of the problem came from Zimbabwe, internal maize stackburn appears to be widespread with reports from Ghana, Mozambique, Tanzania, Malawi and Zambia in both white and yellow maize (Tyler 1992; Kutukwa 1994). It has affected national stocks as well as food aid grain. In many of these countries the increasing incidence of maize stackburn has also been linked to a change from jute to WPP sacks.

Stackburn is a discoloration of the pericarp (the outer layer of cereal grains) and embryo of stored maize. It ranges from light through to very dark brown. The discoloration often starts around the embryo (fungal hyphae are often seen) and progresses over the kernel's surface. It is both temperature and moisture dependent, and occurs in the presence or absence of mould growth. Discoloration can occur at low moisture contents (12–13%) and high temperatures (> 40°C) but occurs more rapidly when high moisture contents are combined with high temperatures. Non-enzymic browning reactions and, to a lesser extent, polyphenol oxidase activity have been suggested as the processes causing the discoloration (Devereau 1995).

Apart from the obvious colour changes, stackburnt maize is brittle and smells as if it has been roasted. It cannot germinate and has a lower nutritive value (Sefa-Dedeh & Senanu 1996). It may have altered starch and protein, and lowered oil yield, test weight and non-reducing sugars. The financial consequences of stackburn are significant. For example, in Zimbabwe in 1993 downgrading of imported maize due to stackburn caused an estimated loss of $US23m; in Zambia in the same year downgrading of imported yellow maize resulted in an estimated loss of $US30m based on the landed value of the maize (Giga & Mushingahande 1996).

Top stackburn appears to be caused by high temperature and high moisture content, and length of storage period. The black tarpaulins used in Zimbabwe to cover outdoor stacks absorb radiant heat during the day and the temperature under them may rise to 60°C. Tarpaulins cool at night so that moisture forms on their inner surfaces. This moisture is deposited on grain and over time the maize becomes discoloured.

Internal stackburn in Zimbabwe is caused by a characteristic temperature rise to a maximum of 43°C over a period of about 100 days, followed by cooling. The exact cause of heating remains unclear. The change from jute to WPP sacks was linked to the increase in internal stackburn and this has been investigated as a possible cause. Jute sacks were found to allow substantially higher air flows than WPP sacks when two layers of the sack material were held together under pressure, but transmission of water vapour was similar. WPP bags could therefore accelerate the discoloration process by reducing heat loss from the interiors of maize stacks due to air movement (New 1995). Trials in Zimbabwe using bag stacks made from WPP and jute sacks showed that higher temperatures and discoloration occurred in the WPP stack (Conway 1996; Odamtten & Clerk 1996). Measures to counter internal heating in WPP bag stacks were developed and are described in the following section.

Commodity and pest management regimes

Commodity management

Good maize storage is dependent on the procurement of maize grain of good quality. To this end, a Crop Variety Release Committee comprising plant breeders, farmers, extension staff, GMB and millers meets to consider the release of maize and other crop varieties. Farmers who grow maize receive technical backup from government agricultural extension experts and representatives of agro-industries. Each depot has a depot liaison committee that meets with farmers and traders and advises them on how to present their maize for marketing through GMB.

When the maize is finally delivered (or imported) to a GMB storage depot, the acceptance standards are usually adhered to rigorously, especially with respect to moisture content. Old and new crops as well as different grades are not mixed. Stores are cleaned and proofed against rodents, birds and rain and groundwater ingress. Handling equipment, machinery and packaging materials are disinfested. Grass within and at the periphery of the storage site is kept short so as to deny pests hiding places. All grading and handling equipment is checked to ensure that it is in a good state of repair. In particular, moisture meters, scales and test weight apparatus are calibrated.

Since stackburn is an important source of quality decline, a number of actions can be taken to reduce it.

Building channels and chimneys into stacks

Building channels and chimneys (Fig. 2.16) can provide passive ventilation of the stack centre. The normal building pattern for bag stacks – alternate layers of sacks placed lengthways then widthways with a row of 'keys' at the edges – is modified by incorporating channels. The channels are arranged such that they intersect to form vertical chimneys running from the bottom to the top of the stack while maintaining its stability. The building pattern is shown in Fig. 2.17. The first layer of the stack is built without channels, but on top of that is built a layer with channels running in one direction (Fig. 2.17a). The next layer is constructed with the channels running at right angles to the layer below (Fig. 2.17b). The next two layers have channels in both directions (Fig. 2.17c, d). Further layers are added using the same four patterns in sequence. The result is a bag stack with horizontal channels and vertical chimneys, and no bag is more than 2 m from a ventilation channel.

Fig. 2.16 Bag stack of maize built in Zimbabwe showing channels and chimneys for passive ventilation, and dunnage in foreground.

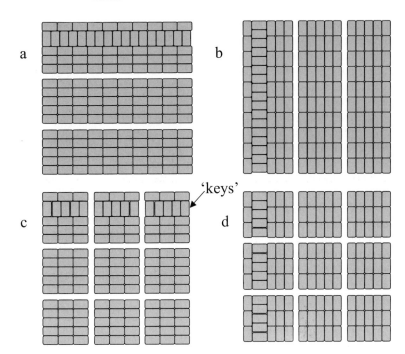

Fig. 2.17 Successive layers in building a bag stack with channels and chimneys to limit 'stackburn'.

Removing covers to air the stock

Tarpaulins are pulled back in strips, weather permitting. During the dry season (May–September) stacks are normally not covered and during the wet season they are opened at least once a week. The tarpaulins may be arranged in such a way as to encourage billowing. This allows hot, moist air out of the top of the stack.

Placing an absorbant layer between grain stock and tarpaulins

A layer of hessian, sawdust or groundnut shells is placed between the tarpaulins and bags to absorb any condensation.

Using tall dunnage

Tall dunnage can encourage better ventilation of bottom bags. Hardstands of concrete should be used (in preference to compacted soil) as these exclude groundwater and prevent sinking of stacks.

Reducing the stack size

The size of stacks has been reduced from 5000 to 1000 t to increase the surface area to volume ratio. This has the positive effect of enhancing passive ventilation (cooling) and therefore reducing grain discoloration. Another positive effect is that multiple fumigations of the same stack of maize may be avoided, as 5000-t stacks are less likely to be completely discharged leaving fragments that will need to be refumigated.

Pest management

Pest management regimes recognise depot hygiene as being of paramount importance and to this end stores, stacking areas, silos and surrounding areas are kept clean. Grass is cut short, spillage is collected and disposed of properly and machinery, plant and equipment are cleaned. It is recognised, however, that despite good hygiene, infestation will still develop and this is controlled by fumigation using either methyl bromide or phosphine. Most pre-shipment and quarantine fumigations are undertaken using methyl bromide at the rate of 30 g t[-1] for an exposure period of 48 h. Pre-shipment fumigations are applied even if no live infestation is detected and, in the case of exports, phytosanitary regulations of the importing countries are brought to bear.

The decision to fumigate is based on visual inspection every two weeks of bag stacks and sampling inspections of silos. Quality controllers receive the infestation reports and use this as the basis for deploying pest control.

The quality controllers also visit all depots within their jurisdiction to check on the accuracy of the reports and also to prioritise fumigation schedules. Bag stacks are fumigated by well-trained and equipped mobile pest control teams using methyl bromide or phosphine, while depot personnel are responsible for all silo fumigations.

All maize handled in bulk is fumigated using phosphine at the rate of 11 pellets per tonne (2.2 g phosphine per tonne) over a two-week exposure period. The exception to this is Chiweshe and Mukwichi bulk handling facilities, where Actellic Super is sprayed on to the maize as it is loaded into the silos.

When the infestation pressure on a depot is high, stack surfaces are sprayed with a contact insecticide to lessen cross-infestation. Empty stores, packaging material, dunnage and tarpaulins are also disinfested. A stack of maize remaining at a depot for 12 months typically receives an average of three methyl bromide fumigations, although the target is two fumigations per year.

Economics of operation of the system

Storage costs vary with the type of storage chosen. CAP storage is the cheapest, followed by bulk storage and shed storage. The major cost elements are labour, fuel, electricity, pesticides, spares and maintenance, transport, dunnage, fumigation sheets, tarpaulins, rope and wire.

According to the balance sheet extracted from the 1998 *Annual Report* of the Grain Marketing Board, the Board was capitalised at $US6.2m (Table 2.7).

As the length of the storage period increases the storage costs will also increase, given that longer storage periods are associated with infestation pressure and therefore greater use of pesticides, replacement of fumigation equipment, increased spillage losses and quality deterioration mainly due to grain discoloration in CAP storage. The costs associated with handling and storage as shown in Table 2.8 but exclude interest charges levied on the finance for procurement.

Assets	$US '000
Land	17 382
Building, storage installations, roads, drainage	2 821 491
Fencing and water supply	463 091
Rail facilities	27 927
Plant, machinery and equipment	1 913 636
Furniture and office equipment	219 455
Motor vehicles	567 509
Work in progress	142 109
Total	6 172 600

Table 2.7 Capitalisation of GMB Zimbabwe as at 31 December 1998.

	GMB costs	Third-party charges†
Storage type		
Cover and plinth (CAP)	3.48/tonne/annum	4.32/tonne/annum
Shed	6.96/m²/annum	8.76/m²/annum
Silo, pooled	5.28/tonne/annum	6.60/tonne/annum
Silo, full	4.92/tonne/annum	6.12/tonne/annum
Handling		
Bag in bag out	0.73/tonne	0.91/tonne
Bulk in bulk out	0.73/tonne	0.91/tonne
Bulk in bag out	0.95/tonne	1.18/tonne
Pest control		
Fumigation	1.08/tonne/annum	1.35/tonne/annum
Transport	0.73/km	0.91/km

*Exchange rate: $US1 = $Z55.00.
†Charges to millers, brewers and commodity brokers for handling maize.

Table 2.8 GMB costs/charges* for handling and storage of maize ($US) (1998) in Zimbabwe.

Future developments

Depot network

Since the advent of market reforms and the liberalisation of grain trading, in particular in 1993, the intention of GMB has been to reduce the number of storage sites by closure or leasing out. However, to date this has not been possible for social and/or political reasons. Most of the storage sites now operate on a seasonal basis and are only fully manned during intake and despatch to parent depots. Collection points operate in more or less the same way.

The liberalisation of grain trading has allowed commercial players into the storage business. Bulk and bag handling facilities have been set up in the maize belt of the country. Parallel to this, improved storage facilities are also being encouraged and promoted at the village level. These developments mean that the national maize output will be handled by a number of different concerns so that expansion of the GMB network is not envisaged.

Strategic developments

For the import or export of maize, the port of Beira in Mozambique presents the most economic route. The development of bulk handling facilities at Beira or Mutare (the third largest city in Zimbabwe and situated on the border with Mozambique on the Beira route) is considered by the writer to be viable and of national interest. Such a development would reduce overall costs associated with maize import and export. Similarly, in drought-prone southern Zimbabwe the development of bulk handling facilities in Gweru and Masvingo would improve food distribution.

Operational/technical developments

Silo fumigations

Currently, phosphine fumigations are done by dispensing pellets of aluminium phosphide into grain that is being moved from one silo to another. This grain transfer is a considerable expense in power consumption and results in grain breakage, thereby reducing flow properties, increasing the angle of repose and increasing the rates of reinfestation. To facilitate fumigation without the need for grain transfer, a fan system (The J-system) is being adopted that will draw phosphine through the grain bulk and recirculate it.

Alternatives to methyl bromide

Perhaps the major challenge facing pest management regimes is the planned phasing out of the fumigant methyl bromide under the terms of the Montreal Protocol. Demonstration projects, funded by the United Nations Development Programme, and implemented at selected GMB depots, have promoted phosphine as the immediate replacement for methyl bromide. The changeover to phosphine means that the GMB will have to invest in some new fumigation equipment, judiciously plan stock movements and ensure that the relevant staff are thoroughly trained. Besides the use of phosphine fumigation, airtight storage may also have a role to play at distribution/collection points. At these locations it may be possible to use airtight grain cocoons holding up to 150 t. GMB has experimented with these cocoons flushed with nitrogen.

Argentina

Historical perspectives

The fertile plains of the Pampas provide 26.5m ha of agricultural land, enabling Argentina to be a major grain-producing country since the nineteenth century. By 1890, the railway system was widespread throughout the humid Pampas, transporting wheat and maize from the countryside to the port at Buenos Aires. The mechanisation of farming, the introduction of tractors, row planters, combine harvesters, the improvement of the transportation system, and the rise in export demand during two world wars, made Argentina one of the most important grain-producing countries in the first half of the twentieth century. In the early 1950s, the adoption of hybrid seed maize caused a revolution in the grain market. In the 1970s, a second generation of maize hybrids and the widespread adoption of soyabean production increased Argentine grain production. During the 1990s, the massive use of fertiliser, pesticides and irrigation yielded another significant increase in grain production and positioned Argentina among the top five grain producers–exporters (Table 2.9), with the US, European Union, Canada and Australia. The primary and agro-industrial sectors have produced about 60–80% of the wealth of the country, and remain Argentina's most important economic sector.

| Grain | Production | Exports | | | World ranking |
		Grain	Oil	Pellets	
Maize	16.8	10.8	—	—	2
Soyabean	20.2	4.1	3.1	13.6	1
Sunflower	6.1	0.2	1.6	1.8	
Wheat	15.3	10.8	—	—	5
Sorghum	3.3	0.8	—	—	—
Total	61.7	26.7	4.7	15.4	

Table 2.9 Grain production and export volume in Argentina (millions of tonnes) (SAGPyA 2000).

Argentine farmers have a strong maize-producing tradition. Maize is one of the most important crops based on the number of farmers involved, the land utilised and the volume of grain generated. During the last few years, maize production in Argentina remained the same due to the expansion of soyabeans. Today, Argentina produces about 16.8m t of maize. The domestic consumption of maize in Argentina oscillates around 6m t, with the balance exported. An increase in domestic maize consumption is not likely in the near future. Thus, if Argentina increased its maize·production, the resulting excess of grain will have to be sold on the international market.

Up to beginning of the 1990s, the Argentine National Grain Board (JNG) built and controlled a large network of country, terminal and export grain elevators, which facilitated the flow of maize from the farm to the end user. Private entrepreneurs and farmer co-operatives built a large number of additional facilities tied into the government system. During a privatisation phase at the beginning of the 1990s, the JNG was abolished and government-owned facilities were sold to private investors and farmer co-operatives. As a result, the Argentine grain industry has undergone a significant reorganisation, with the addition of new, and modernisation of existing, export elevators and processing plants. In the light of ongoing free market policies and increased global competition many less efficient country and terminal elevators are being consolidated.

Physical facilities

The storage capacity of Argentina has increased in line with increases in grain production (Table 2.10), and was also influenced by fluctuations in the country's economy. In 1977, storage capacity deficit was 54% of production, 16% in 1982, 14% in 1986, almost zero in 1991 and 30% in 2000. This deficit is generally not important because grain passes through the system very quickly and harvesting is spread across the year; April–June for maize and soya and December for wheat. When the harvest starts, most of the grain from the previous harvest has already been marketed and just a little carryover remains in storage. However, in the period February–August there may be insufficient capacity for some crops such as sorghum and sunflower seed, and these may have to be stored in the open air (Hajnal 1999).

There has never been an official survey of storage capacity in Argentina and the best industry estimate of current capacity is around 43m t, which can be segregated into five categories; private elevators, co-operative elevators, ports, processors and on-farm facilities (Table 2.11). In addition, during the harvest a considerable amount of grain is held on trucks and trailers while grain is transported to the ports but the actual amount of grain that remains in these is unknown (Hajnal 1999).

	1977	1982	1986	1991	2000
Grain production	28.0	32.0	36.0	36.0	61.7
Storage capacity	12.8	27.0	31.1	36.4	43.0

Table 2.10 Evolution of grain production and storage capacity in Argentina (million of tonnes) since 1977. Source: SAGPyA (1993, 2000), Bolsa de Cereales (1997).

Table 2.11 Storage capacity in Argentina (industry estimate for 2000).

Storage facility	Capacity (million tonnes)
Private elevators	14
Cooperative elevators	6
Ports	4
Processors	6
On farm	13
Total	43

Fig. 2.19 Hydraulic platform for unloading grain (Argentina).

Private and co-operative elevators

Private and co-operative elevators are the backbone of the modern Argentine grain handling and storage system (Fig. 2.18). Facilities are distributed throughout the entire grain production area. A typical Argentine elevator has one or more receiving dump pits. Truck unloading is by gravity and may be aided by a hydraulic platform, which is very efficient (Fig. 2.19). Grain handling is done with bucket elevators for vertical movement, and screw augers, belts and chain conveyors for horizontal movement. Facilities have a wide range of storage capacities (from 15 000 t to more than 100 000 t) and utilise an assortment of technologies. Storage structure types include concrete silos, metal bins and tanks and horizontal sheds. Hopper bottom construction is very popular in Argentina. Often corrugated metal bins are constructed on top of in-ground concrete hoppers. Typical hopper-bottom steel bins have up to 1500 t of capacity, and flat-bottom bins up to 7500 t. Many horizontal storage structures have V-shaped bottoms and range in capacity from 10 000 to 250 000 t.

Elevators are typically equipped with aeration and temperature monitoring systems, and many of them have automatic aeration control systems. Drying capacity is based on the tonnage of maize that the facility receives. All elevators have continuous-flow dryers. Some operators prefer cross-flow dryers (column dryers) and others mixed-flow dryers (cascade or rack-type dryers). The primary business objective is to buy grain from farmers, condition it to the standard market moisture content and sell it to the domestic processing industry and to exporters.

Ports

Most of the maize and soyabean export terminals are located on the Paraná River for shipping maize on seagoing vessels (Fig. 2.20). There are also several important export terminals on the Atlantic coast, which are mainly for wheat. Maize is received from private and co-operative elevators, and also directly from larger farmers. The exporting facilities have large storage capacities and high-capacity grain handling systems to receive, grade and ship maize.

Grain processors

The grain handling system of grain processors varies according to the end use of the grain. Processing industries such as flour milling, dry maize milling, oil extraction and popcorn processing have grain handling and storage facilities set up with the latest technology to preserve their high-value maize. Normally they have significant storage capacity. However, some of them buy grain in

Fig. 2.18 Typical private elevator storage facility (Argentina).

Fig. 2.20 Port facility on the Paraná River (terminal 6), Argentina.

Fig. 2.21 Typical small-farm grain silos without an aeration system, temperature monitoring equipment or grain drying system (Argentina).

the market whenever they need it. Thus, they do not require large storage capacities. Processors receive grain from private and co-operative elevators and also have special contracts with farmers for supplying high-quality maize. Industries that produce eggs, poultry and pork also have their own storage facilities, but frequently buy grain from elevators when they need it. Beef and dairy production industries are set up on farms, and they also produce their own grain. These facilities have maize management practices less complex than those in the maize processing industry and so do not use the latest technology for storage.

On-farm storage

The level of technology employed in on-farm storage is related to the size of the farm. Relatively few farms have their own storage facilities. Smaller farms may have three to five bins (Fig. 2.21) ranging in capacity from 100 to 400 t without aeration systems or temperature monitoring equipment. These facilities usually do not have a drying system. Larger farms often have newer facilities equipped with the technology for successful grain storage. They typically have corrugated steel bins equipped with aeration systems and temperature monitoring devices. Some of these facilities have continuous-flow dryer systems and others in-bin drying systems (Fig. 2.22).

Temporary storage structures

Many farmers and some elevators are short of storage capacity. The investment required to increase their perma-

Fig. 2.22 Typical large-farm grain silos with an aeration system, temperature monitoring equipment and a continuous-flow grain dryer (Argentina).

nent storage capacity is very high, so temporary storage structures are a good option. Additionally, these structures allow farmers and elevators to be more flexible because they can increase or decrease total storage capacity more easily depending on their needs. In the past, elevators used mostly Australian bunkers (see Chapter 3), while farmers often used round wire-mesh silos with cloth sides to meet temporary storage needs. Today, the silobag (Fig. 2.23) has become the most popular system for the temporary storage of whole grains in Argentina. It consists of a white plastic tube of 2.7 m diameter and up to 60 m long with a capacity of 200 t. Farmers place these bags along the edges of their fields and use special equipment for loading and unloading them. The bags are airtight so that grain respiration results in a CO_2-rich,

Fig. 2.23 Maize grain being loaded into a plastic silobag (Argentina).

low-oxygen atmosphere. This decreases all biological activity so allowing dry grain to be stored safely for three to four months. The use of silobags for temporary storage has increased significantly since they were first introduced into Argentine during the mid-1990s. In 2001, more than 1.5m t of maize, soyabeans, wheat, sorghum and sunflower were stored in silobags and this is expected to increase to around 7m t in coming years (Bartosik *et al.* 2003).

Objectives of storage

The main objective of the maize post-harvest system in Argentina is to concentrate and condition the grain for export. Argentina has three grades for maize (Table 2.12) the grade applied is the lowest indicated by any of the five assessment criteria. Most of the grain in Argentina is assessed to be grade 1.

Each component of the grain post-harvest system in Argentina has particular objectives for grain storage.

On-farm storage

Farms that grow maize and also have beef, dairy, pork, poultry or egg production will condition and store their own grain. Farms that produce grain to sell to the market will operate according to the facility they have available. As maize is harvested at moisture contents higher than recommended for safe storage, most Argentine farms without dryers have to send their maize to the elevator at harvest, when the price of grain is lowest. Farms with dryers can dry all of their own maize and then keep it stored until prices are more favourable. Later in the year they can also take advantage of lower transportation costs. Maize has to compete for storage space with soyabeans. Depending on the expected price rise of each grain and their financial needs, farmers will keep more maize or more soyabeans. The on-farm storage period of maize and soyabeans is typically not more than six to seven months because farmers need to free up their storage space for the wheat harvest in December. Farms sell their maize to private and co-operative elevators, but they can have special contracts with processors or exporters.

Private and co-operative elevators

The objectives of these elevators are to buy grain from farmers and other elevators, concentrate the grain, grade, condition (dry), store, blend and sell it to processors and exporters.

Exporters

Argentina exports around 10m t of maize per year. Port facilities receive maize from private and co-operative elevators, and also from large farmers. The main goal of the export facilities is to handle grain (receive, grade, segregate and ship) at a very high rate, rather than store it for a long time.

Processors

Argentina has an important grain processing industry, which is mainly related to soyabean processing. The

Table 2.12 Grades and tolerance limits for maize in Argentina.

Grade	Moisture content (%)	Test weight (kg/hL)	Damaged kernels (%)	Broken kernels (%)	Foreign material (%)
1	14.5	75	3	2	1
2	14.5	72	5	3	1.5
3	14.5	69	8	5	2

maize industry requires 5–6.5m t per year. This amount is broken down as follows:

(1)   The wet milling industry processes 1m t (this segment has been booming for the last three years, because Argentina started to export fructose to neighbouring countries);
(2)   The dry milling industry uses about 0.25m t of maize primarily for breakfast cereals and snack foods;
(3)   The poultry industry consumes 1m t;
(4)   Egg production consumes 0.7m t;
(5)   Pork production uses around 0.3–0.4 m t; and
(6)   The dairy industry consumes 0.5m t.

Poultry, eggs, dairy and pork producers are interested in maize with good nutritional properties free of mycotoxins. Argentina has an important beef industry, and although Argentine farmers prefer to feed their cattle with grass rather than grain, they use about 1.5m t in their feedlots. Farmers are the owners of these feedlots and they produce and store their own grain to supply them. Processors have their own quality standards and they may offer higher prices for high-quality maize. The wet and dry milling industries prefer maize with low levels of stress cracks, which is directly related to the effect of drying. Private and co-operative elevators have high-temperature drying systems (cross-flow and mixed-flow dryers), which make it difficult for them to reach the quality requirements of those industry segments. Farmers equipped with low-temperature drying systems, such as in-bin dryers, have a good opportunity to sell their high-quality maize directly to the dry and wet milling industry at higher prices. Additionally, some processors are interested in grain that is not genetically modified (GMO-free grain). They offer special contracts to farmers to grow non-GMO maize. In order to store high-quality maize throughout the year, processors require well-designed grain handling facilities.

Major sources of quality decline

The major sources of quality decline for maize are moulds, insects and incorrect drying practices that cause stress cracking. Maize is harvested at higher moisture content than recommended for safe storage and, even though most maize is dried before storage, sometimes the grain is still binned at unsafe moisture contents. Around 80% of the maize is harvested with a moisture content greater than 17%. Private and co-operative elevators are set up with high-temperature drying systems,

so the stress crack level in maize can reach significant levels. Movement of the grain after drying increases the percentage of broken kernels, lowering the grade and value of the stock. Most aeration systems in Argentina are set up to push air, which can cause some condensation problems on the underside of the bin roof and rewetting of maize at the surface of the bin. If the grain remains wet for a long period there will be crusting or spoilage in the bin. Ambient temperatures during maize storage season are generally not favourable for insect infestation, which can in any case be avoided easily with correct grain management procedures.

Commodity and pest management practices

This section focuses on the maize handling and pest management operations in commercial elevators. The maize harvest starts in April and can last into June. Trucks, trailers and wagons transport maize from the field to the elevator. Once at the elevator, the first step is to weigh and sample the grain. To receive grain, each elevator must have a trained operator who follows the official Argentine grain standard that specifies the number of samples per load that must be taken, the procedure as to how those samples must be taken, the handling and treatment of each sample and the grading procedures.

The official inspector will grade the grain for its physical characteristics, determine its moisture content and check for the presence of live insects. Argentina has a zero tolerance policy for live insects in the grain trade and so if a single live insect is detected the entire grain mass must be disinfested. If the grain is free of insects it will be stored, segregated by grade. If the moisture content of the grain is 14.5% or lower, the manager will send the maize directly to storage. If the moisture content of the maize exceeds 14.5%, the elevator will discount the moisture shrinkage and will also charge the farmer for drying. The manager then decides to either blend the wet maize with dry maize, or pass it through the dryer. Some elevators have wet holding bins and can segregate the wet maize according to moisture content before passing it through the dryer.

Once dried, the maize is stored in vertical or horizontal storage structures. Before storage, or even before drying, maize is pre-cleaned and most of the broken kernels and foreign material is segregated from the grain for safer storage. Commercial storage structures have aeration and temperature monitoring systems. Large elevators utilise automatic control of their aeration fans. High-value grain, such as popcorn, high-oil maize or non-GMO maize receives closer management attention.

Argentina penalises the marketing of maize with more than 3% insect damage. Even though only 2% of the maize marketed exceeds this maximum limit, insect control is an important issue for farmers and elevator managers. Until the 1970s, weevils (*Sitophilus* spp) were the main pests, but with the use of specific insecticides this pest was easily controlled. During the 1980s, the lesser grain borer (*Rhyzopertha dominica*) appears to have emerged as an important pest. This insect produces large amounts of dust, which supports the development of other insects and moulds. Today, the most important grain feeding pests are weevils, lesser grain borer and mites (Yanucci 1996).

The pest management regime consists of cleaning, aeration and chemical control. Hygiene practices involve cleaning the grain, the storage facilities and the grain handling equipment. The use of aeration to cool the grain and reduce the risk of insect infestation is a common practice. Chemical pesticides are the main insect control methods, but these are far from efficient and adequate. Pesticides can be used for facility treatment, preventive treatment of the grain, treatment of the grain during shipping and curative treatment. Around 98% of grain managers treat their grain by phosphine fumigation and/or admixture of insecticide when grain is binned. Effective pest management at elevators is commonly hindered by:

(1) Managers with insufficient knowledge of grain and grain pests to be able to decide on appropriate pest management action;

(2) The rules for effective pest control, especially phosphine fumigation, are not respected;

(3) Elevators lack grain temperature monitoring systems and inefficient grain cleaning systems, and

(4) There is a shortage of time during the harvesting season to operate the system and managers are not usually aware of how long their grain will be in store.

The most important source of insect infestation is the presence of insects in the store on residues of the previous harvest. Infestation of the crop in the field and insects flying into the store appear to play a more minor role. The key to successful pest control is the thorough cleaning of facilities and grain, proper chemical treatment of facilities and correct use of aeration.

Future needs

Argentine farmers are very efficient maize producers although they face unfair competition from farmers in the US and the European Union who receive government subsidies. To make maize farming more profitable, Argentina has to increase the value of its grain and improve its post-harvest grain systems. The grain processing industry has a good opportunity to add value to its products and receive better prices. Argentina has to increase the internal consumption of its grains and, instead of exporting grain, it should export value-added goods such as beef, milk or baking products. Argentina can also offer to the international market speciality crops such as hard wheat, flint maize or non-GM soyabeans. These speciality grains are already produced, but there are difficulties in segregating them for identity-preserved handling, and so on many occasions these speciality crops do not receive the price premium they deserve.

References

Adams, J.M. & Harman, G.W. (1977) The evaluation of losses in maize stored on a selection of small farms in Zambia with particular reference to the development of methodology, *G109*. Tropical Products Institute, Slough, UK.

Bartosik, R.E., Rodriguez, J.C., Malinarich, H.E. & Maier, D.E. (2003) Silobags: evaluation of a new technique for temporary storage of wheat in the field. In: P.F. Credland, D.M. Armitage, C.H. Bell, P.M. Cogan and E. Highley (eds) *Advances in Stored Product Protection, Proceedings of the 8th International Working Conference on Stored Products Protection*, York, UK, 22–26 July 2002. CABI Publishing, Wallingford, UK.

Belmain, S.R. (2002) Botanicals. In: P. Golob, G. Farrell & J.E. Orchard (eds) *Crop Post-Harvest: Science and Technology. Volume 1 Principles and Practice.* Blackwell Publishing, Oxford, UK.

Birkinshaw, L.A. (2002) Insect control. In: P. Golob, G. Farrell & J.E. Orchard (eds) *Crop Post-Harvest: Science and Technology. Volume 1 Principles and Practice.* Blackwell Publishing, Oxford, UK.

Birkinshaw, L.A., Hodges, R.J., Addo, S. & Riwa, W. (2002) Can 'bad' years for damage by *Prostephanus truncatus* be predicted? *Crop Protection* **21**, 783–791.

Bolsa de Cereales (1997) Número estadístico 1994/95–1995/96 de la Bolsa de Cereales de Buenos Aires. Bolsa de Cereales de Buenos Aires, Argentina.

Boxall, R.A. (2002) Storage structures. In: P. Golob, G. Farrell & J.E. Orchard (eds) *Crop Post-Harvest: Science and Technology. Volume 1 Principles and Practice.* Blackwell Publishing, Oxford, UK.

Brice, J.R. (2002) Small-scale farm storage in the developing world. In: P. Golob, G. Farrell & J.E. Orchard (eds) *Crop*

Post-Harvest: Science and Technology. Volume 1 Principles and Practice. Blackwell Publishing, Oxford, UK.

Chibwe, J.K.S. & Kalinde, G. (1999) Post production food loss reduction practices in Malawi. In: S. Nahdy & A. Agona (eds) *Proc. 1st Conference on Stored Product Insect Pests: Their Status, Coping Strategies and Control in Eastern, Central and Southern Africa*. 29 November–1 December 1999, Kampala, Uganda.

CODA (1996) *Post-Harvest Loss Reduction Study in Malawi*. Draft Final Report, CODA and Partners, Lilongwe, Malawi.

Compton, J.A.F., Tyler, P.S., Hindmarsh, P.S., Golob, P., Boxall, R.A. & Haines, C.P. (1993) Reducing losses in small farm grain storage in the tropics. *Tropical Science,* 33, 283–318.

Conway, J.A. (1996) Investigation on stackburn in national stocks of stored maize in sub-saharan Africa and its association with increased levels of insect infestation and the introduction of woven polypropylene sacks. *Annual Technical Report*, ECDGXII Project, Year 3. Natural Resources Institute, University of Greenwich, Chatham, UK.

Coulter, J. (1994) Liberalisation of cereals marketing in sub-Saharan Africa: lessons from experience. *Marketing Series No. 9*. Natural Resources Institute, University of Greenwich, Chatham, UK.

Coulter, J. & Golob, P. (1992) Cereal market liberalisation in Tanzania. *Food Policy. December 1992*, 420–430.

De Lima, C.P.F. (1979) The assessment of losses due to insects and rodents in maize stored for subsistence in Kenya. *Tropical Stored Product Information*, 38, 21–26.

Devereau, A.D. (1995) Investigation into the causes and prevention of heating and discolouration ('stackburn') in bag stored maize. *Report R2231(S)*. Natural Resources Institute, University of Greenwich, Chatham, UK.

Dick, K. (1988) A review of insect infestation of maize in farm storage in Africa with special reference to the ecology and control of *Prostephanus truncatus*. Bulletin No. 18. Natural Resources Institute, Chatham UK.

Farrell, G. (2000) Dispersal, Phrenology and predicted abundance of the larger grain borer in different environments. *African Crop Science Journal*, 8, 337–343.

Giga, D.P. & Mushingahande, M. (1996) The problem and extent of maize stackburn in Zimbabwe. In: J.A. Conway (ed.) Investigation on stackburn in national stocks of stored maize in sub-saharan Africa and its association with increased levels of insect infestation and the introduction of woven polypropylene sacks. *Annual Technical Report*, ECDGXII Project, Year 3.

Giles, P.H. & Ashman, F. (1971) A study of pre-harvest infestation of maize by *Sitophilus zeamais* Motsch. (Coleoptera: Curculionidae) in the Kenya highlands. *Journal of Stored Products Research*, 7, 69–83.

Golob, P. (1981a) A practical appraisal of on-farm storage losses and loss assessment methods in Malawi. 1. The Shire Valley Agricultural Development Area. *Tropical Stored Products Information*, 40, 5–13.

Golob, P. (1981b) A practical appraisal of on-farm storage losses and loss assessment methods in Malawi. 2. The Lilongwe Land Development Area. *Tropical Stored Products Information*, 41, 5–12.

Golob, P. (1984) Improvements in maize storage for the smallholder farmer. *Tropical Stored Products Information*, 50, 14–19.

Golob, P. (1991) Evaluation of the campaign to control the larger grain borer, *Prostephanus truncatus*, in western Tanzania. *FAO Plant Protection Bulletin*, 39, 65–71.

Golob, P. (2002) Safety. In: P. Golob, G. Farrell & J.E. Orchard (eds) *Crop Post-Harvest: Science and Technology. Volume 1 Principles and Practice*. Blackwell Publishing, Oxford, UK.

Golob, P., Marsland, N., Nyambo, B. *et al*. (1999) Coping strategies employed by farmers against the Larger Grain Borer in East Africa. In: Jin Zuxun, Liang Quan, Liang Yongsheng, Tan Xianchang & Guan Lianghua (eds) *Proceedings of the 7th International Working Conference on Stored-Product Protection, Volume II*, 14–19 October 1998, Beijing, China. Sichuan Publishing House of Science & Technology, Chengdu, Sichuan, China.

Golob, P. & Webley, D.J. (1980) The use of plants and minerals as traditional protectants of stored products. **G138**. *Tropical Products Institute*, Slough, UK.

GTZ (1994) *Storage losses of maize under smallholder conditions part 1: Karonga ADD, Northern Region*. Malawi German Biocontrol and Post-Harvest Project, Lunyangwa Agricultural Research Station Crop Storage Unit, Malawi.

Hajnal, R.D. (1999) Grain handling in Argentina. *World Grain*, 17, 26–33.

Henckes, C. (1994) Dividing the harvest: an approach to integrated pest management in family stores in Africa. In: E. Highley, E.J. Wright, H.J. Banks & B.R. Champ (eds) *Proceedings of the 6th Int. Working Conf. Stored-Product Protection, 17–23* April 1994, Canberrra, Australia.

Hindmarsh, P.S. (2002) The case for food security reserves. In: P. Golob, G. Farrell & J.E. Orchard (eds) *Crop Post-Harvest: Science and Technology. Volume 1 Principles and Practice*. Blackwell Publishing, Oxford, UK.

Hindmarsh, P.S. & MacDonald, I.A. (1980) Field trials to control insect pests of farm-stored maize in Zambia. *Journal of Stored Products Research*, 16, 9–18.

Hodges, R.J. (2002) Pests of durable crops – insects and arachnids. In: P. Golob, G. Farrell & J.E. Orchard (eds) *Crop Post-Harvest: Science and Technology. Volume 1 Principles and Practice*. Blackwell Publishing, Oxford, UK.

Hodges, R.J., Addo, S. & Birkinshaw, L.A. (2003) Can observation of climatic factors be used to predict the flight dispersal rates of *Prostephanus truncatus*? *Agricultural and Forest Entomology*, 5, 123–135.

Hodges, R.J., Dunstan, W.R. Magazini, I. & Golob, P. (1983) An outbreak of *Prostephanus truncatus* (Horn) in East Africa. *Protection Ecology*, 5, 183–194.

Keil, H. (1988) Losses caused by the larger grain borer in farm stored maize. In: G.G.M. Schulten & A.J.Toet (eds) Workshop on the containment and control of the larger grain borer, Arusha, Tanzania, 16–21 May 1988. Pest Control Services, Ministry of Agriculture and Livestock Development [Tanzania] and Food and Agriculture Organization of the United Nations [Rome, Italy].

Kutukwa, N. (1994) Maize stackburn in Zimbabwe, review and current status. In: Investigation on stackburn in national stocks of stored maize in sub-saharan Africa and its association with increased levels of insect infestation and the introduction of woven polypropylene sacks. In: S.I. Phillips (ed.) *Annual Technical Report to the EC for Year 1–1993/94.* Natural Resources Institute, Chatham, UK.

Marsland, N. & Golob, P. (1999) Participatory and rapid rural appraisal for addressing post-harvest problems: a case study in Malawi. In: Jin Zuxun, Liang Quan, Liang Yongsheng, Tan Xianchang & Guan Lianghua (eds) *Proceedings of the 7th International Working Conference on Stored-Product Protection, Volume II,* 14–19 October 1998, Beijing, China. Sichuan Publishing House of Science & Technology, Chengdu, Sichuan, China.

Meikle, W.G., Holst, N., Degbey, P. & Oussou, R. (2000) Evaluation of sequential sampling plans for the larger grain borer (Coleoptera: Bostrichidae) and the maize weevil (Coleoptera: Curculionidae) and of visual grain assessment in West Africa. *Journal of Economic Entomology*, **93**, 1822–1831.

Mkoga, Z.J & Shetto, R.M. (1999) Southern highland farm level post-harvest maize systems research highlights and future trends. In: *Proceedings of the Workshop on Farmer Coping Strategies for Post Harvest Problems with Particular Emphasis on the Larger Grain Borer,* August 1999, MAC.

Morris, M., Butterworth, J., Lamboll, M. *et al.* (2001) Understanding household coping strategies in semi-arid Tanzania: Annex 1 Household livelihood strategies in semi-arid Tanzania: synthesis of findings. NRI unpublished report. Natural Resources Systems Programme of the Department for International Development, UK.

New, J. (1995) Sack material properties. In: G.T. Odamtten & G.C. Clerk (eds) Mitigation of stackburn in woven polypropylene bag-stacks for improved food security in sub-Saharan Africa. European Union STD3 Project ERBT3, CT920097. European Union, Brussels, Belgium.

Odamtten, G.T. & Clerk, G.C. (1996) Mitigation of stackburn in woven polypropylene bag-stacks for improved food security in sub-Saharan Africa. European Union STD3 Project ERBT3, CT920097. European Union, Brussels, Belgium.

Parathasarathy, N.S. (2000) Tanzania Soil Fertility and Agricultural Intensification Project, Project Preparation Mission March – May 2000, Working Document on Marketing.

SADC (2002) SADC food security quarterly bulletin Tanzania. http://www.sadc-fanr.org.zw/rewu/qfsb/Qfsbtz.htm.

SAGPyA (1993) Almacenamiento de granos: analisis de la capacidad instalada en la Republica Argentina. Secretaria de Agricultura, Ganaderia, Pesca y Alimentacion, Buenos Aires, Argentina.

SAGPyA (2000) Estimaciones Agrícolas de la Secretaría de Agricultura, Ganaderia, Pesca y Alimentación. http://www.siiap.sagyp.mecon.ar.

Sefa-Dedeh, S. & Senanu, E. (1996) Physical, chemical and functional properties of stackburnt maize. In: G.T. Odamtten & G.C. Clerk (eds) Mitigation of stackburn in woven polypropylene bag-stacks for improved food security in sub-Saharan Africa. European Union STD3 Project ERBT3, CT920097. European Union, Brussels, Belgium.

Semple, R.L. & Kirenga, G.I. (1994) Facilitating regional trade of agricultural commodities in Eastern, Central and Southern Africa. FAO, http://www.fao.org/inpho/vlibrary/new_fao/x5417e/x5417e00.htm#Contents.

Taylor, R.W.D. (1995) Initiation of a survey to locate plants or other traditional materials used in Malawi as insecticides for stored cereals and pulses. *R2267(S).* Natural Resources Institute, Chatham, UK.

Taylor, R.W.D. (2002) Fumigation. In: P. Golob, G. Farrell & J.E. Orchard (eds) *Crop Post-Harvest: Science and Technology. Volume 1 Principles and Practice.* Blackwell Publishing, Oxford, UK.

Temu, P.E. (1977) Marketing board pricing and storage policy: the case of maize in Tanzania. East African Literature Bureau, Dar es Salaam, Tanzania.

Tyler, P. (1992) Heating and discoloration of bagged maize. *World Grain,* September 1992, 14–16.

Tyler, P.S. & Bennett, C. (1993) Grain market liberalisation in southern Africa: Opportunities for support to the small-scale sector. *R1971(S).* Natural Resources Institute, Chatham, UK.

Tyler, P.S. & Boxall, R.A. (1984) Post-harvest loss reduction programmes: a decade of activities; what consequences. *Tropical stored Products Information,* **50,** 4–13.

Urono, B. (1999) Evaluation of Actellic Super dust efficacy in the control of storage insect pests, Larger Grain Borer (LGB), *Prostephanus truncatus* (Horn) and Maize Weevil, *Sitophilus spp.* in northern Tanzania. In: *Proceedings of the Workshop on Farmer Coping Strategies for Post Harvest Problems with Particular Emphasis on the Larger Grain Borer,* August 1999, MAC.

World Bank (2000) *Tanzania. Agriculture: Performance and Strategies for Sustainable Growth.* World Bank, Washington DC, USA.

Yanucci, D. (1996) Evolución del control de plagas de granos almacenados en Argentina. United Nations Food and Agriculture Organization, Rome, Italy.

Chapter 3
Wheat

H. K. Shamsher, D. M. Armitage, R. T. Noyes, N. D. Barker and J. van S. Graver

Wheat (Plate 3) is believed to have evolved in the Middle East through repeated hybridisations of *Triticum* spp with members of a closely related grass genus, *Aegilops*. This evolution was accelerated by an expanding geographical range of cultivation and by human selection, and had produced wheat for bread making by about 8000 BP. The Egyptians discovered how to make yeast-leavened breads about 4000 BP. Since wheat is the only grain with sufficient of the protein gluten to make leavened bread, it quickly became favoured over other grains grown at the time, such as oats, millet, rice and barley. Nowadays, more foods are made with wheat than with any other cereal grain. Wheat contributes between 10–20% of the daily calorie intake of people in over 60 countries and there are more than a thousand varieties of bread prepared worldwide.

The physical and chemical characteristics of wheat have a strong influence on its end use. For this reason wheat is commonly subject to grade criteria not necessarily applied to other grains, such as test weight, protein content and Hagburg falling number (HFN). Test weight is a measure of the weight of grain in a specific volume, values of 74 kg hL^{-1} or more are given by healthy, well-filled grains and are associated with high milling yields. Flour containing approximately 11% or more protein is usually considered suitable for yeast-leavened breads whereas wheat with lower protein is generally more suitable for other baked foods such as chemically leavened cakes, biscuits, and so on.

The grain type affects gluten content – a strong, elastic gluten is required. This is measured indirectly by nitrogen quantification after chemical treatment or by near infra-red (NIR) which can also estimate moisture content, oil content and grain hardness from the same sample. Low HFN values are associated with grain that has a high concentration of the enzyme α-amylase that converts starch to sugars during germination, and indicate that the wheat would be unsuitable for bread making. HFN is in effect a measure of the amount of sprouting grain present in a sample and is measured by testing the viscosity of ground grain in water by recording the time in seconds taken for a plunger to descend through the suspension. Grain harvested and maintained under dry conditions will not sprout and so would have high HFN values. HFN values will vary from year to year according to harvesting conditions.

More usual grain quality criteria include moisture content and impurities. There are two classes of impurity in wheat: dockage and foreign matter. Dockage is weed seeds, weed stems, chaff, straw or grain other than wheat, plus underdeveloped, shrivelled and small pieces of wheat kernels. Foreign matter is the material that is not dockage or wheat; usually dust or stones. Agronomic and harvesting techniques greatly influence dockage levels. Producers faced with an unusual weed problem or unfavourable harvesting conditions will likely experience higher levels of dockage.

This chapter deals with wheat storage and handling in Pakistan and Afghanistan, UK, the USA and Australia. Wheat is the major food grain of both Pakistan and Afghanistan: they share a similar wheat culture and unlike the other countries mentioned in this chapter rely mostly on smallholder farmers for production. Annual per capita wheat consumption in Pakistan is about 120 kg and in Afghanistan 160–180 kg, making these populations among the largest wheat consumers in the world. In both countries, farm storage facilities are typically traditional structures. In Afghanistan, before 1979 there was substantial large-scale storage capacity within the public and private sector but almost two decades of strife has severely damaged the agricultural marketing and storage infrastructure and there is no officially controlled wheat market. Reliance is now placed on temporary storage facilities such as old school buildings and so on. In recent years, the country has been dependent on food aid imports and trade inflows from Pakistan but agriculture is recovering, helped by a bumper cereal harvest in 2003.

The public sector in Pakistan is responsible for an official grain market and has about 4.3m t capacity of covered wheat storage including bag warehouses, concrete silos and bunkers. In addition, there is seasonal, temporary outdoor storage. However, some of the better storage facilities are under-utilised owing to technical and operational problems; this increases the use of temporary outdoor storage that is associated with greater quality decline. The private sector, primarily traders and flour millers, has storage facilities that are usually sufficient for one week of operation.

Public bodies procure 70% of wheat grain from farmers or through intermediaries at a government-fixed 'floor' price and ensure adequate supplies to deficit provinces; the private sector also moves wheat from surplus to deficit areas. Wheat in Pakistan is harvested dry, usually about 9% m.c., which minimises quality deterioration in storage. The wheat quality standard 'fair average quality' (FAQ) for buying/selling has been in practice since the 1930s. However, there is no objective testing of quality at the time of buying, selling or during storage and FAQ has no legal status. Wheat is purchased from the government by agencies such as the World Food Programme (WFP) and USAID for supply to Afghanistan as food or seed grain. A sizeable quantity is apparently sold through open trade to Afghanistan. The main concerns for Pakistan include minimising post-harvest grain losses, reducing handling and storage costs, establishing grain quality standards, mechanising storage facilities and the resumption of training and research activities in grain storage.

In the UK, wheat is marketed against a range of quality standards depending on the intended end use such as bread making, animal feed, seed, and so on. On-farm storage capacity for wheat is about 14m t mostly stored in bins or silos holding 20–100 t, while commercial storage has capacity for about 3m t in large 'floor' stores. Both store types are usually equipped with under-floor aeration systems. The main quality problems concern grain protein content, specific weight, moisture content and attack by pests. Some of these are related to the UK's humid, maritime climate that accounts for the use of storage strategies that rely on drying and cooling. These are essential for preventing fungal growth, associated ochratoxin production and mite population increase.

Mites are of special concern because they are a potential source of allergens. Temperatures at harvest and after discharge from dryers are ideal for the development of insects and make cooling an essential part of the storage process. Although the average running costs of drying and cooling are relatively modest at about $US4.5 and $0.03 t^{-1}, respectively, storage running costs for farms are more likely to be about $6 t^{-1} and for co-operatives $12 t^{-1}. Future concerns include the need for product traceability, the development of associated recording and decision support computer software, and the need to replace current residual insecticides with other more natural products or physical control measures that maintain equivalent quality standards.

Wheat marketed in the USA is subject to long established standards for quantity and grades specifying quality. The marketing system is supported by a network of silo complexes (grain elevators). Some farmers may store their grain on-farm in steel bins but most move it directly to small country elevators from where it is transported to larger terminal elevators. The majority of elevators built before 1975 are of concrete construction; those built subsequently are generally bolted-steel galvanised bins. The steel bins are often technically superior to the concrete elevators, being equipped with facilities for aeration and temperature monitoring, but many grain managers do not understand how to interpret temperature data and this precludes timely management action and has perpetuated a dependency on residual insecticides.

Pest management problems are greater in the more southerly area of the USA when high grain temperatures following harvest promote rapid development of insects, although mites are not found to be a problem. The number of pesticides available for use on wheat is steadily declining as their registrations are withdrawn by the US Environmental Protection Agency (EPA). Malathion is still used widely on grain, mainly by farmers, even though there is extensive insect resistance to it. A new and effective residual insecticide, Storcide, was approved for use on grain by the EPA in 2003. However, there is growing demand for wheat that is free of both insects and insecticide residues.

There is also a demand for food safety that necessitates traceability of grain to source. This has called for improvements to the application and integration of existing options for quality preservation which is being pursued through a package focusing on sanitation, loading, aeration and monitoring (SLAM). The adoption of SLAM, especially better aeration practices coupled with the introduction of closed loop phosphine fumigation, is leading to significant improvements in grain quality.

Australia has a unique wheat storage and handling system that has evolved over several decades. It consists of a range of storages found across the wheat growing areas. These include permanent sheds, bins and silos and semi-permanent bunker storage. The majority of these stores are gas-tight and so can be fumigated. They are able to receive an entire year's production within a few weeks. Thereafter, most of the crop is transported by rail from country storage sites to export terminals, which have high rail receival rates, and ship loading capacities of up to 5000 t per hour. The costs of the semi-permanent bunker stores are considerably lower than for other store types; for this reason few of the replacement stores constructed in recent times have been of the permanent type.

Australian wheat is usually received in storage at low m.c.; this minimises spoilage caused by moisture migration and mould growth. However, insect infestation still poses a threat to grain quality preservation. The grain storage and handling industry has invested significantly in systems that ensure wheat quality is carefully monitored on intake, is segregated and stored according to very specific receival standards, then delivered to end users in accordance with their quality requirements. All export wheat is subject to inspection for insect pests by the Australian Quarantine Inspection Service that operates a policy of 'nil tolerance'. The application of this measure over the last 30 years has enabled the Australian wheat industry to maintain a reputation for exporting largely insect-free grain.

The state-owned Australian grain storage organisations, which were established early in the twentieth century, have now been privatised. Economic decisions made by these publicly owned companies have driven a number of changes, including a reduction in the number of country receival sites and greater rationalisation of rail transport systems. These companies are now merging and diversifying into other areas of business such as grain marketing, rural merchandising and raw material processing.

Pakistan and Afghanistan

Historical perspectives

In Pakistan the major demand is for hard, *chapatti* quality wheat. About 80% of the wheat grown by the four provinces of Pakistan comes from the Punjab that

harvests at least 15m t from an area of over 6m ha. This production makes up a shortfall in the other three provinces, Sindh, Balochistan and North West Frontier Province (NWFP).

> 'Pakistan has achieved remarkable progress in the development of its wheat-producing sector during the past. National wheat production increased from 4m t in 1965–66 to over 7m t by 1968–69, making Pakistan the first developing country in Asia to achieve self-sufficiency in wheat production; the period was marked as the Green Revolution.'
>
> *Borlaug (1989)*

The Punjab is blessed with a 100-year-old canal irrigation system, although there is a large area of the western side dependent on rainfall where the introduction of improved wheat varieties is central to raising productivity. In Afghanistan, 69% of wheat comes from irrigated land; the remainder is rain-fed. Here yields are about 1.66 t ha^{-1}, giving a production of 1.99m t in 2000–01. In remote villages harvesting is still done by manual cutting and threshing but elsewhere tractor-driven threshers or combine harvesters are used.

Grain storage is not new in Pakistan and archaeological evidence suggests that the ancient civilisation in the Indus valley established granaries for grain storage. However, British India was a pioneer in the development of organised agricultural research. In 1905, it created the Punjab Department of Agriculture and the Punjab Department of Agriculture College and the research institute at Lyallpur (now Faisalabad) was established in 1906. Since Pakistan independence in 1947, the Provincial Food Departments (PFDs) have been responsible for regulating the food grains' business including purchases, storage, sales, transfer, milling and so on, under the Foodstuff (Control) Act 1958. These departments also established ration shops where people could purchase wheat flour at subsidised prices; however, these were discontinued in 1986.

Since the 1960s considerable technical assistance has been given by the international community for the design, construction and mechanisation of the Pakistani grain storage system. Specific assistance from international donors was extended to Pakistan for improving efficiencies in grain storage management. With the assistance of many donor agencies, especially WFP, Pakistan has supplied wheat to Afghanistan since the 1960s. There has also been a silent movement of substantial quantities of wheat from Pakistan to Afghanistan over many years. Wheat is taken along the transit route to

Afghanistan by truckers as grain is easy to sell and there are growing numbers of mills in the Afghan-neighbouring provinces of NWFP and Balochistan to supply flour. PFDs often enforce laws (section 144) to prevent inter-district and inter-province movement of wheat and to check smuggling of wheat out of the country. However, such bans do not prohibit movement of baked bread or other products across the border.

Facilities and practices at storage sites

Storage capacity

Grain in both Pakistan and Afghanistan may be stored by government organisations, flour and feed mills, seed corporations or by farmers. On-farm storage facilities are simple and include mud bins, metal drums (Fig. 3.1), *kothis* (small compartments; Fig. 3.2) and *bokharies* (straw structures). Farmers also store their grain in bags (Fig. 3.3). In both countries wheat is harvested in the dry summer months of April–June. In Afghanistan and remoter areas of Pakistan the crop is harvested manually, tied into small bundles and stacked until it is threshed; elsewhere harvesting is mechanised. Wheat is transported from farms mainly by animal driven carts or carried on camelback. Large-scale farmers use tractor-driven trolleys and trucks. Whatever the means of transport, grain is moved in sacks.

In Afghanistan, after wheat is delivered by farmers to the village market or to a government food corporation it may be cleaned, usually manually by labourers or oc-

Fig. 3.2 A grain storage compartment, *kothi*, constructed in a house (Pakistan).

Fig. 3.3 Wheat stored in polypropylene bags on a small farm (Pakistan).

Fig. 3.1 Half-tonne capacity metal drums for use as grain stores, for sale at a village shop (Pakistan).

casionally by machine. Flour mills hold small amounts of grains for milling purposes in sheds or conventional bag warehouses, while large-scale storage facilities are generally lacking. In 1989–90 a survey of 11 provinces in Afghanistan by the United Nations High Commissioner for Refugees (UNCHR) included information on the availability of grain storage facilities. This showed that at that time in Herat there was one silo complex of 39 000 t, other facilities of 6000 t with the co-operatives and in Kandahar an aircraft hangar that could house 1000 t. All other facilities were of a temporary nature, such as school buildings. At the time of writing the situation appears similar, farmers continue to use traditional storage and permanent large-scale facilities have still to be restored around the country. The International Centre for Agricultural Research in the Dry Areas (ICARDA) is assisting construction of farm-level storage facilities in Afghanistan and food aid agencies such as the WFP have established temporary warehouse facilities in Pakistan and Afghanistan for emergency food distribution.

In Pakistan, large-scale wheat storage is undertaken by the public sector to ensure national food security. The main public sector wheat handling agencies are the Provincial Food Departments (PFDs), the Pakistan Agricultural Storage and Services Corporation (PASSCO), the Federal Food Directorate, the Defense Department, the National Logistic Cell (NLC) and the seed corporations.

Most storage capacity in Pakistan is found in the Punjab (Table 3.1). There are six different types of grain storage facility (Table 3.2). Five are permanent or semi-permanent types – conventional warehouses (Fig. 3.4) and binishells (Fig. 3.5) for the storage of bagged grain, and hexagonal bins, circular bins, silos and bunkers for bulk grain. The sixth type, cover and plinth storage, is a temporary outdoor method. Here date mats and polythene sheets are laid on hardstanding, usually raised on a plinth, bag stacks are built on the mats that are then covered with tarpaulins or polythene (Fig. 3.6). These stores are called *gunjies* and may be of 150–200 t capacity.

There is a substantial shortage of permanent storage in both the public and private sector and about 40% of wheat stocks are held in *gunjies*. Temporary outdoor storage, for periods of six to twelve months, may adversely affect wheat quality and substantial quantities are reported damaged at public and private sector facilities. The deficit provinces prefer to receive wheat from

Table 3.1 Public sector storage capacity* in Pakistan by province in 2003 (Ministry of Food Agriculture and Livestock, Islamabad).

Provinces/PASSCO	Capacity in tonnes ('000)
Punjab	2 483
Sindh	709
NWFP	365
Balochistan	223
PASSCO†	441
Total	4 221

*Excluding temporary outdoor storage.
†PASSCO (Pakistan Agricultural Storage and Services Corporation, under federal government control).

Table 3.2 Large-scale grain stores used in Pakistan by public and private sector.

Store type	% permanent storage capacity	Description of store type
Conventional warehouse	69	Warehouses vary in design and capacity from 500 to 1600 tonnes, constructed in different parts of the country to suit local climate and specially bag handling requirements
Hexagonal bins	6.9	Mostly constructed before 1947. Hexagonal bins are structures of beehive shape with per bin capacity of 35.7 tonnes and the total single site capacity ranging between 500 and 3000 tonnes. The bins are supported on columns 2.6 m above ground level
Binishells	14.4	Dome-shaped structures made of reinforced concrete, built recently to increase covered storage capacity on an emergency basis. The height of the dome in the centre is about 10 m while floor area has diameter of about 32 m. Each binishell has a storage capacity of 1500 tonnes of bagged grain
Bunkers	6.0	Of Australian design and built in the 1990s with World Bank funding. Mostly held by PASSCO in Punjab, Sindh and Balochistan. Not being utilised as bulk storage due to non-availability of PVC sheets and other technical reasons
Concrete/steel silos	3.7	Concrete silo complexes have grain storage capacities of about 50 000 tonnes. Built in the 1990s by National Logistic Cell for Ministry of Food Agriculture and Livestock, these facilities are located in Punjab (3), Sindh (1) and Balochistan (1). Steel silos with Punjab Food Department in Multan (1) 30 000 tonnes capacity. Imported steel silos are mostly used by the feed industry at various locations.
Temporary cover and plinth storage	—	In Pakistan about 60% of wheat stocks are stored in bags outdoors. Bag stacks are built in pyramid shape on date mats and polythene and covered with tarpaulins/polyethylene. There are concerns about the deterioration of quality of grain in these stacks

Fig. 3.4 Conventional warehouses holding wheat grain in either jute or polypropylene bags (Pakistan).

Fig. 3.5 A binishell made of reinforced concrete with capacity for the storage of about 1500 tonnes of bagged wheat (Pakistan).

Fig. 3.6 Typical cover and plinth open-air storage, *gungi*, a bag stack of wheat built in a pyramid shape on hardstanding and covered with a tarpaulin. Seen here between two bag warehouses (Pakistan).

the issuing agencies from permanent covered storage since wheat held in *gunjies* is more likely to be habouring insect infestation, and wheat in the outer layer of bags may be badly affected by heat and increased moisture.

Until recently, bag storage in Pakistan has relied entirely on the use of jute sacks of 100 kg capacity. A large industry is devoted to weaving these bags using jute imported from Bangladesh. In 2000, the PFDs of the Punjab and Balochistan purchased substantial quantities of woven polypropylene (WPP) bags for wheat storage. It is expected that the use of WPP bags will become widespread; an unpopular move as it threatens the jute bag industry.

In both the public and private sectors, about 10% of capacity of the more sophisticated storage facilities, especially hexagonal bins (Fig. 3.7), bunkers (Figs 3.8 and 3.9) and grain silos (Fig. 3.10), is under-utilised due to technical or managerial constraints. The main problem

Fig. 3.7 Hexagonal (hex) bin storage facilities in the Punjab (Pakistan).

Fig. 3.8 Bunker storage facilities with covers off (Pakistan).

Fig. 3.9 Bunker storage facilities with covers on (Pakistan).

Fig. 3.10 Concrete silos in Faisalabad (Pakistan).

is a lack of bulk handling equipment for the operation of hexagonal bins and both bulk handling equipment and suitable cover sheeting for bunkers.

Objective of storage

There is no officially organised wheat marketing system in place in Afghanistan. Marketing intermediaries or stockholders purchase wheat at the time of harvest and sell directly to flour millers or other traders. Public sector bodies in Pakistan, the PFDs and PASSCO, procure 70% of wheat grain from farmers or through intermediaries. The government fixes a support price for wheat well before each annual harvest. In 2002–03, this price was R7500 t⁻¹ (about \$US125) which was little different from the issue price. The price at which wheat is issued to the flour mills does not cover the total cost, i.e. the price at which wheat is procured plus the incidental costs. The

issue price fixed by the government for wheat flour was at a subsidised rate of R8130 t⁻¹ (about \$134). Farmers can sell their wheat at temporary procurement centres established by the government. They can also sell to flour mills that buy grain in May and June until wheat is released from public sector storage from July onward.

Private sector traders buy wheat from the open market in the Punjab and transport it to distant locations (300–600 km). These consignments are sold to flour mills in the deficit provinces, which desperately need wheat to make flour for the open market. The PFD of the Punjab and PASSCO supply wheat to deficit provinces at prices fixed by the federal government. Wheat is also purchased from the government by agencies such as WFP and USAID for supply to Afghanistan as food or seed grain. A sizeable quantity is reported sold through open trade to Afghanistan.

Pakistani PFDs have to meet procurement targets to stock food security reserves and supply wheat to flour mills at issue price. In order to meet these needs, inter-district and inter-provincial wheat movement is often restricted under administrative orders (section 144) that also control movement of wheat out of Pakistan by trade or smuggling. The enforcement of such restrictions causes producer prices to fall in the surplus areas, to the disadvantage of growers.

In 2002 Pakistan exported wheat for the first time since 1981, about 600 000 t, mostly to the Middle East and Africa. In 2003 PASSCO sold 300 000 t of Punjab wheat from the 2002 harvest for export from its 508 000 t stockpile. In order to ensure that the grain did not end up on the local market, it was sold to local exporters at a raised rupee price and a refund offered after the grain had been shipped. In previous contracts tenders were only invited where exporters had confirmed international orders; this change in practice has helped exporters by giving them more time to take positions according to market conditions.

The Pakistan wheat market is well established. Buyers and sellers normally know the characteristics of grain they require in order to prepare *atta, maida* and other flours, and no grain testing is done at procurement, during storage or during distribution. It is left to the visual inspection and good judgement of the buyer or seller. There is an expectation that grain reaches 'fair average quality'. This concept originated in the grain exchanges of Pakistan that were common in Lahore, Lyallpur, Okara, Multan and other cities in the 1930s. FAQ was developed for each exchange and adjusted to the quality of each harvest. The FAQ specifies m.c., shrunken, bro-

Table 3.3 Specifications of fair average quality (FAQ) for wheat procurement by the Pakistan Agricultural Storage and Services Corporation (PASSCO wheat policy 2003, unpublished).

Quality variable	Tolerance limit
Moisture content	10% (max)
Test weight	76 kg/hL (min)
Foreign matter (dirt/dust and other non-edible matter)	0.5% (max)
Broken and shrivelled grains	3% (max)
Other food grains	3% (max)
Damaged grains	0.5% (max)
Wet gluten	26% (min)
Protein (dry basis)	10% (min)

Table 3.4 Specifications of fair average quality for wheat procurement by the Punjab Food Wheat Procurement Scheme, 2002–03.

Foreign matter	Acceptance limits
Inorganic matter	0.25%
Organic matter (barley straw, weed sand and other edible food grains)	Up to 2.0%
Shrunken and broken grains	Up to 3%
Damaged grains	
Insect damaged	Nil
Heat damaged sprouted	Up to 0.5%

Note: FAQ buying variations at wheat harvest – no mention of moisture limits by the Provincial Food Department in wheat procurement policy.

ken and insect-damaged grain (Table 3.3) but is checked mainly by visual inspection. PFDs may define their own FAQ; that used in the Punjab is shown in Table 3.4. Wheat produced is typically superior to FAQ because harvesting is now largely mechanised and wheat varieties improved.

A set of wheat quality standards (Table 3.5) have been proposed for Pakistan (Maxon *et al.* 1994) but these have yet to be adopted. Since 1972, when it became necessary to test the chemical composition of wheat products, use has been made of the quality standards for wheat referred to in the 'Food Composition Table for use in East Asia' (FAO 1972).

Table 3.5 Proposed grading standard for wheat in Pakistan. Source: Maxon *et al.* (1994).

Grade	A	B	C
Moisture contents %	> 9.1	9.1–10.0	10.1–12.0
Foreign matter %	0–0.5	0.6–1.5	1.6–2.5
Broken and shrivelled %	0–1.5	1.6–2.0	2.1–5.0
Other food grains %	0–1.0	1.1–2.0	2.1–5.0
Damaged grains %	0–0.5	0.6–1.0	1.1–2.0

The following variables are important considerations for wheat quality.

Moisture content

Under hot conditions in Pakistan and Afghanistan m.c. values above 12% encourage the development of insect infestation and above 14% encourage mould growth. Harvested wheat in Pakistan is typically dry and grain would be expected to be at a m.c. of not more than 9%. In Afghanistan it is not uncommon for newly harvested wheat in humid areas to be at 15% m.c. so that it must be dried immediately to protect it against mould. At 14% m.c. grain can be safely stored for two to three months but for longer periods (4–12 months) it must be reduced to 12% or below.

Protein

Wheat protein content varies from 6% to 20%, depending on cultivar, class and environmental conditions during growth, although Pakistani wheats typically have 9–11% protein.

Hagberg falling number

In a study of 30 different wheat cultivars harvested in the 1994 and 1995 seasons, the former had an average HFN value of 701 s and the latter 684 s, which proved to be statistically significant. Nevertheless both were well above the minimum acceptable value of 400 s, indicating that both had the very low α-amylase contents typical of the dry harvested Pakistan wheat (Butt *et al.* 1997).

Seed quality

Good seed is an important contributing factor to a high-yielding, good-quality harvest while poor seed that may be diseased and/or mixed with weed seeds will result in a poor yield and may have repercussions for future harvests. The germination rate of seed may be reduced by poor storage conditions, especially high temperatures, high moisture conditions and/or insect attack. Most farmers use their own seed, which in unsanitary crop production conditions becomes infested with off-type wheat or weed seeds and, over time, accumulates seed-borne diseases. The use of such seed may result in poor crop yields. For improved production farmers should raise their seed in separate fields, using low sowing rates and weed control measures. The supply of good-quality seed has been a crucial component in efforts to restore Afghan agriculture. In 2003 USAID provided 7000 t of wheat seed that was estimated to increase the area under harvest by about 70 000 ha, so decreasing Afghanistan's dependency on food aid imports.

Impurities

In the remote areas of the Punjab, Sindh and Balochistan high levels of impurities such as sand or dust result from manual harvesting. Mechanised harvesting produces cleaner wheat although dockage, shrivelled and blackened grains are still prevalent. Similarly in Afghanistan traditional grain handling techniques lead to high levels of impurities.

Major causes of quality decline

Grain temperature and moisture affect the rate of quality decline of stored wheat. Low temperature and moisture are associated with good quality preservation, high temperature and moisture with attack by insect pests and micro-organisms. For long-term storage, wheat should be at 12% m.c. or less. However, before milling the grain is conditioned using wet cleaning that results in the m.c. rising from 13.5% to about 15%. This could adversely affect the shelf life of the flour but usually does not as the product is consumed within a few days. An exact measure of grain m.c. at the time of purchase is important since prices should be reduced if grain weight is increased by excess moisture. Moisture meters are not used at the time of harvest, storage or milling in Pakistan although those working in the grain industry are well practised in recognising appropriate moisture levels by the feel and hardness of the grain.

Pests

In a study of farm storage in irrigated and rain-fed areas of the Punjab, wheat losses averaged 5.2% and showed the greatest rise during the monsoon period (Baloch *et al.* 1994); the monsoon similarly affects stored sorghum in South India (see Chapter 5). In 1984–85, a preliminary review of public sector grain storage facilities in Pakistan confirmed that losses due to insect infestation, mould growth and the activities of birds and rodents are important. Insect pests are the most serious (Table 3.6) and substantial losses of grain caused by insects have been seen at flour mills in Pakistan. The most important storage insect pests are the beetles *Sitophilus oryzae*, *Rhyzopertha dominica*, *Trogoderma granarium*, *Tribolium castaneum* and *Oryzaephilus surinamenisis*, and the moths *Sitotroga cerealella*, *Ephestia cautella* and *Corcyra cephalonica* (Vol. 1 p. 100, Hodges 2002). In a study of farm and public facilities the most frequently occurring pests were *R. dominica*, *S. oryzae* and *T. castaneum* with little difference between farm and public storage.

In Afghanistan it is estimated that over 200 000 ha of irrigated wheat production in 2002 was rendered unusable after being infested by Sunn pest (*Eurygaster integriceps*). This insect is a pre-harvest pest that injects chemicals into the developing grain, causing the gluten to break down. If as little as 2 or 3% of the grain in a crop has been affected, the grain is unusable for baking.

Surveys for vertebrate pest infestations at 349 locations at provincial grain storage centres in Pakistan showed that about 29% were infested by one or more rodent species and about 78% were infested by several bird species (Brooks & Ijaz 1988). The overall loss from these vertebrate pests was estimated to be 0.2 to 0.5% of all stocks. The most important contributing factors were poor storage sanitation and poor structural maintenance.

Table 3.6 Estimates of public sector wheat storage weight losses 1984–85 in various provinces of Pakistan. Source: Baloch *et al.* (1994).

Province	Mean months in storage	Mean % weight loss			
		Insects		Moulds	Total
		Pre-storage	Storage		
Balochistan	2.6	0.5	1.2	0.5	2.2
North West Frontier Province	6.5	2.9	2.6	0.7	6.2
Punjab	6.3	0.1	1.8	0.3	2.2
Sind	6.4	0.1	2.9	0.3	3.3

The common rodent pests are the house mouse (*Mus musculus*) and the roof rat (*Rattus rattus*), and major bird species found around grain storages are house sparrow (*Passer domesticus*), domestic or wild rock pigeon (*Columba livia*), collared dove (*Streptopelia decaocto*) and house crow (*Corvus splendens*). Sparrows can damage bags by pecking holes in jute, increasing spillage and contaminating the grain with their copious droppings. In Afghanistan rats and *Bandicota bengalensis* are serious pests and in upland valleys of both wet and dry mountains losses are reported in the range 1–6%.

Fungi and mycotoxins

Wheat if dampened will support fungal growth, leading to mouldy and blackened grains. While such wheat is rendered unfit for consumption it may also be dangerous for humans or other animals to consume as moulds may produce mycotoxins including aflatoxin and fumonisin (Vol. 1 p. 128, Wareing 2002). To date aflatoxin has not been detected on wheat in Pakistan (Shah & Abdul 1992) but tests have not been made for other toxins.

Incorrect handling

In order to move wheat grain from the farm to market, farmers pack their grain into jute or WPP bags of 100 kg capacity. The bags are handled manually using hooks and may be thrown from shoulder height by manual labourers. This causes damage to the bags and grain, and poor stitching of bags allows spillage. These problems are particularly evident at temporary procurement centres and in Port Qasim when loading and offloading vessels.

Commodity and pest management regimes

Farm storage

In farm storage, traditional methods are employed for the protection of wheat. These include sun drying which is simple and cheap but involves risks such as grains getting wetted or stolen or the grain becoming contaminated with foreign materials such as dirt and stones. Protectants are also admixed. A mixture of mercury and sand is a tradition in South Asia, especially in the Punjab provinces of both Pakistan and India and nearby districts, despite its potential hazards. The neem tree (*Azadirachta indica*) is native to the Indo-Pakistan subcontinent and grows abundantly in this region; farmers sometimes mix dried neem leaves with food grain stored in jute bags. Those who store wheat in mud bins rub fresh neem leaves on the inside walls. In the districts of Nawabshah and Khairpur of Pakistan, straw storage structures (*bokharies*) are in common usage. In Rahim Yar Khan District, neem extract is sprinkled on the wheat straw packed at the bottom of *bokharies* before loading the grain. Considerable research has been undertaken on the properties of neem as a grain protectant (Ghluam 1992), however it is unlikely to be adopted commercially for direct application to food due to its bitter flavour.

The use of pesticides (insecticides, herbicides, fungicides) has risen steadily in Pakistan over the past 20 years. The total consumption of pesticides reached 45 680 t, in 1999 worth R7324m (about $121m). The use of pesticides is more common in the irrigated areas where 13% of farmers use insecticides and fumigants and 41% treat the grain with mercury. Although some degree of control seems to have been achieved, most chemical treatments are unsatisfactory and can be dangerous to human health. Moreover, the widespread and uncontrolled use of pesticides wastes scarce resources when treatments are ineffective. The exposure of insect pests to sublethal doses may promote the development of resistant strains of insect pests.

Large-scale storage

Hygiene and monitoring

Store hygiene is an essential element of commodity management. This should involve pre-storage inspection of buildings, cleaning and treatment of the surfaces of empty stores with insecticide. These procedures are generally not implemented, resulting in the multiplication of insects in public and private sector stores. In flour mills cross-infestation to finished products from infested residues in the mills is a serious problem.

Ensuring that grain entering storage is of good quality is an important element of commodity management. At the time of harvest wheat is delivered in sacks to the procurement and storage sites. Sacks are sampled randomly; for each 100–150 bags, three or four bags are selected and a sample removed from them with a grain probe. The grain is given a visual inspection before being accepted.

Insecticide and fumigant application

Both chemical (insecticides and fumigants) and non-chemical techniques are used for insect control in large-scale wheat storage in Pakistan. Public sector grain managers use contact insecticides (malathion, lindane, pirimiphos methyl, fenitrothion, dichlorovos or bioresmethrin) to spray the surfaces of bags stacked inside or outside warehouses.

Phosphine is used for fumigation but as normal tarpaulins employed in open storage (*gunjies*) are insufficiently gas-tight, gas concentrations are inadequate and such treatments frequently fail. In silos the current practice is to apply aluminium phosphide at the rate of two 3-g tablets per tonne. All tablets are placed on the wheat surface. The manhole lid is closed and sealing is supplemented by mud plaster around the top lid and the bottom spout opening. A critical study of this methodology has shown it to be unsatisfactory as the required gas concentration is not achieved for the minimum exposure period; this results in fumigation failure (Tariq *et al.* 1994). Such failures not only increase grain protection costs but also encourage the development of phosphine resistance in

insects. To improve fumigation performance a technique has been developed to seal bag stacks into a plastic envelope prior to releasing the fumigant. The envelope is airtight and so ensures a good retention of the fumigant. It can be kept in place once the fumigation is completed, so preventing reinfestation; the same technique is used in rice and coffee storage (see Chapters 1 and 11).

Strains of insect pests with pesticide resistance are quite common in Pakistan, particularly in large-scale premises, due to continuous use and misuse of pesticides. In a survey during the 1980s, resistance to both insecticides and fumigants was documented in storage insects pests. Of 33 strains of *S. oryzae,* 13 were resistant to malathion, and none of 40 strains of *T. castaneum* was susceptible to this insecticide. At the time of testing malathion had been in use in the public storage system for about 20 years. Five strains of *T. castaneum* showed some resistance to phosphine (Baloch *et al.* 1994).

Economics of operation of the system

The food and fibre sector is a dominant force in the Pakistani national economy, provides revenue for both the federal and provincial governments and is the source of raw materials to other developing sectors. From the point of view of wheat producers, the government support price and issue price policy have a negative effect on the wheat market. Wheat issued to flour mills is subsidised, i.e. costs less than the procurement, storage and handling costs. This distorts the workings of the free market since private traders are unable to compete for the supply of wheat and farm gate prices remain low. On the other hand, urban consumers have access to cheap food products. A completely free market may enable entrepreneurs to make excessive profits at the expense of society at large while the government monopoly works to the disadvantage of the producer. It is clear that a more equitable balance between the producer and consumer is required. This will result in a more effective market – one in which there is more incentive for the producer, better information flow on prices and consumer quality demands, sufficient financial surplus for investment in better storage systems and an assurance of food security for the urban consumer.

The average public sector handling and storage expenses for wheat in the year 2002–03 ranged between R1600 and R1800 t^{-1} (about $27–$28). Much could be done to improve the cost performance of the system (Table 3.7). In the public storage system there is an accounting convention known as the 'no loss policy'.

Table 3.7 Possible approaches to improving the cost performance of large-scale storage in Pakistan.

Cost item	Reasons for increased costs	Suggested measures
Jute bags	Rise in costs of imported raw jute and of local manufacture	Bags should be saved for reuse at next harvest. This requires that labour should avoid excessive use of hooks that can damage the bag
Prolonged storage	Late discharge of grain leading to deterioration of the stocks and higher bank loan repayments	Priority to be given to discharge of stocks in temporary storage to avoid use of excessive dunnage material. Faster turnover of stocks can reduce interest payment on loans taken out to finance procurement and handling
Use of insecticide	Frequent reapplication of insecticide to bag and store surfaces to reduce infestation problems	More rational use of pest management, better fumigation performance will reduce the need for repeat spraying with insecticide
Transportation	Transportation of wheat from distant locations, 200–500 km	To avoid long-distance transport, procurement agencies should be allocated permanent operational areas
Efficient utilisation of storage	Under-utilisation of covered storage facilities	To make full use of existing covered storage so minimising outdoor, temporary storage
Losses of grain and material	Grain losses and quality decline of wheat in the storage system	Training for better management and operational techniques at public and private sector storage sites
Labour costs	Labour costs are increasing due to repeated handling and extra activity for cleaning and reconditioning of open stack	Efficient use of existing covered storage and use of mechanised grain storage facilities can reduce extra labour costs incurred in various operations
Reconditioning of deteriorated wheat	Stocks need to be reconditioned if grain is infested or damaged	Correct timing and application of pest management and appropriate storage practices will reduce the need for costly reconditioning

This states that once grain enters a government store no changes in weight can be made. This convention has been put in place to prevent pilfering from stores but, while individuals are aware that losses occur, the official policy is a disincentive to improvement in storage performance since the system is officially perfect. Grain storage losses are a serious cost to the economy, especially when they result in a need for food grain imports.

The milling sector is also facing enormous challenges; about 20% of newly built flour mills are inoperative due to technical problems. The main concern of the industry is to procure good-quality wheat to make good-quality flour. A change from bag to bulk grain handling, from shift washing (wet cleaning) to dry cleaning and development of appropriate pest management practices will do much to improve milling operations.

Future developments

For Afghanistan, future concerns include the revival of the economy and rebuilding of the infrastructure for agriculture. In the past the country was a net exporter of a wide range of high-quality agricultural commodities, especially dried fruit and nuts. Reconstruction of grain storage and marketing infrastructure will help farmers to grow more wheat and reduce Afghanistan's dependency on the international community.

The post-harvest grain sector in Pakistan faces numerous problems. There has been little or no technological improvement in recent years and the difficulties are exacerbated by inadequate management strategies in both the public and private sector. The inadequate use of grain storage and milling facilities in Pakistan places a burden on the national economy.

Wheat will continue to be the largest crop in the future. Strong demand for wheat means that it has eclipsed other crops that are essential food items, especially oilseeds and pulses. There is a growing deficit in these crops, resulting in large imports; the edible oil import bill was expected to be over R30bn ($50m) in 2002. The country has been importing an average of 2m t of wheat per annum during the decade ending in 2000, although there was a surplus produced during 2002–03. The Punjab is the breadbasket of the country and can continue in this role due to its network of irrigation canals, possibly raising production to over 22m t per annum by the year

2015, over and above its own 14m t annual requirement for wheat.

The government has given an assurance that it will deregulate industry including the export and import of agricultural products and grains, and has recently allowed private sector flour mills to import limited quantities of wheat. The private sector is willing to assume a larger role in grain storage and distribution, but requires technical guidance in the development of grain handling and storage. The long-term intention is to remove all subsidies and taxes, including those on imports and exports. If the agribusiness sector is to develop further it is essential that:

- a complete and reliable market information system is established;
- there are policy changes that give a greater role to the private sector in agricultural marketing;
- government services are expanded for the development of efficient agricultural marketing and storage systems;
- higher quality standards are set for marketed products; and
- new and improved grain storage and handling systems are introduced.

The goal of any development strategy must be to encourage farmers and the private sector to participate fully in the market. In order to preserve valuable food and feed grains a modern storage and handling system has to be established in the country. The first steps towards this will be to reduce handling charges by mechanising the marketing and storage system using low-cost, locally fabricated equipment and to work towards an efficient utilisation of facilities to reduce storage and transportation expenses.

United Kingdom

Historical perspectives

There is evidence that the Romans built granaries with under-floor ventilation (Nash 1978) and today air flow, used for cooling and drying, remains the favoured strategy in the UK for grain preservation. More recently, several changes in agricultural practice and market conditions, that demand consignments free of contaminants, have dictated advances in storage strategies.

During the latter half of the twentieth century, the advent of mechanical harvesting using combines and of hot-air drying have probably had a marked affect on the pest status of species infesting grain in the UK. Previously, the grain weevil *Sitophilus granarius* had been the most common pest because it could penetrate undamaged grain and prosper in the relatively cool maritime northern European climate, developing at temperatures as low as 10–12°C. Mechanical combining damages the grain sufficiently for penetration by species such as the saw-toothed grain beetle *Oryzaephilus surinamensis* and the red-rust grain beetle *Cryptolestes ferrugineus* (Armstrong & Howe 1963). Since most stored product insects are of tropical origin they would normally be able to increase only very slowly, or not at all, at British post-harvest temperatures of around 20°C. However, grain harvested in damp weather often needs to be dried and the swiftest way of doing this is in a hot-air dryer in which layers of grain are exposed to air at high temperatures for a short time. This heats grain several degrees above ambient, which is ideal for the rapid reproduction of these stored grain insects.

In recent times the trend has been away from prolonged on-farm storage towards storage by processors, co-operatives and commercial concerns. The motives for this include the advantage of buying grain at harvest when the grain price is low, rather than when there is demand or when supply may be unsure and the price relatively high. Another advantage for processors is that they will have more control of wheat quality during storage. For farmers the development of co-operatives may save individuals from having to invest in expensive machinery such as dryers and/or storage facilities and to provide the time needed to manage them. The European guaranteed price for grain, which offers the prospect for realising more than the market price at a particular time, also encourages the development of very large stores so that more grain can be stored for longer periods. Thus the size of storage bulks has expanded from 10–20 t bins on-farm, to commercial bins of thousands of tonnes, to 'floor' or 'flat' stores of tens of thousands of tonnes. However, with falling grain prices, the most recent development is for processors to consider reducing their costs by putting the onus for storage back on producers. Today there is no longer a guarantee that the costs of long-term storage will be justified by higher prices later in the season. There is a fine balance between the cost of storage and the added value resulting from storage.

The consequences of storage in larger and larger bulks include slower natural cooling, since grain is such a good insulator. This increases the risk of infestation and quality loss. The storage of grain in such large bulks was only possible because of the development of efficient and cheap cooling strategies and the practice of pesticide admixture to the grain.

Facilities and practices at storage sites

Storage capacity

Grain in the UK may be stored by the growers, by farming co-operatives, at commercial sites or in government-sponsored intervention (designed to stabilise markets under the European Union's Common Agricultural Policy), by food manufacturers at animal feed and cereal premises or mills, or at ports prior to export. It is tempting to think of a neat flow from farm and commercial stores to food manufacturers and export, but there is some flow in all directions. The UK storage system was most recently surveyed between 1987 and 1989 when data were collected from 742 farms and all commercial sites holding over 1000 t of grain (Prickett 1988; Prickett & Muggleton 1991). Some of the findings are presented in Table 3.8.

Table 3.8 Features of the UK grain storage system as surveyed in 1987 and 1989. Source: Prickett (1988), Prickett & Muggleton (1991).

Storage capacity				
On-farms	11.4 million tonnes	72% floor storage* 8% external bins 20% internal bins		
Central stores	2.4 million tonnes	81% floor storage 15% external bins 4% internal bins		
Grain cleaners				
On-farms	32% with cleaners			
Central stores	59% with cleaners	93% hot air dryers 7% ambient air		
Grain coolers				
On-farms	60% with coolers			
Central stores	91% with coolers	3% with refrigeration 8% portable aeration spears 31% ventilated floor 84% above floor ducts		
Pesticides				
	Applied to store fabric	Applied direct to grain		
On-farms	52%	10%		
Central stores	86%	72%		
Sampling for physical factors and insects				
	Temperature	Moisture	Insect monitoring	
On-farms	37%	6%	31%	
Central stores	81%	97%	83%	
Advice				
	Chemical companies	Agricultural companies	Private consultants	Government advisory service
On-farms	10%	63%	3%	34%
Central stores	26%	25%	9%	48%

*Of these, 85% were overloaded with grain above the grain walling.

Farm stores are typically indoor bins or outdoor silos of 20–100 t that are 3 m deep. In contrast, co-operative or commercial operators are likely to keep grain in a pile on a floor, held in place by concrete or wooden walls (Fig. 3.11). Such stores usually hold about twice the depth of farm stores with a capacity of 1000–3000 t. Intervention floor stores often hold tens of thousands of tonnes. Commercial stores may have 'jumbo' outside bins that are up to 10 m deep and hold 1000 t. In most cases the floor stores or bins are fitted with ducts to allow ventilation at rates that lead to grain cooling (Figs 3.11 and 3.12).

Fig. 3.11 Floor store with the grain pile held in place by wooden walls. For aeration of the grain, fans outside the store (Fig. 3.12) are used to force air through the pipes (right-hand side) into under-floor ducts (UK).

Fig. 3.12 External view of a floor store showing a row of aeration fans used for cooling, each fan serves one under-floor duct (UK) (see Fig. 3.11).

Objectives of storage

Intended fate of grain

The UK storage infrastructure survey indicated that 95% of farms intend to sell some grain (Prickett 1988; Prickett & Muggleton 1991). Of the commercial sites 67% intended to sell some grain for export, 30% stored grain for intervention while 19% intended to sell into intervention, 65% intended to sell some to animal feed mills, 41% to flour mills and 18% sold to food manufacturers. At the time of writing, wheat prices in the UK are at an all-time low of about $90 t^{-1} and a premium of $15–22.5 is available for wheat destined for milling. Consequently, if producers are to operate profitably it is important that quality standards for different end-uses are maintained or enhanced by storage.

Quality standards

The standards used in the grading of UK wheat quality are shown in Table 3.9 (Anon. 1986a).

For bread making the protein content needs to be high (10–11%). A safe m.c. is in equilibrium with relative humidity below 65%, which for cereals is 14–15%. A minimum HFN of 220 s is required for bread-making wheat, or common wheat for intervention. A test weight of 76 kg hL^{-1} is expected for bread-making wheat, compared with 68 kg hL^{-1} for standard wheat for feed. Seed grain must show shoot and root development within a given time when provided with adequate water, and germination needs to be 85% for seed. Broken grains, straw, dust, foreign seeds, insects, stones and all other non-grain material can be separated by sieves of different sizes and weighed to give the proportion of impurities. Intervention standards may further define the nature and quantity of the impurities; broken grain, sprouted grain, shrivelled grain, miscellaneous impurities and so on, as may standards for seed and futures markets, defining numbers of specified non-cereal seeds. Acceptable levels are listed in Table 3.9.

Major causes of quality decline

In 1999, a major trader analysed the number of rejections on wheat moved in the last week of November (Allen-Stevens 2000). Of 900 loads there were 162 complaints, 48 for test weight, 46 for moisture content, 35 for protein, 11 for HFN and 10 for insects. Although protein and HFN may be altered by careless high-temperature drying

Table 3.9 Wheat quality standards. Source: *Schering Grain Quality Guide*.

	Minimum values				Maximum values	
	% protein	HFN	% germination	Specific weight kg hL^{-1}	% impurities	% moisture content
Milling						
Bread	10.0–11.0	220		76	2	16
Biscuit	10.0	140		72.5	2	16
Other	9.2	180		74	2	16
Feed wheat						
Denaturable				70	2	16
Standard				68	12	16
Intervention						
Premium wheat	14.0	240		72	10	14.5
Common wheat	11.5	220		72	10	14.5
Feed wheat	—	—		72	12	14.5
Seed						
F1 – HVS			85		1	17
F1 – LVS			85		2	17
F2 – HVS			85		1	17
F2 – LVS			85		2	17
GAFTA futures						
EU wheat				72.5	2	16
Export						
Milling wheat	11.0–11.5	225–250		76	2	16
Feed wheat	(9.8–10.0)	(220) (150*)		70	2	16

These data are only for guidance; all market values need to be checked with customers.
F1 = certified 1st generation seed, F2 = certified 2nd generation seed, HFN = Hagberg falling number, HVS = higher voluntary standard, LVS = lower voluntary standard.
Values in parentheses = general trading standards for low grade milling wheat. *Biscuit milling wheat.

and the latter improved by cleaning strategies, most of the qualities listed are due to growing conditions and do not change in storage. The important issue in such cases is that representative sampling is used to obtain an accurate picture of each load. Moisture content and insects are particular post-harvest problems while temperature is a major problem for grain entering EU intervention storage. Maximum temperatures are laid down for reception according to time of year. In November, grain is acceptable at 18°C, thereafter it should be not more than 15°C but unsubstantiated claims are often made that the temperature rises during conveying, or that low temperatures are unobtainable due to unseasonal conditions.

Excess moisture content

As biological deterioration usually begins at 65–70% r.h., the m.c. in equilibrium with this is usually taken as the 'safe' m.c. for storage and is the target for ambient air and continuous drying operations. The 'safe' m.c. for wheat under UK conditions is in the range 14–15% and is the reason the intervention m.c. is set at 14.5% (Pixton & Warburton 1971).

The measurement of m.c. is crucial and therefore often an object of contention in the trade. Simple meters used by growers depend on electrical qualities that are indirectly linked to moisture, such as resistance and capacitance, and can rarely be more accurate than ± 0.5%. Their accuracy also depends on seasonal calibration as the electrical properties of the grain appear to vary from season to season. Moisture content is more accurately measured by standard oven methods such as ISO 712 (Vol. 1 p. 84, Devereau 2002). Perhaps as important as the method used to determine m.c. is the selection of representative samples. If loads from different sources are mixed within a bulk, if there is moisture stratification due to ambient air drying or a hot-spot, or if the surface is sampled during the winter, it may be difficult to decide which samples represent the bulk.

The m.c. of the bulk may change during storage for a variety of reasons. If the grain is not cooled, then warm air rising from the bulk may condense on the grain surface as ambient temperatures fall in autumn, raising the m.c. there to such an extent that mould growth and sprouting may result. The same may occur when insect activity, usually pre-pupal *S. granarius,* leads to the development of a hot spot. In the UK, however for well-dried grain, the m.c. at the surface will increase as it gains equilibrium with the high r.h. prevalent in winter. The surface of a bulk at 13% m.c. is likely to reach at least 16% m.c. while a bulk at 16% m.c. may increase to 19–21% (Armitage & Cook 1999).

The m.c. of a bulk can, of course, be changed if sufficient air is blown through it and this is the principle of the ambient-air drying technique. In this case, air at $0.05 \text{ m}^3 \text{ s}^{-1} \text{ t}^{-1}$ is blown through the grain and drying should be completed in about ten days. It is often alleged that cooling aeration rates, which are one-twentieth those required for drying, will achieve drying or dampening but, with certain exceptions, this is not the case. Regardless of the ambient r.h., cooling will not redampen the grain because (i) cool air can carry little water, (ii) fans heat the directed air flow by a few degrees, thus lowering the r.h., and (iii) as the cool air meets the warm grain, its r.h. falls. In practice, cooling usually results in a reduction of only 0.2–0.5% m.c. over a storage season. Moisture reductions of 0.2–0.25% for each 5°C reduction in grain temperature occur by a process termed 'dry aeration' (Lasseran 1977). Some aeration systems operate by suction, and in this case there have been instances of a dampening front being pulled through the grain, probably exacerbated by the phenomenon of winter moisture uptake at the grain surface.

Occasionally, after drying, the same high air flow rates are used to cool the grain and in this case it should be remembered that only one-twentieth the time normally required for drying is required. Under such conditions, cooling the grain to ambient temperature may be achieved in a single night. Continued aeration using drying rates after the drying front has passed through the grain could result in redampening of the bulk.

Elevated temperatures

Temperature is crucial to storage deterioration and affects the survival and reproduction of insects, mites and fungi, as well as factors such as germination rate and baking qualities. Less obviously, it affects the relationship between m.c. and r.h. so that, for a given m.c., the lower the temperature, the lower the r.h. The temperature of dryers is also critically controlled for different end-uses to avoid lowering quality (see the section on drying).

In UK stores there is often a requirement for record keeping, and storage temperatures may be measured by an array of sensors connected to a central data logger or a hand probe. The records may be printed out or transferred to permanent records in a book or to record-keeping facilities of a decision support system such as the Integrated Grain Storage Manager (Knight *et al.* 1997). Sensors are normally thermistors, since alcohol or mercury in glass thermometers cannot be used in grain stores because breakage could lead to contamination of the grain. Temperature monitoring is important because it indicates the effectiveness of the aeration system and highlights problems areas. If rapid temperature rises occur it also indicates hot spot formation due to the activity of insects or fungi.

Grain in the UK is normally harvested at temperatures around 25°C, even though the daily average ambient temperature in the UK rarely exceeds 20°C. Grain discharged from hot-air dryers is at much higher temperatures because the cooling sections of the dryer are frequently not efficient enough to reduce the grain temperature to ambient. It is always possible to cool the grain to below 20°C within a fortnight by aeration (see below) and, although it is possible to cool to near 0°C by December, most grain is only cooled to 10–15° for the storage season.

Pests

Most of the serious stored crop pests in the UK are assumed to be of tropical origin and distributed through trade. These include insects (beetles, moths, psocids) and mites (Vol. 1 p. 100, Hodges 2002).

Beetles

The storage beetle pests, *Oryzaephilus surinamensis,* *Cryptolestes* spp and *Sitophilus granarius,* have great potential for population increase and are important for the direct damage they cause and the transmission of micro-organisms. As a result there is no tolerance for insect presence in trading. In the UK survey (Prickett 1988; Prickett & Muggleton 1991) insects were observed in 10% of farm stores and 40% of commercial sites, the most frequently occurring beetles in farm stores being the adventitious species, *Typhaea stercorea* and *Ahas-*

Table 3.10 Proportion (%) of farm stores and commercial sites where beetles were detected (after Prickett 1988; Prickett & Muggleton 1991).

	Oryzaephilus	*Cryptolestes*	*Sitophilus*	*Ahasverus*	*Typhaea*	*Cryptophagus*	Ptinids
Farm	5	5	4	6	20		
Commercial	20	17	17	9	15	20	50

verus advena. Similarly, the most frequent commercial store species were ptinids, principally *Ptinus fur*, and the fungus-feeding *Cryptophagus* spp (Table 3.10).

Cryptophagus spp, *A. advena* and *T. stercorea* are species that would not be expected to reproduce under UK grain storage conditions (e.g. Jacob 1996), requiring a much higher m.c. than that found in stored grain, but they occur widely in association with mould. It is assumed that they invade stores from nearby haystacks and so on, outside the store. *Ptinus fur* is a species that builds up only very slowly as it is of low productivity and diapauses as a pupa in August. Adults are usually only found between October and June and require free water and a diet high in protein. Nevertheless, high populations have occurred in intervention wheat stored for over three years (Armitage *et al.* 1999a). All the adventitious species can survive for some time in cool grain without reproducing and may damage the grain directly.

The most important pest species are all adapted to survive in dry commodities and may live for several months at low temperatures. While the grain weevil, *S. granarius*, can penetrate undamaged grain and completes its life cycle protected therein, *Cryptolestes* and *Oryzaephilus* depend on damaged grain and the larvae are free-living in the space between the grains.

The activity of insects can cause grain heating or insect-induced 'hot-spots' and in the case of *S. granarius* most of the heating is caused by the last instar larvae (Howe 1962). For instance, their activity results in ten times more gaseous exchange than that of half-grown larvae and five times more than that of adults. Heating of grain by insects is thus affected by the age structure of a population. Very high densities of insects are necessary to cause a noticeable increase in temperature, but nevertheless temperature rises of 10°C in two weeks are possible since populations can increase 1000-fold in 16 weeks at initial temperatures of 23°C. The spread of the hot spot is due to mass movement of insects migrating from hot areas caused by insects breeding in small pockets. Grain as near as 50 cm to the hot spot can remain cool as steep temperature gradients occur between the heated pocket

and uninfested grain nearby. Most heat is produced near the edge of the hot spot.

Hot spots caused by insects that can develop at low temperatures (e.g. *S. granarius*) can allow a succession of infestation, enabling more active insects that require higher temperatures, such as *O. surinamensis* and *C. ferrugineus*, to increase in number explosively. As hot air rising from these hot spots condenses on cool grain at the surface, mould growth may take over and allow thermophilic fungi to eventually push grain temperatures toward 50°C, beyond the lethal threshold of insects.

Moths

Moths tend to be a problem only on the grain surface, as their soft-bodied larvae, which cause the damage, cannot easily squeeze between the spaces of compacted grain. They cause direct damage to grain but their webbing and faeces may also foul machinery and so on. *Ephestia* spp were present in 3% of farm stores and 2% of commercial sites while *Endrosis sarcitrella* occurred on 27% of farms and 6% of commercial sites. *Hofmannophila pseudospretella* was found on 8% of commercial sites and moths, as a whole, occurred on 65% of commercial sites (Prickett 1988; Prickett & Muggleton 1991).

Psocids

Psocids are almost omnipresent but, because they require high humidity, they are largely associated with the grain surface in winter. They have been found on 52% of farms and 87% of commercial sites, the most common species being *Lepinotus patruelis* and *Lachesilla pedicularia* (Prickett 1988; Prickett & Muggleton 1991).

Mites

Mites are ideally suited to the cool, maritime climate of the UK, where they have received a considerable amount of attention as their populations are often large. They damage grain directly, often attacking wheat germ and so the cavity they create in the germ may be found packed with, for example, *Acarus siro*. In experiments it may appear that mite-infested grain has a higher germination rate when compared with uninfested grain as mites inhibit fungal growth and hence the associated damage (Armitage & George 1986). However, when a comparison is made with fungicide-treated grain it is clear that the mite damage itself reduces the germination rate (Zdarkova 1996). They impart a taint to the grain, often imaginatively described as 'minty'. More seriously, they produce respiratory allergies and occasionally have been suspected of causing anaphylactic shock after ingestion (Blanco *et al.* 1997) or even of being involved in the transmission of prions (Wisniewiski *et al.* 1996). The biology of a wide number of UK storage and domestic species is summarised by Hughes (1976).

Mites have been found on 72% of farm and 87% of commercial sites where the commonest species were *A. siro*, *Tyrophagus longior* or *Tyrophagus putrescentiae* and *Lepidoglyphus (Glycyphagus) destructor* (Table 3.11).

The predatory mite, *Cheyletus eruditus*, is frequently present but requires higher temperatures to reproduce than the pest species although it tolerates lower r.h. and so usually occurs in the spring. Other predators frequently encountered include *Androlaelaps* spp and *Haemogamasus* spp. Although predatory mites have potential control benefits in an integrated pest management strategy, they are unfortunately no more tolerated in trade than the pest species.

Fungi and mycotoxins

Under the dry, cool conditions recommended for UK storage there should be no growth of fungi yet, in 1997, 20% of samples of cereals contained ochratoxin A (OA) in excess of 1 μg kg^{-1} (Scudamore *et al.* 1999). Under UK conditions, OA is produced by certain strains of *Penicillium verrucosum* that require 80% r.h. for growth (Northolt & Bullerman 1982) and 85% r.h. for the production of the toxin, corresponding to a wheat m.c. of about 17% and 18.5%, respectively (Henderson 1987). However, in laboratory tests, OA has been found in samples at between 16.3% and 18.3% m.c. (Hetmanski 1997). Slow drying, such as near-ambient air drying, appears to risk

the production of OA as the fungus can grow in undried grain above the drying front. If this happens then the engineering process or method needs refinement. Another hazard is the harvest backlog, when undried grain may be held for several days, even weeks, before it can be passed through a continuous-flow dryer. Normally though, the fungi most common on UK grain are those tolerant of cool, dry conditions, such as *Aspergillus restrictus* and *Aspergillus glaucus* (perfect stage *Eurotium* spp) species groups and *Wallemia sebi*. *Aspergillus* species may cause germination loss and grain heating (Christensen & Kaufman 1969), while *Wallemia* is responsible for the production of its own toxins (Wood *et al.* 1990) but is often overlooked because of its small colony size and media requirements during isolation.

Rodents and birds

Vertebrate pests cause some direct damage to grain but are mainly important because of their indirect damage to structures, faecal contamination and spread of pathogens and invertebrate pests during their movement between stores. About 72% of farms and 59% of commercial sites suffered from bird pests – house sparrows, feral pigeons and collared doves. Rats and mice were found on 70% and 79% of farm and commercial premises, respectively (Prickett 1988; Prickett & Muggleton 1991). Exclusion is the most important control strategy, supplemented for rodents with use of rodenticides following baiting regimes to aid assessment of the population size and to habituate the rodents to baits (Vol. 1 p. 284, Meyer & Belmain 2002).

Grain properties

Seed senescence and dormancy

During storage grain will slowly break dormancy and then the ability to germinate will decline at a rate dependent on grain m.c. and temperature. Pre-storage damage such as that caused by combine harvesting and excessive heating during drying will have an effect on the initial condition of the grain. Although for most markets wheat germination is not as important as for malting barley

	Acarus	*Lepidoglyphus*	*Tyrophagus*
Farm	55	47	26
Commercial	68	64	32

Table 3.11 Proportion (%) of farm stores and commercial sites where mites were detected (after Prickett 1988; Prickett & Muggleton 1991).

(Armitage & Woods 1999), it is obviously crucial for seed wheat.

Incorrect handling

Damage to grain caused by incorrect handling, for instance by allowing augers to run partially empty so the screw rattles against the casing, should be avoided as it increases the amount of damaged grain. This renders the consignment more susceptible to attack from insects, mites and fungi and impedes control measures such as cooling, fumigation and pesticide admixture (Armitage 1994; Armitage *et al.* 1996). Most markets have limits for broken grain in grain samples.

Commodity and pest management regimes

The current UK recommended strategy for grain storage aims to minimise pesticide usage. It is therefore based on reducing pest inocula by scrupulous hygiene and preventing and controlling pest development by grain drying and cooling. Good advice on this subject is increasingly hard to obtain. The government advisory service was privatised in 1997, the staff drastically reduced and grain storage expertise largely lost. Extension work is now mainly undertaken by the Home-Grown Cereals Authority, a body that funds research and technology transfer through a small levy on grain trading. Current advice is summarised in a decision support system, Integrated Grain Storage Manager (Knight *et al.* 1997) and *The Grain Storage Guide* (Armitage *et al.* 1999b).

Good pest and commodity management systems require routine sampling for insects and recording of grain temperature and moisture conditions. The extent to which this is undertaken is documented in the UK storage survey (Prickett 1988; Prickett & Muggleton 1991). Temperature was measured in 37% of farm stores and 81% of commercial stores (Table 3.8) where 35% used fixed sensors and 52% took spot readings. A total of 32% measured it daily, 38% weekly and 19% monthly. Moisture contents were measured on 63% of farm stores and 97% of commercial stores where 75% used a meter and 23% used both oven methods and a meter. In commercial stores, 35% measured moistures either weekly or monthly, 17% measured monthly and, surprisingly, 11% never measured after intake. To detect insects, samples, mainly by spear (trier), were taken on 31% of farms and in 83% of commercial stores (Table 3.8). Insect traps were used on 3% of farms and 34% of commercial sites but visual inspection was relied upon by 74% of farms

and 52% of commercial sites. The frequency of checking for insect pests varied: 42% checked weekly, 35% monthly and 19% less than monthly. The use of insect traps to monitor insect populations has recently become mandatory under some crop assurance schemes.

Cleaning, hygiene and fabric treatments

Store preparation begins after the previous harvest has been discharged, usually between December and May, after 4–10 months' storage. The techniques applied include cleaning out residues by vacuum or brush. Certain servicing companies also offer steam cleaning that is a form of thermal disinfestation, although no study has been made of its effectiveness. After cleaning it is common practice to treat the fabric of stores with insecticides, in particular, inaccessible dead spaces such as those that exist between grain walling and the external walls of storage buildings or underfloor ducts. There is no documented evidence that a fabric treatment has ever completely eliminated residual infestation of insects or mites in a building.

Drying

In a climate such as that found in the UK much of the harvested grain has to be dried artificially. About half is dried in bulk over a period of weeks, using ambient air, or perhaps air heated by up to 5°C to lower the r.h. sufficiently to ensure that fans do not have to be switched off during more humid periods. The other half is dried with air above 40°C over a period of hours.

Near-ambient air drying

The equipment for ambient air drying is similar to that for cooling but at least twenty times the air flow rate is required to remove moisture, compared to that required to remove heat, and so larger fans and ducts are required to provide and carry the air. Drying from 20% m.c. in ten days requires an air flow of at least $0.05 \text{ m}^3 \text{ s}^{-1} \text{ t}^{-1}$ and the bed depth should be reduced from a maximum of 2.8 m by 0.5 m for each percentage point increase in grain m.c. above 20%. Normally a drying 'front' passes through the grain and grain above this front remains largely unchanged in m.c., or may even increase and so biological deterioration continues until the front breaks through the grain surface. Strategies to maximise the efficiency of the system include switching fans on and off depending on ambient r.h., but this does not prevent deterioration when the air flow is off (Nellist 1988). Another strategy heats the air, depending on ambient humidity, but this often causes over-drying of grain near the air inlet. A recent development evens out m.c. stratification by ensuring grain above and below the drying front is mixed by grain stirrers (McLean 1993).

Hot-air drying

This is performed in batch dryers where the grain is emptied between batches and the final m.c. depends on the time in the dryer. Thus the cooling time can be independent of the drying time. Sometimes the grain may be recirculated and mixed during the process. However, in continuous dryers the moisture removed is dependent on the time taken to pass through the heated air section, which is determined by the grain discharge rate. Extra cooling may therefore be required after continuous drying. To avoid grain damage during hot-air drying of wheat for milling it is normally recommended that a maximum of 65°C is used at 20% m.c., decreasing by 1°C for every 1% increase in m.c. For feed grain, temperatures of 100°C for 3 h or 120°C for 1 h are tolerated (Nellist 1979; Nellist & Bruce 1987).

Cooling

The cool climate of the UK is ideal for ambient aeration of commodities, to eliminate temperature gradients, reduce the temperature below the breeding thresholds of insects and mites, and holding the temperature below 5°C sufficiently long to kill insect populations within the grain. It is often forgotten that the original aim of aeration was not to eliminate pests but to equalise temperature gradients within the grain and prevent moisture migration. The rates of air flow recommended for installations are based on the premise that the first cooling front must lower the temperature below the insects' breeding thresholds before they can complete their life cycle. Examination of 20-year average records and extreme years has indicated that an air flow of 10 m^3 h^{-1} t^{-1} is sufficient to achieve this in the British climate (Armitage *et al.* 1991). In the UK survey of grain storage, control of ventilation was largely manual; 84% mostly switched the fans on and off by hand, whereas 29% used automatic control by set thermostat, normal thermostat, thermostat and humidistat combined, or by time; the majority of these (17%) used a combination of thermostat and humidistat (Prickett 1988; Prickett & Muggleton 1991).

In conventional aeration systems the air is distributed via perforated ducts beneath the grain or through a per-

forated floor (Figs 3.11 and 3.12). However, an increasing number of installations use vertical aeration systems (Bartlett *et al.* 2003). In these, duct columns are laid out in a floor store at intervals determined by the depth of grain and the tonnage to be cooled and fans suck or blow air down the columns (Fig. 3.13). These systems are often cheaper to install than conventional horizontal ducting and have the advantage over above-floor ducts that they are unlikely to be damaged by tractors unloading the flat store. Another vertical application of aeration is the 'hot-spot cooler' (Armitage & Burrell 1978). A duct with a screw end can be turned into the grain to a depth of about 2 m and a fan sucks air (and insects) from areas of heating grain. As the fans supplied usually produce an air flow of 3–4 m^3 min^{-1}, and the aeration rate is considerably higher than that required for prophylactic cooling, the core of a hot spot can be usually cooled in a day or two.

In the UK, even during the warm months of summer, sufficient cool night air is available to reduce grain temperatures rapidly below insect development thresholds. This contrasts with many countries where cooling only reduces insect breeding rates so that chemical control methods, such as fumigation, need to be applied immediately after harvest until sufficiently cool weather is available later in the storage season.

Insecticide admixture

Until recently only three insecticides, all organophosphorous compounds (OPs), were cleared for admixture to

Fig. 3.13 Martin Lishman pile dry pedestal and fan in grain (left) and whole assembly out of grain (right).

grain in the UK as dusts or emulsifiable concentrates: pirimiphos-methyl, chlorpyrifos methyl and etrimfos. They were applied to all the bulk or as a top dressing combined with cooling as part of an integrated pest management (IPM) strategy (Armitage *et al.* 1994). The UK storage survey (Prickett 1988; Prickett & Muggleton 1991) reveals details of insecticide usage in UK wheat stores. Of those admixing insecticide, 40% did so as prophylaxis and 85% did so to combat a known infestation; some admixed for both reasons. A total of 71% used pirimiphos-methyl, 23% used chlorpyrifos-methyl and 8% used etrimfos. When insecticide was admixed, about 50% was applied as an emulsion using a sprayer and about 10% as a dilute dust formulation. A sizeable proportion (18%) applied the insecticide to the surface of the bulk alone. Each year, about 11–12 t of active ingredient pirimiphos-methyl was applied to storage structures and 4–5 t to the grain itself. At the time of the survey only 0.3% of farms and 1.5% of commercial sites were reported to fumigate their grain, usually with phosphine. Rodenticides were used on 78% of farm and 98% of commercial sites. More recent pesticide usage data are shown in Table 3.12 (Fox *et al.* 2001, Garthwaite *et al.* 2001).

The use of etrimfos was revoked in 2000 and the final use date of the dust formulation of pirimiphos-methyl was 2003. Replacements for OPs are urgently required and at the time of writing alternatives are not widely available on the UK market (see section on alternatives below). Recommended application rates are 4–5 mg kg^{-1} but, as within the EU, maximum residue limits may be progressively lowered, and this will have the effect of reducing the period of protection against insect pests (Collins & Cook 1999). There is a further problem in that many mite populations can already tolerate twice the recommended application rates (Prickett & Muggleton 1991).

Records must be maintained of any pesticide applications to a particular parcel of grain and this 'pesticide passport' accompanies the grain to avoid repeat treatments that may result in pesticide levels exceeding the maximum residue limit (MRL).

Economics of operation of the system

The calculated cost components for grain storage structures with ventilation facilities are shown in Tables 3.13 and 3.14.

Table 3.12 Use of pesticides on wheat in farm and commercial stores 1998/99. Source: Garthwaite *et al.* (2001), Fox *et al.* (2001).

	Store type	Wheat stored ('000 tonnes)			
		Farm		Commercial	
	Flat	9 256		925	
	Silo/bin	2 712		2 085	
	Active ingredient applied (kg)				
		To bulk	To surface	To bulk	To surface
Dust	Etrimfos	44	619	0	6
	Pirimiphos methyl	789	817	21	110
Spray	Etrimfos	0		66	0
	Pirimiphos methyl	1 788		653	38
	Chlorpyrifos methyl	0		6	0
Fumigant	Phosphine	421		6	
	% stores receiving a treatment*				
By farmer/ storekeeper	Fabric	52		41	
	Admixture	1		7	
	Fabric and admixture	4		26	
By a contractor	Fabric	14.1†		50.1 (37.1‡)	
	Grain	6.5		8.0‡	

*Based on all commodities. †Aluminium phosphide. ‡Methyl bromide.

Table 3.13 Capital and running costs for a 500-tonne bin or floor store in the UK. Source: Chadwick (1996).

	US\$ '000	US\$/tonne
Bins	15.0	30.0
Low-volume fans	2.7	5.4
Circular bins with drying floor (total)	98.5	197.0
On-floor storage	56.0	112.0
Low-volume ventilation	4.5	9.0
On-floor drying (underground ducts add US\$19/tonne)	130.0	260.0

Table 3.14 Annual operating costs for 500 tonne of grain including drying and cooling ventilation in the UK. Source: Chadwick (1996).

	Bins		On-floor	
	\$ '000	\$/tonne	US\$ '000	\$/tonne
Depreciation (10 y)	10.0		13.2	
Interest (6%)	5.9		7.8	
Electricity (\$2/tonne)	1.0		1.0	
Repairs, etc.	1.5		2.0	
Total	18.4	36.8	24.0	48

The running costs for a 4000 t on-farm storage facility have been compared with the price charged by a co-operative or 'grain pool' (Thearle 2000). Farm costs, which include cleaning, repairs, insecticides, drying and handling, monitoring and discharge were estimated to be about $5.25 t^{-1}$, compared to $11.6 t^{-1}$ for the co-operative handling charge and marketing fee. Typical co-operative prices include an interest-free qualification loan of $45 t^{-1}$ (refunded in full after ten years), $8.1 y^{-1} t^{-1}$ storage charges, and handling charges of about $12.8 t^{-1}$. However, this comparison does not take into account the capital costs of a farmer establishing a storage facility. For example, capital costs of storing 1500 t on-floor with under-floor drying are about $172 t^{-1}$, with a continuous-flow dryer $15 t^{-1}$, or in outdoor bins with in-bin drying $135 t^{-1}$. Even a Dutch barn conversion for 700 t including proofing and walling would cost $43.5 t^{-1}$. This compares to a co-operative's qualification loan and storage charge of $126 t^{-1}$, making co-operative storage financially attractive.

The advantages of a co-operative were less clear when the costs of a 750 t floor unit, a continuous-flow dryer with bin storage and membership of a co-operative were compared (McLean 1989). Annual costs ($ t^{-1}$) were found to be about 24, 42 and 28.5, respectively. The first two options included depreciation (amortisation) over ten years at 12% per annum, energy to remove 5% moisture and maintenance. The co-operative membership included interest foregone on an interest-free loan of $60 t^{-1}$ and a transport charge of $4.5 t^{-1}$. In addition, greater yield and higher quality resulting from early harvesting can offset the increased fuel costs of removing more moisture from the crop. Mixed-flow dryers with dryer throughputs of 10–30 t h^{-1} require 12–29 kW while cross-flow dryers vary from 28 to 9 kW. The cost per tonne per hour varies with throughput, for instance falling from $2650 at 10 t h^{-1} to $1650 at 40 t h^{-1}. The cost of ambient-air drying in energy and over-drying using a number of strategies and fuels and tariffs varies from $4.05 to $5.25 t^{-1} based on tariffs for on- and off-peak electricity and propane of $0.08, 0.028 and 0.026 kWh^{-1} and grain valued at $150 t^{-1} (Nellist & Bartlett 1988).

For pest management, the power costs for grain cooling can be as little as $0.03 t^{-1} using off-peak electricity compared to insecticide admixture costs of $0.75–1.00, depending on pesticide and formulation. Costs of commercial fumigation with phosphine in the UK vary from $1.5 to $4.5 t^{-1}, increasing with decreasing grain bulk. Prophylactic fumigations at $0.75–1.20 are also offered shortly after harvest where it is feared that cooling cannot occur swiftly enough or where there has been a history of infestation. The cost per tonne of cleaning varies from about $4.5 based on cleaning 1000 t over 50 h to $0.75 for cleaning 15 000 t over 750 h (McLean 1989).

Future developments

The current issues for grain storage in the UK include the fall in grain prices, the demand for traceability and increased quality standards, including absence of insecticide residues and mycotoxins, and the pressures against the use of synthetic pesticides on stored food.

Traceability

The UK Assured Combinable Crops Scheme (ACCS) was devised largely to improve standards as a reaction to the need to increase public confidence in food safety issues relating to agriculture. This scheme demands proper record keeping, for instance of temperature and moisture and pest control measures such as drying, cooling, pesticide use and measures taken against rodent and bird access to grain stores. It implies a greater dependence on instruments used to monitor changes during storage and requires evidence that steps are taken to deal with any untoward changes that may be detected.

Climate change

We have already seen how changes in agricultural technology and storage scale have affected the ecology of wheat storage. The possibility in the future of warmer summers and milder winters in the UK may challenge some of the underlying principles of physical control of grain and allow the establishment of a wider range of storage pests. Warmer weather, decreasing the availability of cool air immediately after harvest, may prolong the time taken to cool grain so that a pest generation may be complete before a cooling front passes through the grain. Higher minimum mid-winter temperatures may enhance the overwintering of storage insects. Both these factors may allow the establishment of destructive pests such as *Rhyzopertha dominica*, hitherto not a major problem in the UK. Solutions to this potential threat may include an upgrading of cooling recommendations, particularly increased aeration rates, already proposed experimentally in the US (Harner & Hagstrum 1990). Another approach may be to store wheat at lower moisture contents which would slow the development rate of many storage insects (Fleming & Armitage 2003).

Alternatives to organophosphates

The worldwide search for alternatives to OPs has led to the registration of diatomaceous earths (DEs) as grain protectants or for treating grain storage structures in Australia, Canada, China, Croatia, Germany and the USA (Golob 1997; Vol. 1 p. 271, Stathers 2002). The active ingredient, silicon dioxide, is already present as an additive to foods so there should be little concern about its addition to grain. It is regarded as having a physical mode of action, damaging the cuticular waxes of arthropod pests, resulting in death by desiccation. This being so, it is evident that the efficacy of DE is related to r.h., but in grain in which the recommendation is to store at an m.c. below 14.5% (65% e.r.h.) this should not be a concern. However, one of the challenges to storage in the UK is the uptake of moisture by surface grain in the winter; the grain m.c. may increase to 20% and mite populations can develop under these conditions. Nevertheless recent work in the UK has established the efficacy of DEs against mites and insects in the UK climate (Cook *et al.* 1999). Their use as a top dressing combined with aeration has proved satisfactory in both Australia and the UK. In 2001, DEs were marketed in the UK for the first time, partly because they were outside the scope of the Control of Pesticides regulations, since they act by physical means. This timely introduction coincided with the withdrawal of certain OPs from the market.

Fumigant replacements

The two fumigants in common usage, methyl bromide and phosphine, both present problems. Methyl bromide is listed as an ozone-depleting compound under the terms of the Montreal Protocol and there are deadlines for a phase-out in the countries of the European Union (60% reduction by January 2001, 75% reduction by January 2003 and no production, except for essential quarantine uses, by 2005). For phosphine, storage pests are gradually developing resistance and its release into the atmosphere is under increasingly stringent control in some countries, such as the USA and Germany, perhaps motivated by reports that it may cause transient chromosomal rearrangements. Consequently, the storage community is looking for alternatives, ways of minimising use and ways of increasing the efficiency of existing fumigants. Studies on methyl bromide include recirculation, recovery and destruction while, in the case of phosphine, the quest has been for more effective formulations, such as mixtures with carbon dioxide or nitrogen in cylinders (Ecofume® and Frisin®) or with more efficient circulatory systems (SIROFLO®, SIROCIRC®, Closed Loop Fumigation or J-systems) (Winks & Russell 1994; Noyes *et al.* 1997).

Other experimental approaches have included the use of controlled (CA) or modified (MA) atmospheres, with nitrogen to replace air, or low-oxygen atmospheres enhanced with some carbon dioxide generated by the combustion of propane (Conyers *et al.* 1996). As with conventional fumigants, the efficacy of these applications requires long treatment times under the low-temperature conditions characteristic of stored grain in the UK. MA and CA also take considerably longer to achieve complete control, compared with conventional fumigants, at all typical grain storage temperatures.

Remote sensing, automation and decision support

The demise of a free-of-charge government advisory service and demand from agriculture for accessible research outputs has led to the development of a number of decision support systems (DSS). One of the earliest, *Grain Pest Adviser*, was developed in the UK and specifically aimed at grain storage (Denne 1988). This later evolved into *Integrated Grain Storage Manager* (Knight *et al.* 1997). The requirement for greater traceability and ACCS also necessitated the design of record-keeping modules for DSS. This often requires that temperature or humidity data recorded automatically be transferred laboriously to the computer. It follows that automatic integration of record-keeping sections of a DSS with separate monitoring systems is urgently required to save duplication of effort. Further, since differences in temperature between ambient and grain are often used, for example, for efficient cooling, the record-keeping and monitoring systems could be further integrated with fan control for cooling or drying. These could all be controlled by a central computer that, with remote logging, could manage a number of widely distributed stores (Gibbs 1994).

Biological control

Biological control as part of an IPM strategy is now established for many fields of application but is still regarded with suspicion and is largely experimental for grain storage. It is likely that the chief application will be as a fabric treatment in empty stores. Protozoans, fungi or bacteria usually require registration but native species of parasitoids and predators often do not. The subject has been reviewed recently by Cox and Wilkin (1998). Potential predators include the mite, *Cheyletus eruditus,* that may feed on insect larvae as well as mite pests, and the bug, *Xylocoris flavipes,* that may eat the

larvae and eggs of moths and beetles. Parasitoids that could be applied in the storage environment include the wasps *Anisopteromalus calandrae* and *Choetospila elegans* that attack *Sitophilus* spp while *Bracon hebetor, Trichogramma pretiosum* and *Venturia canescens* consume moths. The protozoans *Mattesia trogodermae* and *Nosema whitei* have been used against beetles, as have the fungi *Beauveria bassiana* and *Metarhizium anisopliae.* Fungal isolates have been obtained from UK stores and several have shown good efficacy in laboratory studies (Wildey *et al.* 2002). Dose response data for the most promising isolates has been produced and pilot-scale studies will be undertaken shortly. Variants of the toxigenic bacterium, *Bacillus thuringiensis,* have been found to have efficacy against moths and some beetles and may have a place in future IPM strategies. However, this will have to take account of evidence that insect resistance to it can develop in the field after only a few generations.

Sampling

An ability to sample grain accurately is becoming more important to achieve compliance with various assurance schemes and to satisfy regulation for maximum level of contaminants. In order to reduce some of the disagreements that inevitably occur between farmers, merchants and end-users, a unified method of sampling every trailer-load of freshly harvested grain for quality has been proposed. In addition, a uniform, validated sampling regime for static grain bulks that will serve for a number of contaminants such as mycotoxins, arthropod pests, heavy metals, *Salmonella,* pesticides and GM contamination is required. These sampling regimes will be dependent on the accuracy and uniformity of laboratory analysis.

United States of America

Historical perspectives

Wheat was introduced to North America from several sources. Spanish and French explorers brought wheat into the southern and southwest regions during their early explorations of southern North America. Christopher Columbus, on his second voyage from Spain in 1493, brought wheat seed to the Spanish colonies in Central America and Mexico. By 1532, the Spanish Government ordered all ships sailing to America to carry seeds, plants and domesticated animals (Whitaker 1929) and as early as 1731 Spanish wheat was being grown in the San Antonio missions, where it spread to southern and eastern Texas over the next 40 years (Koehnke 1986).

During the early 1800s, wheat was the primary crop that supported farmers in the westward expansion into the Indiana Territory wilderness, being used for both baking bread and as livestock feed (USDA 1979). By 1840, the main wheat growing states were New York, Pennsylvania, Ohio and Virginia. However, from 1845 to 1855, as maize production began to increase in these states, wheat production shifted to newer, cheaper land to the west. By 1860–65, major wheat production had shifted to Wisconsin, Indiana and Illinois, and by 1870–75 the wheat belt had moved across the Mississippi River to river delta and high plains land.

From 1830 to 1890, soft red winter (SRW) wheat spread through northwestern Texas and the Texas Panhandle (Koehnke 1986). During the early 1900s wheat production had reached all sections of the Great Plains. The US wheat belt reached north–south from Texas through North Dakota, and east–west from western Minnesota through Montana to eastern Washington and northeastern Oregon. Winter temperatures in northern states are too severe for winter wheat, so soft and hard spring wheat is grown there. Minimum temperatures in central and southern high plains states are suitable for winter wheat. Texas and Oklahoma farmers grow hard red winter (HRW) wheat for pasture and grain. Milder winter temperatures are followed by hotter summer temperatures, making the southern plains and coastal states a high-risk zone for wheat storage.

Facilities and practices at storage sites

On-farm wheat stores built during the late 1800s and early 1900s were generally rectangular wooden bins inside wooden granaries. Farmers began to store wheat in steel bins in the mid-1930s to 1940s. Today, the majority of farmers in the high plains states from Texas to North Dakota harvest their wheat and haul much of it directly from the field to the local country elevators, usually within ten miles of the farm. Although most farmers store some seed wheat for the following year, less than 10% of Texas farmers and about 15–20% of Oklahoma and Kansas farmers store commercial wheat for market, similarly in the other high plains US wheat-belt states. The US Government has also used small galvanised steel bins, of 30–100 t capacity and constructed during the depression of the 1930s in connection with a market board operation to support declining grain prices (Tontz 1955).

The first commercial grain elevator was built in the US in 1842 by Joseph Dart (Lee 1937). Although there is no indication of the type of construction, it probably consisted of rectangular wooden frame bins similar to farm granaries of the time but on a much larger scale with thick, rigid, well-insulated structural walls. In 1848, the Chicago Board of Trade (CBT) was organised to accommodate the rapid expansion of cash grain crops in the US, creating the first grain futures market (Mehl 1954) and the Illinois Warehouse Act of 1871 was the first important law regulating grain elevators (Larson 1926). The New York Produce Exchange established a uniform system for grain inspection and grading in 1874, initiating early grain marketing standards in eastern US markets and shipping ports (Fornari 1973).

Grain elevator facilities are normally distinguished as either country elevators, that receive grain directly from the producer, or terminal elevators that receive from the country elevators and usually have a much large capacity for cleaning, drying, storing and conditioning grain. In the 1910s–1920s, corrugated, galvanised sheet-metal clad elevators appeared although these still had wooden, rectangular grain bins. Concrete elevator construction began in the late 1920s and early 1930s. Many concrete country elevators were built in stages, starting with the concrete head-house with a few silos in the 1930s, where all silos were gravity filled by separate downspouts or fill spouts to each silo. As more capacity was justified, concrete elevator storage volumes were expanded in stages by adding new silo annexes at 10–20-year intervals into the 1970s. Each multi-silo addition or annex was typically a more advanced design with individual silo cells becoming taller and wider. Larger silos were especially important where land became scarce at elevators in towns and cities.

Most wheat storage systems built during the 1940–1970s were steel-reinforced concrete elevators with either multiple silo cells or a large single concrete cell; walls were 20–25 cm thick. Early concrete elevator storage designs typically used two rows of circular main silos (Fig. 3.14) with smaller four-pointed 'star'-shaped cells in the interstices. This configuration of three silos across used one load conveyor belt overhead to fill the silos and one unload conveyor belt fed by rectangular discharge spouts from the sloped concrete hoppers at the bottom of each silo. Rack and pinion slide gates on each discharge spout allowed grain flow to be carefully controlled from one or more silos onto the belt, for blending or to unload silos (Fig. 3.15).

Fig 3.14 Typical concrete, country, wheat elevator built in the USA during 1940s–1960s. The covered conveyor belt gallery is called a 'Texas House'. The headhouse near the middle of the elevator with the larger annex to the right was added on several years after the original elevator was built.

Fig 3.15 Typical concrete elevator basement showing the conveyor tunnel with rack and pinion slide gates on the sloped discharge spouts of two main silos and, in the background, the vertical discharge spout of a star cell (USA).

Terminal elevators

From the mid-1960s to mid-1980s there was a change in design of the larger silo complexes that typically held hundreds of thousands of tonnes of grain. Multiple rows of large round silo cells with small star cells, which are difficult to access for service, were replaced by hexagon-shaped cells of uniform size with half-silos (four-sided) spaced alternately to form flat end walls on the annexes

(Fig. 3.16). Thus, instead of the two rows of main cells with a row of star cells in between, terminal elevator annexes with hex-shaped bins were built three, six, nine or twelve silos wide (Fig. 3.17).

Many of these huge elevator annexes were built in sections with a flex joint between sections along their length, which allows each section to move and shift independently of the adjacent sections. For example, a 60-cell length annex would likely have flex joints at quarter- or third-points with 15 to 20 cells per section. To conserve power, simplify design and reduce costs, many terminal elevators used a continuous belt to fill and

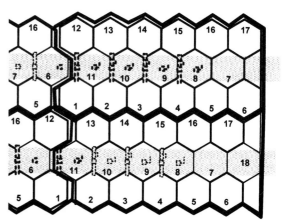

Fig. 3.16 Diagram of a portion of a 'hex' bin grain elevator showing hexagonal cells, a flattened end wall with four-sided half silos (left) and the conveyor belt serving three rows of silo cells (shaded).

Fig. 3.17 Grain terminal elevator constructed with hexagonal silo cells (see Fig. 3.16) used to eliminate inconvenient and hard-to-access star cells formed in the interstices of conventional round silo cells (USA).

unload each three rows of cells rather than have separate belts for loading and unloading. This typically decreases the length of the conveyors by about 40%, requires only one drive motor, not two, and, with no idler support rollers to carry the return belt directly under the loaded belt, the conveyor frames are simpler and require less maintenance.

From the 1950s to early 1970s wheat farms expanded and farming equipment became larger and required equal or less manpower. At the same time production efficiencies improved by 20–25% through increased use of commercial fertiliser, herbicide applications and new wheat varieties. These technological improvements increased wheat yields by 50–75% in all regions of the USA. To cope with large carryover grain stocks during this period, farmer-owned co-operative grain companies and privately owned elevators built thousands of new, larger concrete grain elevators or added concrete silo annexes with increased grain handling speed to existing elevator storage capacities. Over 10 000 grain elevators were built from the early 1900s through the mid-1960s, mostly since 1940. By the 1980s, over 14 000 commercial grain elevator facilities were operating in the country. However, during the past 15 years the number of grain elevators in the USA has declined, to less than 10 000 in the year 2000, but the total storage volume for wheat and other cereal grains has gradually grown as production yields increased. Likewise, the number of grain elevator companies in Oklahoma has dropped from a peak of about 410 in 1985 to about 250 in 2000 through consolidations and buyouts. New construction of large steel bins and independent concrete silos have replaced some older, inferior quality flat storage, so the total storage volume of Oklahoma grain elevators, and in most other major grain states, has remained more or less constant during the past quarter of a century.

Country elevators

As wheat yields increased in the 1950s and 1960s, and more storage capacity was needed, many country elevators that typically had a concrete head-house (the main elevator structure containing the main vertical cup and belt elevator leg system) and concrete silos fed and unloaded by gravity spouts, added concrete silo annexes built a few feet from the main structure. These annexes were loaded by horizontal belt conveyors mounted on the silo roof deck, covered by a galley structure, and unloaded by a belt conveyor in a basement built beneath the centre. Almost all concrete silo cells had concrete floors

sloped at 35–45°, which allowed gravity discharge. A few (5–10%) concrete head-house and annex silos had aeration systems installed.

The older concrete elevators, built in the middle of the twentieth century, have gradually deteriorated. Since 1975 most commercial grain elevators have increased their storage capacity by constructing large low-cost steel bins; these operate as independent cells connected only by conveyors. The change to large steel bins has resulted in more difficult grain quality management due to ease of access by grain insects. The concrete silos allowed only limited access of insects to grain through the exterior vents under the roof decks. In contrast, steel bins are open to insects at the base of the structure through aeration fans, unload conveyors, sidewall access doors and so on. Since bolted-steel corrugated bins have relatively lower sidewalls (typically 10–15 m compared to about 40 m in concrete silos), grain insects can more easily access the roof headspace of steel bins due to lower surrounding wind velocities. Therefore, permanent sealing of grain structure base and sidewall openings and semi-permanent sealing of equipment (such as aeration fans and unload conveyors), when not in use, is an important measure to limit insect infestation.

Objectives of storage

US Government support of grain programmes

In 1916, Congress passed the US Grain Standards Act authorising official grading standards for grain and requiring inspection of grain sold by grade in interstate or international export markets (Brown 1918). The Agricultural Marketing Service (AMS) was established in the US Department of Agriculture (USDA) in 1939 to help farmers market their grain and other commodities internationally (USDA 1940) and became a separate division of the USDA in 1953. The role of the AMS was strengthened by the Agricultural Marketing Act of 1946, which provided for integrated administration of marketing programmes and scientific approaches to marketing problems. It also included basic authority for major AMS functions including market research, federal standard grading and inspection services, market news services, market expansion and consumer education (Banfield 1949). The Act also established the organisational structure and programme outlines that were the basis of the Federal Grain Inspection Service (FGIS), set up later as a separate service in USDA in 1976 through an amendment of the original US Grain Standards Act.

The FGIS is responsible for inspection, sampling and official grading of all US grain exports. FGIS either performs this service directly or delegates inspection responsibilities to the state where export terminals are located (Manis 1992).

Grain standards for quantity and quality have evolved over an extended period and an account of this is given below, relying heavily on Hill (1990).

Development of US grain standards – quantity

The standards used for marketing grain have evolved since the late 1700s in parallel with the growth of interstate commerce. The need to develop standards for grain quantity was a much earlier issue than grain quality. Initial measurements were based on English standards, using the Winchester bushel (2150.42 cubic inches and based on the English wine gallon). On 14 July 1836 both houses of Congress passed a resolution adopting the Winchester bushel, or its derivative, as the official measurement volume for US grain, but by 1850 Chicago grain traders were buying grain based on weight, not volume. However, continued discrepancies and disputes over grain marketing uniformity prompted Congress to establish import standards of weight per bushel, or test weight; this federal legislation covered only import standards. While the Chicago Board of Trade (CBT) urged all states to use 56 pounds as the standard weight per bushel, data from the Baltimore Corn Exchange showed that although all 38 states were consistent for wheat (marketed at 60 pounds per bushel), the legal weight for other market grains varied by state through the 1930s. In December 1975 Congress passed the Metric Conversion Act, promoting voluntary use of metric measurements in the US grain industry. However, the US grain industry still uses Imperial units for internal trading, with metric units used almost exclusively by exporters.

Development of US grain standards – quality

Establishing consensus between states on standard grain quality was even more difficult than for quantity. Initial quality analysis for grain trading during the 1800s was done by visual inspection of representative grain samples. As trading volume and distances grew it became difficult and impractical for buyers and sellers to meet and inspect submitted samples visually. Terminology used to describe grain quality evolved out of trade practices over many decades. By the early 1840s grain trad-

ers began to recognise the need for a uniform descriptive standard for each grain. Instead of merely listing classes of wheat, such as winter or spring wheat, subclass descriptions were developed based on wheat origin. For example, '4900 bu. Massillon wheat, a beautiful sample, the berry being plump and a bright yellow hue' or '1100 bu. Wisconsin wheat, in bags, not properly cleaned'. Grain quality language began to evolve terms such as sound, bright, common, extra, choice, merchantable, clean, fair, hot and unsound.

The CBT was the first grain board to develop a formal grain grading system for description and control of grain quality attributes. In 1856 the CBT passed a resolution establishing standard classifications for wheat as White Winter, Red Winter and Spring – prime quality. The resolution stated that 'variations from prime quality should be specified'. In 1858, CBT expanded its wheat grade classifications to include Club Spring, No. 1 Spring (the standard), No. 2 Spring and Rejected.

By 1859 CBT added test weight, a measure of weight per unit volume (grain density), as a quality grade factor and in 1860 all grain bought and sold on the Chicago market was inspected officially and fees were established for inspecting grain moved on canal barges. The CBT found that although red winter wheat usually sold for 2 to 3 cents per bushel more, stringent inspection and grading for spring wheat increased its demand and value, making it competitive with red winter wheat.

Due to the success of the CBT wheat grading and inspection system, grain exchanges in Detroit, Toledo, Cleveland, St Louis, Indianapolis, New York City, Philadelphia and Montreal adopted the CBT grading and inspection system, making slight modifications to represent regional grain production, milling characteristics and grain varieties. However, adoption of grades at major US ports led to local, national and international trading problems. Uniform methods to determine test weight measurements and moisture content were lacking. The primary obstacle to creating uniformity in grades at all markets was the lack of objective, standard measures and instrumentation. Grain standards and terminology were too general and subjective, leaving wide latitude for interpretation by inspectors. Major complaints and dissatisfaction with grain producers, millers and foreign grain buyers led to the Illinois Railroad and Warehouse Act of 1871. This law removed control of grain inspection from the CBT and gave it to the three-man Illinois Railroad and Warehouse Commission. The Commission appointed a chief grain inspector and other inspectors.

Even with diligence and experimentation with grading, the industry was still dissatisfied half a century later. This led to a push for national uniformity in grain grading standards, which resulted in the US Grain Standards Act of 1916. Since most grain quality losses were identified as the direct or indirect result of excessive grain moisture, moisture was a factor in the Grain Standards Act. The act specified the need for more objective, measurable standards of grain quality. Objective grading factors for wheat included test weight and the following grade factors as percentages: damaged kernels, shrunken and broken kernels, foreign material, wheat of other classes, stones and other materials. Grades are given values of 1 to 5 where grade 1 is the highest quality (Table 3.15). There is also a sample grade which is wheat that does not meet the criteria specified for US wheat grades or has a sour, musty or commercially objectionable odour, or is heating or is of distinctively low quality.

In addition, the act provided for standardised sampling and measurement methods and required the development of official measurement instruments and equipment, such as official segmented grain probes or triers, grain dividers to provide uniform subdividing of samples, test weight equipment, moisture testers, and so on. The USDA's official standards for wheat (Table 3.15) became effective on 1 July 1917 (Post & Murphy 1954) and remains virtually unchanged in 2002.

Major causes of quality decline

Excess moisture, leading to moulding and insect damage, is the major source of quality losses in US wheat. This is compounded by lack of management skill at many farm and commercial grain elevator facilities.

The primary grain storage insect pests in US wheat are the beetles *Rhyzopertha dominica* and *Sitophilus oryzae* which cause significant damage to kernels, and the moth *Plodia interpunctella*, the larvae of which spread silken threads forming a webbing over grain surfaces, so causing significant contamination and blockage of aeration

Table 3.15 Grades and grade requirements for all classes of wheat, except mixed wheat. Source: USDA (1988, 1999).

| | Minimum limits | | Maximum limits | | | | | | Wheat of other classes¶ | |
| | Test weight per bushel | | Damaged kernels | | | | | | | |
Grade	Hard Red Spring Wheat or White Club Wheat† (pounds)	All other classes and subclasses (pounds)	Heat damaged kernels (%)	Total‡ (%)	Foreign material (%)	Shrunken and broken kernels (%)	Defects§ (%)		Contrasting classes (%)	Total** (%)
US No. 1	58.0	60.0	0.2	2.0	0.5[0.4]	3.0	3.0		1.0	3.0
US No. 2	57.0	58.0	0.2	4.0	1.0[0.7]	5.0	5.0		2.0	5.0
US No. 3	55.0	56.0	0.5	7.0	2.0[1.3]	8.0	8.0		3.0	10.0
US No. 4	53.0	54.0	1.0	10.0	3.0[3.0]	12.0	12.0		10.0	10.0
US No. 5	50.0	51.0	3.0	15.0	5.0[5.0]	20.0	20.0		10.0	10.0
US Sample grade*										

*US Sample grade is wheat that: (a) does not meet the requirements for the grades US Nos 1, 2, 3, 4, or 5; or (b) contains 32 or more insect damaged kernels per 100 grams of wheat; or (c) contains 8 or more stones, 2 or more pieces of glass, 3 or more crotalaria seeds (*Crotalaria* spp), 3 or more castor beans (*Ricinus communis* L), 4 or more particles of an unknown foreign substance(s) or a commonly recognised harmful or toxic substance(s), or 2 or more rodent pellets, bird droppings, or an equivalent quantity of other animal filth per 1000 g of wheat; or (d) has a musty, sour, or commercially objectionable foreign odour (except smut or garlic odour); or (e) is heating or otherwise of distinctively low quality.
†These requirements also apply when Hard Red Spring wheat or White Club wheat predominate in a sample of Mixed wheat.
‡Includes heat-damaged kernels.
§Defects (total) include damaged kernels (total), foreign material, and shrunken and broken kernels. The sum of these 3 factors may not exceed the limit for defects (for each numerical grade).
¶Unclassed wheat of any grade may contain not more than 10% of wheat of other classes.
**Includes contrasting classes (USDA 1988).

airflow (Vol. 1 p. 95, Hodges 2002). Secondary stored grain insects, although less damaging to grain, are nevertheless serious because the presence of any live insects in wheat reduces the official grade and requires the grain to be fumigated before it is acceptable to millers and other buyers. Important secondary insect pests in US wheat are the beetles *Tribolium castaneum*, *Tribolium confusum*, *Oryzaephilus surinamensis* and *Cryptolestes ferrugineus* (Krischik & Burkholder 1995). Mites are a relatively insignificant problem in US wheat.

The USA is divided into grain storage risk zones (Fig. 3.18), based on inherent grain storage ecosystem conditions in each zone. For example, HRW wheat grown in the southern high plains states of Texas, New Mexico, Oklahoma, southern Kansas and southern Missouri has a higher risk from insect damage due to earlier harvest, followed by a long period of hot storage weather before grain can be cooled in the autumn. HRW wheat grown in northern Kansas, Colorado and Nebraska is in the central risk zone where later harvest and cooler summer nights allow aeration during the summer, so it is at less risk from insect infestation. Also in this central, medium-risk zone is SRW wheat grown in the central and eastern corn belt. In the low-risk storage zones are HRS wheat grown in the northern high plains in South Dakota, North Dakota, Montana, Minnesota, Iowa, Wisconsin and Michigan, soft red spring wheat grown in the northeastern US and SRW and durum wheat produced in the Pacific Northwest, Oregon, Washington, Idaho and Montana.

Due to wheat spoilage at many of the government-controlled stores established in the 1930s, research was undertaken on grain bins holding HRW and hard spring wheat (HRS) stored at sites in Kansas and North Dakota (Schmidt 1955). Most of the experimental bins were constructed from corrugated preformed steel sheets assembled with lead-washer headed bolts for watertightness. All joints were caulked and sealed. Galvanised sheet steel floors were coated on the underside with bitumen paint to prevent rusting due to soil moisture. There was no serious damage to bins from grain pressure or storms during the study. The primary causes of grain spoilage were leakage of rainwater into most of the bins around roof doors and fill hatches, and snow blown into the headspace through vents. The study included an analysis of grain moisture profiles versus side of bin (north vs south exposures) and time in 30, 80 and 150 t bins. The average m.c. was 11.3% for HRW wheat and 12.4% for HRS wheat. Both moisture levels were considered 'safe' for their geographic location. Although the average m.c. did not change significantly during storage, moisture migration (Vol. 1 p. 90, Devereau 2002) occurred in the bins due to temperature variations and location in the bin, affecting grain quality in zones where moisture increased above safe equilibrium relative humidity levels, and moulding occurred. None of the bins in these tests was aerated or ventilated. The results of these studies were used to improve management of government surplus wheat storage throughout the USA. USDA ARS agricultural engineers also began research on management of large bulk grain storage in grain warehouses and large concrete elevators.

An important contribution to grain quality decline is management failure; this will be dealt with in the next section.

Commodity and pest management regimes

Although many grain elevator and farm wheat storage facilities are 30 to 50 years old, wheat storage in the USA is generally adequate. However, grain management skills on farms and at commercial elevators vary widely. Farm-stored wheat often has to be fumigated by the elevator manager upon receipt in all states, resulting in discounts and/or fumigation charges. Many grain managers, elevator superintendents and farmers have not developed sufficient commodity management skills to maintain wheat in good condition without reliance on residual grain storage insecticides, which are being withdrawn gradually from the market by Environmental Protection Agency (EPA).

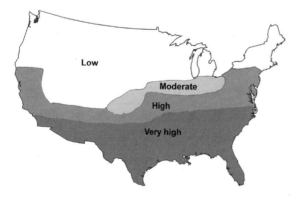

Fig. 3.18 Zones in the USA where there are different risks of wheat becoming infested by insects. Risk is related to the maximum grain temperature during the storage season following harvest.

Aeration systems

During the 1980s and 1990s aeration with ambient air was emphasised as an important means of controlling moisture migration, mould damage and insect pests, while chilled aeration was developed for use in high-value commodities (Noyes *et al.* 1995a; Maier & Navarro 2001; Navarro & Noyes 2001). During the 25 years 1960–85, USDA and university stored product researchers developed aeration systems to cool grain more uniformly. Innovative aeration systems such as cross-flow aeration were developed for concrete silos (Holman 1960; Day & Nelson 1962, 1964). With this system, airflow can be delivered across the diameter of a silo with only 15–20% of the fan power needed to force air vertically through the entire grain depth. This is due to the natural tendency of elongated kernels (wheat, maize, rice, barley, oats, etc.) to align themselves horizontally in silos so that static resistance to horizontal air flow is much lower (Jayas & Muir 1991; Jayas & Mann 1994). However, there are two basic limitations of cross-flow aeration systems: the silos must be full, or a method

devised to block the upper parts of suction and exhaust ducts that are in the silo headspace, to a point a few feet below the grain surface, and the roof sections of the silos must be sealed to avoid short circuiting of air through roof leaks (Noyes *et al.* 1991).

Technical improvements to both the cross-flow and vertical aeration systems of concrete silos are possible. It is believed that the two-duct cross-flow system (silos with full-height vertical inlet and outlet ducts mounted on opposite sides inside the silo) is ineffective while the six-duct system (inside vertical ducts spaced equally with 60° separation around the silo) is too complex and impractical (Noyes *et al.* 2001). To eliminate the dead spots that develop along the sides between ducts in two-duct systems, and the dead centre zone in conventional four-duct systems (Fig. 3.19a), a theoretical four-duct system was developed with one pressure supply duct and three exhaust ducts (Fig. 3.19b). In this system, the supply and exhaust ducts alternate half of the time between opposite ducts. Thus, the original supply duct becomes the centre exhaust duct. The two adjacent ducts are dedi-

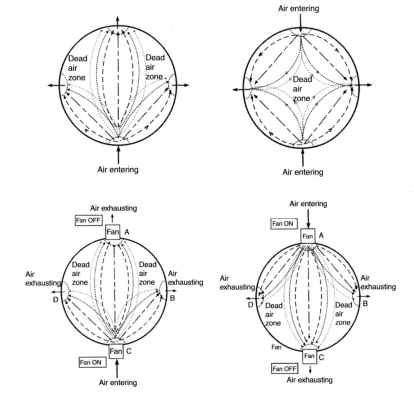

Fig. 3.19(a) Four-duct cross-flow aeration dead air zone with one constant-flow entrance duct (left) and four-duct cross-flow aeration dead air zone with two constant-flow entrance ducts (right).

Fig. 3.19(b) Four-duct cross-flow aeration with two fans timed to operate alternately between ducts A and C to eliminate dead zones in grain by periodically reversing air path from operating fan to two side exhaust ducts B and D. Ducts A and C alternate half time as the third exhaust duct, thus, a 24-hour percentage timer is the primary control element.

cated full-time exhaust ducts. The air supply transition duct is valved at the fan discharge to deliver air to the preferred supply duct while a valve on the opposite duct is opened to allow air to exhaust to the atmosphere.

An alternative to valving the fan discharge and installing an aeration duct around half the silo diameter is to install fans on opposite sides on the two exhaust/inlet vents. Each fan would be timer-controlled, operating half the time at specified intervals, such as 6–12 hours per fan, alternately. This system can be manifolded to a series of in-line concrete silos with supply-exhaust ducts along each side of the silos. With this method of cross-flow aeration there is continuous air flow across the centre and around the sides of the silo providing low-cost, uniform cooling with no dead spots.

Push-aeration typically generates heat of compression so the temperature of cooling air is raised by 1.5° to 5°C as it is forced through wheat against high static pressure resistance. Rigid concrete or steel silo construction allows another innovative aeration system, push-pull aeration, which reduces heat of compression by half. Push-pull systems split the fan system, placing half of the fan power at the silo base to push and half on the silo roof as suction.

Unfortunately, the high cost of constructing concrete elevators and the trend to large-diameter steel bins has reduced interest by the grain industry in refining these innovative low-energy aeration systems. The US grain industry will need to evaluate the future designs of commercial grain elevators as concrete elevators 50–60 years old wear out, and maintenance costs become too high for low-margin grain businesses. The USA and other commercial grain storage industries might do well to study Russian reinforced pre-cast modular concrete grain elevator designs, which have good potential for fast, relatively strong, watertight construction. A small crew can build these elevators rapidly as the modules are shipped by rail and a small low-cost crane used to unload and then stack modules for immediate or later assembly.

Residual pesticides

There is widespread reliance on synthetic residual insecticides although these are gradually being taken off the market. Malathion, an inexpensive pesticide used since the 1950s, is still used widely on grain, even though its use is no longer recommended as insect resistance to it is widespread in the USA (Cuperus 2000; Zettler & Cuperus 1990). Other, more effective residual pesticides,

mostly organophosphates, were developed and used from the 1960s through the mid-1990s, such as chlorpyriphos methyl (Reldan®). This was applied to wheat and other grains at 6 ppm of active ingredient and was effective against all grain insects except the lesser grain borer. However, it is no longer approved by the EPA but research during the past five years has shown that a mixture of chlorpyriphos methyl and the synthetic pyrethroid cyfluthrin is highly effective against all grain insects. This new product, called Storcide™ and applied at a rate of 3 ppm chlorpyriphos methyl and 2 ppm cyfluthrin, was approved by EPA for use on wheat in mid-2003. Besides malathion, Storcide™ is the only approved synthetic pesticide for use on wheat in the USA.

An important alternative to synthetic pesticides is the use of diatomaceous earth (DE) (Vol. 1 p. 271, Stathers 2002). When DE is applied to grain bulks at recommended rates the natural sliding friction of kernels is reduced, which in turn reduces grain bulk density. Bulk density is an important quality parameter (Table 3.15), so companies marketing DE no longer recommend treatment of complete grain bulks. However, DE makes an excellent and economical treatment to store surfaces and is recommended for application to empty silos, flat warehouses and steel bins at the rate of 0.3 kg DE per 100 m² of floor surface area. An alternative strategy is to use it as a barrier on the grain surface, applied as a top layer at the rate of 9.6 kg per 100 m² for all types of stores or in silos applied at 7.5 kg per 30 t to both the top and bottom metre of grain.

Fumigation

Two fumigant gases are currently used for disinfesting wheat in the USA, methyl bromide (MeBr) and phosphine (PH_3). Chloropicrin (tear gas), a very effective fumigant approved for empty bin or silo fumigation for the past half-century, was used until the mid-1990s, but production has been discontinued. Methyl bromide was developed as a recirculated grain fumigant used by grain companies and flour mills from the late 1920s up to 2000 as a fast, effective grain treatment and space treatment for elevators and mills. In the USA, phase-out for all but specific import/export quarantine was scheduled for 2005 but flour millers, strawberry producers and other major lobby groups are urging Congress to extend the phase-out period over another five to ten years.

Phosphine is a cheaper, but slower acting, fumigant than methyl bromide. It is generated from solid aluminum phosphide and has been in use at grain elevators since the 1950s. New formulations of magnesium phosphide granules are being developed for use in phosphine generators that will be tested under EPA experimental use permits for commercial fumigation of wheat and other grains in 2003–04. The general guidance dosage range for US fumigation for all types of structures is 4.5–33 pellets or 1–6 tablets per tonne, based on gastightness (Jeyamkondam *et al.* 2003a).

Sulfuryl fluoride (SF) has just received EPA approval for use on grain. It has been used under the trade name Vikane® for termite control for the past 20–30 years and is now marketed for grain fumigation by Dow AgroScience under the trade name ProFume®. ProFume® is a high pressure, cylinderised liquid that vaporises at a relatively low boiling point. Grain insect control by SF is faster than PH$_3$, but not as fast as MeBr. However, it is thought to be a possible replacement for MeBr in flour mills and warehouses as it does not react with metals such as copper in controls and motors. Active recirculation will be required with ProFume® as it has a gas density of about 3.5, similar to methyl bromide.

The EPA reviewed over 3000 pesticides and chemicals in 1999–2001 and since phosphine is highly toxic it proposed to ban or greatly restrict its use. This was challenged by a broadly based Phosphine Technology Task Force that supplied data to support continued use of phosphine under new label guidelines. The Task Force data indicated that although several fatalities had been attributed to phosphine, all deaths occurred due to non-authorised uses that violated label application restrictions. Consequently, the EPA dropped most of the highly restrictive text on the label and a new phosphine industry label emphasising education, training and certification for both commercial and private fumigators is being finalised in 2003–04. Major additions to the new phosphine label were the requirement for each user to develop and maintain a fumigation management plan that documents phosphine gas concentration levels at key locations at the fumigated site, including detection of phosphine gas concentrations at property boundaries, and the requirement that a licensed fumigator be present on the premises during the application of fumigant.

Since the mid-1960s, very little grain research has been dedicated to concrete elevator grain management systems. The traditional method of concrete silo grain management is to turn the grain to check quality, cool, fumigate and blend grain. The key management tool is the grain unloading/reloading transfer system. Fumigation is done in almost all concrete silos during turning or moving grain from full to empty silos, an expensive and energy-consuming process. Concrete elevator operators in most of the USA use an automatic phosphine pellet dispenser to distribute the dosage uniformly into grain as it is transferred. However, alternative methods are being developed and promoted in Oklahoma and surrounding states to avoid grain turning and these are described below.

Improved phosphine gas technology using closed loop fumigation

In 1980, a phosphine recirculation method using a specialised fan, called the *J-system,* was patented (Cook 1980). This improves phosphine gas fumigation efficacy in sealed stores and several J-systems were installed at commercial grain elevators in south and central Texas and in Kansas. Degesch America purchased the J-system US patent rights in 1985. The principles of the J-system have been expanded through the development of closed loop fumigation (CLF). This has been under continuous development in the USA and recirculates gas using a single, centrifugal fan manifolded to multiple steel bins or concrete silos (Noyes *et al.* 1995b, 2001). The CLF system requires only partial sealing of grain storage structures, an important feature since a good hermetic seal with virtually no gas leakage is not practical in most existing or new stores in the USA. A typical target dose of phosphine for this system is 300 ppm ± 30–60 ppm. CLF systems typically connect two to six steel bins so that these are fumigated simultaneously. In the early 1990s multiple concrete silo CLF systems were developed. Concrete CLF systems can now combine from 2 to 40 or more concrete silos in one elevator annex into a common fumigation unit operated by one CLF fan (Noyes *et al.* 1995b).

Experimental work continues to improve CLF as the fumigation method of choice in the US grain industry. The number of phosphine gas recirculation system installations at elevator and farm grain storage sites is steadily increasing. CLF systems must be designed to fit the specific needs of each grain storage site. Thus during the past decade the adoption rate has been slow. However, several factors, including increased pressure by the US Occupational Health and Safety Agency to restrict confined space entry, current changes by EPA in phosphine labelling and national Grain Elevator and Processing Society educational programmes are expected to result in substantial increases in CLF installation. The introduction of new external gas generating and release technologies will simplify application and enhance CLF adoption at US wheat elevators. The following three commercial methods of supplying phosphine gas to CLF systems through equipment mounted outside the fumigated structures (which eliminates confined space entry) have been identified:

(1) ECO$_2$FUME™ (98% CO$_2$ with 2% PH$_3$), or VaporPhos™ (99.9% pure PH$_3$) delivered from high-pressure cylinders. VaporPhos™ requires the use of a precise controller called the HDS-80 Horn Diluphos System, that controls phosphine gas release into a controlled blend ratio with ambient air. Both ECO$_2$FUME™ and VaporPhos™ can be metered precisely into grain storage structures on an intermittent or continuous basis.

(2) Degesch Phosphine Generator produces phosphine gas continuously at 1200 ppm for fumigating large bulk grain masses.

(3) QuickPhlo-R/C Gas Generator uses a prepackaged, coated granular formulation of magnesium phosphide to deliver specific volume of gas during a 90-min generating period at a concentration of approximately 3000 ppm into tightly sealed fumigated structures. This unit recirculates the gas as it delivers it to provide uniform distribution within two to three cycles. A slightly different piece of equipment, the QuickPhlo/C Gas Generator, is designed to generate and deliver a slow steady flow of low concentration gas through a leaky structure over a much longer time.

Integrating commodity management options

From 1980 to 2000 the EPA cancelled the registrations of many pesticides used for the control of grain insects. With fewer pesticides available, quality preservation has relied on improvements to the application and integration of other existing options. Starting in 1983, grain storage leaders in academia and industry in the USA have facilitated the national Stored Product IPM Training Conference at three- to five-year intervals. The aim of the programme has been to minimise the accumulation of residue pesticides on cereal grain by focusing on four management tasks – sanitation, loading, aeration and monitoring (SLAM) (Maier *et al*. 1994). The four task groups are as follows although there is some overlap between them.

Sanitation

This incorporates housekeeping and is the foundation of effective commodity management. It involves 'sanitis-

ing' every part of the facility from cleaning spilled grain, destroying rodent and insect harbourages in and near the storage complex to not mixing grains of different types.

Sanitation may involve the spraying of residual pesticides where needed, such as into cracks and crevices, inaccessible aeration ducts, and spray-down of bin walls, inside and out. Difficult-to-reach silo bottoms that cannot be cleaned out each year should be sprayed down from the top with an effective residual treatment, such as cyfluthrin or Storcide™.

Another important aspect of sanitation involves sealing the bin or silo base and sidewalls to provide a barrier to insect entry at all locations except the roof eave gaps, under-roof gap or vents, the roof surface vents and spouts where these are needed for aeration. Most non-functional bin or silo openings can be sealed permanently while all functional openings can be sealed when not in use.

Loading

Loading includes all management functions that should occur during transfer of grain into storage. Grain should be handled as gently as possible, consistent with getting the job done on time. Loading and transfer provide excellent opportunities to clean wheat of moist weed seeds, dockage and foreign materials that must be removed for good storage. Cleaning is especially important if wheat is weedy or if there is a high proportion of cheat grass kernels (these are shaped much like wheat seeds but are only about half as heavy). Grain cleaning improves the grain mass ecology by removing moist foreign materials and preventing the interstitial space between kernels from being blocked, allowing good air movement throughout the grain mass micro-environment to stabilise inter-kernel equilibrium relative humidity.

An important sub-function of loading is levelling or spreading the grain to give a uniform distribution of grain fines, dockage and foreign material. Mechanical or gravity spreading of fines will minimise aeration air flow restriction under the spout line or fill conveyor. One simple method of cleaning fines from the grain is called coring (Fig. 3.20). The preferred method of coring is to operate the unload conveyor periodically to draw down the grain peak, so forming an inverted cone at the time of grain loading. For consistency, the inverted cone should be some arbitrary diameter, such as one-fifth to one-third of the bin diameter. Alternatively, coring can involve the removal of a specific volume of grain, such as unloading one or two truckloads, or 25 tonnes after every 500–600

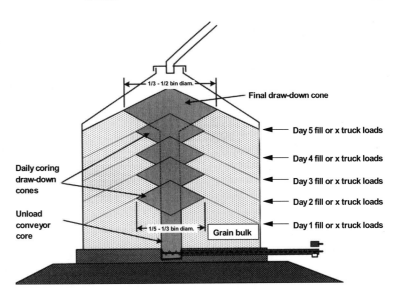

Fig. 3.20 Recommended multiple unload method for 'coring' grain bins removes fines and improves aeration by lowering by a quarter to a half the final grain peak height.

tonnes loaded into the bin. Because most of the fines, weed seeds and small kernels settle in the core at the centre of the bin, periodic coring will remove a significant volume of fines. The high fines-content grain unloaded by coring should be cleaned, then added back to the peak where it will spread over the full surface of the grain.

A second method of coring is to remove the core when the bin has been filled. This is beneficial but removes a lower percentage of fines. It loosens the centre of the grain mass and removes a lower percentage of the dense centre core of material than the multi-unload method (Fig. 3.21). The final inverted cone provides an opportunity to lower the peak so that the average grain depth in the peak is only 50–60% as high as the normal grain surface peak. This coring provides a way to rough-level the grain surface.

Aeration

For effective aeration it is important to ensure that the aeration system cools the grain fast enough, based on available cooling weather, grain temperature and moisture content, to keep it in condition until it is cooled. Roof vents must be designed to provide adequate exhaust or inlet area to minimise roof condensation and static pressure under the roof. The total accumulated roof vent exhaust area should allow 300 m min⁻¹ exhaust air velocity (± 25%). Another approximate rule of thumb is that roof vents should provide about 0.09 m² of vent cross-section area per aeration fan horsepower.

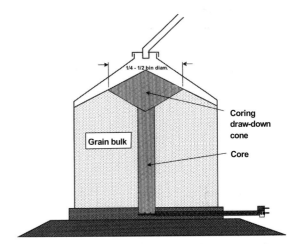

Fig. 3.21 'Coring' a grain silo by a single discharge of the central core after the silo has been filled. This is a less effective method than the multiple unload coring shown in Fig. 3.20.

Aeration should also include automatic fan control by simple mechanical or electromechanical switch operated by a thermostat, preferably with an hour meter to record the amount of accumulated aeration fan operating time. More sophisticated microprocessor or computer-based control may be preferable for large complex, multi-bin or multi-silo aeration systems. Automatic controllers properly operated and co-ordinated with other aeration management such as monitoring of grain temperature or checking exhaust air temperature (fans on suction systems or roof vents on pressure systems) will ensure optimum fan usage.

Suction systems are generally preferred over pressure systems for insect management because they make better use of ambient air for grain cooling. Pressure fans add the heat of compression to the air entering the base of storage structures on upflow aeration systems. If bin or silo base openings are sealed then stored grain insects can only enter at the grain surface. During the first 2–3 months, as insect population pressures build, the insects move slowly downward in the top 1–2 m. In a 10 m deep grain bed, suction aeration can cool the top 1–2 m of grain to 24°C during the summer in at least the northern half of the high-risk storage zone, while rather lower temperatures are possible in the medium-risk storage zone. At an aeration rate of 0.1 m³ min⁻¹ t⁻¹, autumn cooling of clean level grain requires about 100–120 h. Thus in 10–25 h of cooling the initial zone of insect activity can be cooled to the thermostat set-point. By reducing wheat binned at 30°C to 24°C, insect feeding and breeding activity will be reduced to less than half.

Further cooling to 16°C and below will stop reproductive and feeding activity of most stored grain insects. Grain temperature control with pressure cooling is very difficult to achieve in wheat. Thus, in US wheat stores the primary function of suction aeration is insect control. A new method of adding suction aeration to old concrete silos is being tested at two concrete elevators in Oklahoma during the summer 2003. This will demonstrate the feasibility of adding suction aeration to elevators with 14 and 16 concrete cells to provide flow rates of 0.025–0.017 m³ min⁻¹ t⁻¹. Both of these elevators will have automatic aeration control that should cool the top 1.5–3 m of grain, where initial insect activity would be expected, within 50–75 h of fan operation.

For the future, aeration technology should be strengthened. Aeration air flow rates need to be increased from the US conventional rate of 0.1 m³ min⁻¹ t⁻¹ to 0.2–0.5 m³ min⁻¹ t⁻¹, especially in southern, high-risk storage zones where cool weather conditions suitable for cooling grain in summer and early autumn are limited compared to central and northern states. Aeration controllers need to make best use of the available weather conditions. An effective strategy is to start below the grain temperature but above normal autumn grain temperature targets, beginning with thermostat settings of 24°C, falling to 21°C, reducing the set-point a few degrees every 2–3 weeks as average air temperatures drop.

Maintenance of aeration ducts, including removing covers and cleaning dirt, fines and foreign material,

checking for collapsed ducts and perforated screen plugging are very important for the SLAM approach and overlap with the sanitation task group function.

Monitoring

Regular sampling to detect pests and routine measurement of grain temperatures and moistures should be undertaken throughout the storage period.

Insect monitoring is done by hand probing with grain triers, deep cup samplers, power suction probe samplers or static insect traps, called probe traps. Probe traps consist of perforated plastic tubes about 45 cm long and 3.75 cm in diameter with a funnel and trap at the bottom made of very low friction (slick) plastic, up which insects cannot climb (Fig. 3.22). Probe traps are pushed vertically into the grain with the top just below the surface. A cord with flag marker is attached to locate and retrieve the trap. Sampling is usually done at two-week intervals for the first month or two after harvest, but the frequency is increased to weekly intervals as insect populations begin to rise. There is also a newly marketed automatic probe trap that connects to a computer, where

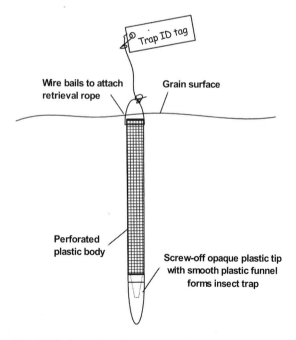

Fig. 3.22 Insect probe trap used to monitor insect activity near the grain surface in all types of storage units. Most of these traps are used just under the grain surface but with a push rod can be positioned at depths down to 2 m.

it indicates not only the number of insects captured but also identifies the species caught (Shuman *et al.* 2003). Sticky traps are used to monitor flying insects inside and outside warehouses or in bin headspaces.

Grain temperature monitoring is done by using sensors such as thermocouples or thermistors; the latter are more accurate, stable and have much simpler transfer wiring. Several hundred thermistor sensors mounted in many grain bins and silos can be monitored automatically by computer using a three- or four-wire signal cable. Computers receiving temperature readings can be used to optimise the operation of aeration systems and give warning of the temperature rises associated with biological deterioration.

Thus, in the USA, the ideal grain management model is based on excellent storage facility sanitation, careful bin loading which may include grain cleaning, cooling the grain using aeration as soon as practical, followed by diligent, continuous monitoring of the stored grain. Fumigation is used as a fifth stage backup to these four major task groups (Maier *et al.* 1994).

Stored grain management using bin boards at elevators

Even though steel bins typically include more aeration and temperature cable monitoring of grain than do concrete silos, the advent of bolted-steel galvanised bins has created more difficult grain management for elevator operators. Convenient handling in concrete elevators, which allowed managers to turn the grain to break up hot spots and blend higher moisture wheat with drier wheat within two to four weeks of harvest, is not an option with wheat stored in stand-alone steel bins.

Grain elevator superintendents and managers with one to five years' experience often do not have the skills to maintain the quality of wheat stored for long periods in steel bins and concrete silos. The wheat is often not segregated according to grade and marketing characteristics, specifically moisture content, test weight, dockage and protein content, so that good and poor quality grain is mixed, especially where most or all of the storage is in a few large steel bins or a flat storage building.

This can be avoided with the use of bin boards, i.e. physical diagrams of the storage units on a chalk board near the truck dump station, a printed sheet of the bin layout, a computer record of all bins and silos or a combination of physical and electronic files. With proper inputs from the elevator superintendent or grain manager, computerised bin boards allow the elevator manager

to print out the volume, condition and status of wheat throughout the elevator at any time. Some electronic bin boards have been developed that allow terminal and large country elevators with concrete silos to input scale ticket data into computer files to document grain characteristics in individual silos. Oklahoma State University researchers are developing an advanced bin board computer program called the Grain Segregation and Blending Model (GBM). This incorporates all US Grain Standard grading factors as well as the moisture and protein content of grain for up to 30 silos or bins. Elevator operators can input all grading factors (informal grades, using in-house grading equipment) at initial receipt, or record preliminary wheat segregating characteristics at initial receipt, then accumulate representative samples for each silo which are graded for all desired factors for input into the GBM. Most wheat elevator managers will record only the grading factors that directly affect their marketing contracts.

Grain managers using the GBM electronic bin board can input grain discount schedules from many terminal elevators and other markets. This will allow grain elevator marketing personnel or grain merchandisers to blend minimum amounts of high-quality wheat to meet market contracts while minimising discounts, thereby optimising their overall wheat stocks. Flour millers can use the GBM to develop or fine-tune blends of protein, moisture and other identified wheat characteristics to improve flour yields. This electronic management tool will allow US elevators to develop optimum blends more quickly than by manual methods. It will also allow elevators to improve their profit by analysing discounts vs transportation costs to establish the best market for their wheat (Jeyamkondan *et al.* 2003b; Noyes *et al.* 2003).

Economics of wheat storage operations

From the early 1950s, the great increase in cereal grain production has led to the construction of many farmer-owned co-operative elevators across the USA. Many of these elevators operated as federally licensed warehouses, receiving US Government storage subsidy fees of about $9–11 t^{-1} y^{-1} to store government grain reserve wheat. Federal grain storage subsidy payments were responsible for much of the new elevator storage construction during the 1960s and 1970s. In 1988, the USDA discontinued subsidy storage payments and many grain elevators, which were dependent on government fees, had serious wheat quality problems when they had to deliver their government grain to market. They often found that government grain that had been stored in the same bins for several years was infested by insects. Much of their wheat had high levels of damage and, where there were more than 32 insect-damaged kernels per 100 g, the wheat was considered sample grade, which reduced its value by 40–50%. As a result many grain elevators filed for bankruptcy and were forced to sell to or consolidate with other, stronger elevator organisations.

Pest management costs vary widely between elevators, based on whether control measures are applied by local crews or by pest control contractors. The cost of phosphine pellets or tablets has remained fairly constant during the past decade, ranging from about $250 to $350 per case of 35 000 pellets or 7000 tablets. Using an average of 10 pellets per tonne, for both concrete and steel stores, actual fumigant costs for a 27 000 tonne elevator would be $2357 for fumigant, or $0.0872 t^{-1}. However, labour is a major component of the cost and commercial fumigators would probably charge $5–7 t^{-1} for a 27 000 t elevator. When fumigating twice a year the cost to an average country elevator can easily add $30 000–40 000 to annual operating expenses.

The costs of residual pesticide treatment with malathion is low, at $7–8 L^{-1}, but this is poor value as this pesticide is relatively ineffective. Storcide®, the only effective residual chemical, costs about $90 L^{-1} (to treat 70 t), or $1.28 t^{-1}. At the time of writing Storcide® does not have Codex maximum residue levels, so international traders would be unlikely to be able to receive Storcide®-treated wheat at this time (Phillips *et al.* 2003).

Wheat elevators must operate on the income margin between buying and selling, plus some storage rental income. Many elevators also sell and apply fertiliser, herbicide and insecticides in the field. Thus trying to analyse typical operating costs of the overall operation is not straightforward. However, considering just the grain operation, the basic variable operating costs are labour, electricity, trucking or rail expenses and elevator maintenance, while primary fixed costs are the cost of interest on property loans and insurance. For a typical 27 000 t elevator, labour costs are about $1.70 t^{-1} and, where the power use averages 150 HP for legs, conveyors, fans, cleaners, etc. operated on average four hours per day, the electricity costs are about $0.27 t^{-1}.

In summary, fixed costs for storage might amount to a total of $13.50 t^{-1}. Variable costs would be for electricity, chemicals (pesticides and fumigants), maintenance, insurance and labour. Variable costs (all per tonne) would be approximately $0.27 for electricity, $1.10 for pesticides, $0.75 for maintenance, $1.10 for insurance

and $1.72 for labour (Anderson & Noyes 1989), giving a total variable cost of about $4.94 t^{-1}. Total storage costs (fixed plus variable costs) could be about $18.44 t^{-1}. Thus, the elevator manager would need to make more than this to generate a profit from the elevator grain storage programme.

Future developments

Since 1975, the USDA has reduced by 75% the number of research scientists and extension specialists working in grain storage management. The USDA has also closed most of its grain research centres such as the National Rice Research Center at Beaumont, Texas (closed in the 1980s), the entomology grain research centre at Savannah, Georgia (closed in 1995) and USDA grain laboratories at the University of Wisconsin and North Carolina State University (closed during the 1990s). These reductions in post-harvest grain scientists, specialists and facilities were due to Congressional cost-cutting mandates of the 1970s and 1980s. Some USDA administrators thought that with the effective control of grain insects by the inexpensive residual insecticides available at the time, there was little need for continued research and education on post-harvest grain storage problems. They did not anticipate the massive pesticide re-registration changes that were being planned by EPA in the mid-1980s and mid-1990s, mandated by the Food Quality Protection Act of 1996.

Primary grain and seed laboratories that are still operating include the US Grain Marketing and Production Research Center (USGMPRC), Manhattan, Kansas, and the National Peanut Research Laboratory, Dawson, Georgia. To counter the decline in federal post-harvest research two new centres have been inaugurated recently; the Post Harvest Education and Research Center (Purdue University) in 1999 and the Stored Product Research and Education Center (Oklahoma State University) in 2001. The post-harvest specialists at these sites work closely with USGMRC staff at Manhattan. Other US universities undertaking post-harvest research and technology transfer include Kansas State, Minnesota, Ohio State, North Dakota State, Kentucky, Georgia, Auburn and Montana.

As the opportunities for use of residual pesticides for grain insect management have declined, so sustainable commodity management has been promoted, particularly through SLAM. However, this approach is often labour intensive and requires more management attention than the use of insecticides and fumigants, with the result that adoption rates have been relatively slow. For the future, applied research and extension development should place emphasis on adding tools to SLAM. The future model grain storage system is expected to be well-sealed silos with multi-functional automatically controlled suction aeration/CLF systems to minimise the need for grain turning, reduce labour and housekeeping and improve elevator sanitation. Full adoption of this sustainable commodity management technology by the country elevators in the USA will greatly improve wheat quality.

Australia

Historical perspectives

Wheat grown in Australia used to be stored and handled in bags (Fig. 3.23). The bags were built into huge stacks in the open and protected from the weather by galvanised iron roofs and hessian curtains hung over their sides (Winterbottom 1922; Callaghan & Millington 1956; Whitwell & Sydenham 1991; van S. Graver & Winks 1994; Ayris 1999).

Prevention and control of insect infestation in depots containing literally millions of stacked bags was difficult. The processes adopted included physical exclusion and sealing, hermetic storage or use of controlled atmospheres and poisonous gases, physical shock, heat treatment and sterilisation, mechanical cleaning, dust treatments, and hygiene practices (Winterbottom 1922). Despite the technological simplicity of these techniques, the actual losses sustained by the stock of bagged wheat

in Australia during World War I were relatively light (Callaghan & Millington 1956).

Bulk handling had been under consideration for some time before World War I. However, the decision to adopt it was, in part, influenced by the losses experienced during the war and subsequently by the cost of jute bags and their handling. In the 1930/31 season in Western Australia the cost of bags amounted to 14% of the total return from the crop and bag handling was three times the cost of handling two years earlier (Callaghan & Millington 1956; Whitwell & Sydenham 1991).

Australia was one of the first countries, after the USA and Canada, to adopt large-scale bulk storage and handling of wheat. The first bulk storage constructed in Australia was a concrete silo, built in 1918, at Peak Hill in New South Wales. This was followed in 1921 by the first shipment of bulk wheat exported from the Sydney terminal. Further expansion of the bulk storage system in New South Wales was disrupted during World War II but continued steadily thereafter. In Western Australia, bulk storage was introduced in 1932 using a basic design derived from horizontal bag storage sheds (Callaghan & Millington 1956; Zekulich 1997; Ayris 1999), and in South Australia using vertical silos during 1952 (Lamshed 1962). Subsequently, a variety of different bulk stores were adopted as was the use of semi-permanent bunker storage that is a unique feature in the development of the Australian grain storage system. These low-cost structures were designed and adapted to cater for the ever-increasing quantities of grain produced in Australia during the late 1970s and early 1980s (Whitwell

Fig. 3.23 In the early twentieth century, before widespread use of bulk handling, Australian wheat was stored in bag stacks in the open and covered with galvanised iron roofs (seen towards rear of photograph).

& Sydenham 1991; Barker 1992). Recent significant improvements in design and construction techniques have led to a strong reliance on this type of storage by all major bulk handling companies.

Australia became a significant exporter of wheat on the international market as the twentieth century progressed (Whitwell & Sydenham 1991; Connell *et al.* 2002). It was therefore in the interests of the industry to develop the most efficient and cost-effective means for receiving, storing and transporting grain from the farm gate to an export terminal. However, it was not until after World War II that bulk handling became the main means of receiving, storing and outloading grain at country locations and delivering it to seaboard terminals or to domestic end-users (Whitwell & Sydenham 1991). At this time, five bulk handling companies (Co-operative Bulk Handling in Western Australia, South Australian Co-operative Bulk Handling, the Grain Elevators Board of Victoria, the New South Wales Grain Elevators Board and the State Wheat Board in Queensland) had been established by each of the mainland state governments to complement the activities of a number of grain marketing boards. These grain marketing boards, including the Australian Wheat Board (AWB), had monopoly powers to buy and sell grain both domestically and on the export market (Callaghan & Millington 1956; Anon. 1987a; Whitwell & Sydenham 1991; Coombs 1994; Zekulich 1997). The structure of the Australian wheat industry and the legislation governing it are summarised by Reeves and Cottingham (1994).

Dramatic changes have taken place in the storage and handling sector of the Australian grains industry in the past decade. A series of mergers and alliances and the privatisation of marketing organisations have taken place so that by 2003 there were four major bulk handling companies operating in Australia: GrainCorp, AWB Ltd and AusBulk in the east, and Co-operative Bulk Handling in the west. In addition to grain handling services these companies now offer grain marketing, farm input merchandising and grain transportation services to their clients. Some companies have also diversified into downstream processing with the acquisition of Allied Mills by GrainCorp in 2002, in partnership with Carghill Australia Ltd, and the acquisition of Joe White Maltings by Ausbulk in 2003.

It is difficult to predict how this industry will evolve in the future, as the amount of change predicted a decade ago has already exceeded expectations. However, it is certain that existing companies will continue to diversify into all areas along the grain supply chain from farm gate to domestic end-user or export terminal and into downstream processing businesses. This will lead to further competition among industry participants, which in turn should lead to greater efficiencies in service delivery and lower costs to growers. However, these processes will need to be achieved without compromising Australia's enviable reputation as a supplier of high-quality grain to the domestic and international markets, a task that will require the participation of all industry players.

While this account concentrates mostly on the central storage system traditionally used to store wheat in Australia, the rapid change and evolution of the grain storage and handling industry has greatly increased the options available to wheat growers to handle and store their crop (Banks 1998; Turner *et al.* 2002). They may now choose to deliver it directly to bulk handling companies, make warehousing arrangements with these companies, deliver it directly to end-users or merchants and/or store it on-farm. This has resulted in a substantial increase in the amount of wheat stored on-farm with a commensurate increase in the number and type of grain storage structures in Australia (Kotzur 1998a; Turner *et al.* 2002).

A wide range of web-based information sources is available on Australian grain storage and marketing system. Some of these can be found in Box 3.1.

Box 3.1   Web-based information sources on grain storage and handling in Australia (all accessed 20 July 2003)

AGIRD – Australian Grain Insect Resistance Database. A national database of pesticide resistance in grain insects: http://www.agric.wa.gov.au/ento/agird1.htm

Agridry (grain drying and aeration): http://www.agridry.com.au/

AusBulk Ltd: http://www.ausbulk.com.au/

Australian Bulk Alliance: http://www.bulkalliance.com.au/

Australian Centre for International Agricultural Research (ACIAR): http://www.aciar.gov.au/

AWB Ltd: http://www.awb.com.au/AWB/user/default.asp

Co-operative Bulk Handling Ltd Western Australia: http://www.cbh.com.au/index.html

CSIRO Stored Grain Research Laboratory: http://sgrl.csiro.au/

Department of Agriculture Western Australia, Entomology Branch: http://www.agric.wa.gov.au/ento/grain1.htm

Grainco Australia Ltd: http://www.grainco.com.au/home/default.asp

GrainCorp Operations Ltd: http://www.graincorp.com.au/docs/home/index.html

GCA (Grains Council of Australia): http://www.grainscouncil.com/IndexGCA.htm

Grains Research and Development Corporation (Search for grain storage): http://www.grdc.com.au/

Kotzur Silos (sealed silos): http://www.kotzur.com/silos.html

New South Wales Agriculture: http://www.agric.nsw.gov.au/

Phosphine. National awareness campaign: http://www.agric.wa.gov.au/ento/publications/national_phosphine_awareness_campaign.htm

Phosphine. Use it responsibly: http://www.agric.wa.gov.au/ento/publications/Phosphine_Use_it_responsibly_or_lose_it.htm

Quality Wheat CRC Limited. Information related to storage can be accessed under the publications section: http://www.wheat-research.com.au/

Queensland Department of Primary Industries. Information related to grain storage is available in the general category of 'broad-acre field crops', under the section on 'crop management principles'. In this section of the website, researchers from the Farming Systems Institute discuss pest management and other grain storage issues. Refer to sections on 'pest management' and 'ready to harvest': http://www.dpi.qld.gov.au/fieldcrops/

Rural Industries Research & Development Corporation (search for 'grain storage'): http://www.rirdc.gov.au/home.html

South Australian Research and Development Corporation. For information related to grain storage look for 'harvest and storage' at: http://www.sardi.sa.gov.au/pages/field_crops/crops/croplinks/general.htm#harvest#harvest

Victoria Department of Primary Industries and Department of Sustainability & Environment. Information related to grain production is available under i) Primary Industries, ii) Farming and Agriculture, iii) Crops and Pastures, iv) Grain Crops, v) Victoria's Grain Industry. At: http://www.nre.vic.gov.au/

All hard copy references cited in this account are obtainable from the Resource Centre of the Stored Grain Research Laboratory, CSIRO Entomology, GPO Box 1700, Canberra, ACT 2601, Australia. Email: ento-sgrlresource@csiro.au; website: http://sgrl.csiro.au/.

Facilities and practices at storage sites

The permanent storage facilities used in Australia vary somewhat according to locality, whereas semi-permanent bunker storage is now an integral feature of wheat storage systems throughout the country (Newman 1988; Barker 1992; Connell *et al.* 1992; Coombs 1994; Ayris 1999; Satterley 2001).

Permanent storage

The design of permanent storage structures varies considerably across Australia. This reflects the various construction and utilisation philosophies adopted by each bulk handling company and the need to accommodate the various segregations of wheat required by the industry (Anon. 1986b; Hinchy 1989; Tutt & Burton 1998). The storage capacities of four larger bulk handling companies varies from 3m t to 20m t (Table 3.16).

The permanent storages include (1) horizontal sheds, (2) squat circular storage bins (silos) with a height to diameter ratio of about 2 : 1, (3) vertical storages consisting of blocks of individual cells (silos) with a height to diameter ratio of 3 : 1 and (4) semi-permanent bunker storages with steel, earth or concrete walls (Prattley 1995).

The silos are of either concrete or steel construction, vary considerably in design and range in capacity from 2000 to 10 000 t (Fig. 3.24). Many smaller silos are of the hopper-bottomed type while larger facilities are emptied by means of sweep augers. The silos are normally (1) sealed to allow effective use of fumigants, (2) may be fitted with gas recirculation facilities (Ripp *et al.* 1984; Green 1986; Newman 1988; AFHB/ACIAR 1989; Burton 1998), to assist in fumigant distribution, and (3)

Table 3.16   Infrastructure and total storage capacity of four of the larger bulk handling companies in Australia.

Company	Area of operation	No. of country receival sites	Number of port terminals	Storage capacity (million tonnes)
GrainCorp Operations Ltd	Queensland New South Wales and Victoria	360	8	20
Cooperative Bulk Handling Ltd	Western Australia	150	4	15
AusBulk Ltd	S. Australia and Victoria	85	6	10
AWB Ltd	Victoria	21	1	3

Fig. 3.24   A silo complex with two rows of four cells with two large free-standing silos either side, at Crystal Brook, South Australia (photograph courtesy of AusBulk Ltd).

Fig. 3.25   Sealed horizontal shed of 20 000 tonnes capacity and fitted with aeration at Roseworthy, South Australia (photograph courtesy of AusBulk Ltd).

are filled using bucket elevator systems with capacities in the range 80–500 th⁻¹ (Scott *et al.* 1998).

Horizontal shed types are usually all-steel structures although some have concrete walls (Newman 1988). They normally range in capacity from 5000 to 20 000 t, although some are considerably larger. Many shed-type storages are fitted with aeration facilities (Newman 1996; Anon. 2002a), and are sealed (Ripp *et al.* 1984; Green 1986; Burton 1998) to allow effective fumigations (Fig. 3.25). They are filled using bucket elevator systems or inclined conveyors and most are emptied using front-end loaders.

A typical country receival site may consist of a number of different storage types which collectively may have capacities in the range 5000–250 000 t (Satterley 2001). Such sites are able to receive 1000–8000 t of grain per day during the harvest period. The network of country receival sites across the Australian wheat belt is closely linked with the state-based rail systems that were constructed to facilitate the transport of wheat to export terminals (Small 1981; Anon. 1987b; Coombs 1994).

As a result, most facilities have the capacity to outload wheat directly to rail as well as to road. Rail outloading rates vary from as little as 70 th⁻¹, in small older-style facilities, up to 1000 th⁻¹ in the larger sub-terminal type facilities (Scott *et al.* 1998).

Almost all export wheat is transported by rail from the country to export terminals at the coast, either directly or through large sub-terminal facilities located at key junctions on the networks. Wheat destined for the domestic market is transported predominantly by road (Whitwell & Sydenham 1991).

Storage capacity at Australian seaboard export terminals varies from 60 000 t to over 1m t (Moore 1980, 1988). One of the latest facilities to be built in Australia, Port Kembla Terminal, uses rising belt conveyors exclusively (rather than the conventional bucket elevator systems) to move grain within the complex. This facility can receive up to 2500 th⁻¹ by rail, can store nearly a quarter of a million tonnes of grain in fully sealed hopper-bottomed bins and can load ships at up to 5000 th⁻¹ (Collins 1996a).

GrainCorp Operations Ltd owns and operates grain storage and handling facilities in New South Wales, Victoria and Queensland. In Queensland the system is capable of storing up to 5m t of grain at any one time. A record 3m t of wheat (and other grains) was received into this system during the 2000/01 harvest season. Many types of vertical (silo) and horizontal (shed) storages of different designs are used in the Queensland bulk handling system (Fig. 3.26).

The AusBulk storage system, which extends into Victoria (Anon. 2002b), is concentrated in South Australia (Anon. 1998; Gentilcore 1998; O'Driscoll 2000; Satterley 2001), and consists predominantly of concrete vertical storage units (Fig. 3.27). Because much of the grain growing areas in South Australia are close to the coast a large proportion of the harvest is delivered by road directly to port facilities during the harvest period. Thus the costs of grain transportation are comparatively low.

Most storages constructed by Co-operative Bulk Handling Ltd in Western Australia are of the steel horizontal 'shed' type, and are between 10 000 and 40 000 t capacity although some are considerably larger (Ayris 1999). In the early 1980s, the decision was made to seal all existing horizontal sheds to permit effective fumigation with phosphine and to build all new storages to specific gas-tight standards (Ripp *et al.* 1984; Green 1986). This policy was adopted to enable the state to export wheat that was both free from insect infestation and from residues of grain protectant pesticides. By the early 1990s all permanent storage in Western Australia's central handling system had been sealed to enable fumigation with phosphine and the use of residual insecticides was phased out altogether (Delmenico 1992; Ayris 1999).

Fig. 3.27 Typical reinforced concrete silo complex at Booleroo in South Australia (photograph courtesy of AusBulk Ltd).

Semi-permanent storage facilities

Bunker storage

Wheat production in Australia is extremely variable because of wide fluctuations in the amount of rainfall received. In seasons of high production, low-cost temporary storages have been used to accommodate the very large quantities of wheat because the erection of permanent facilities is not cost effective (Callaghan & Millington 1956; van S. Graver & Winks 1994). In Australia the term 'bunker', 'bulkhead' or 'pad' storage is used to describe semi-permanent storage structures of up to 50 000 t capacity with earthen or bitumen floors, low-profile earthen, concrete or steel framed walls and PVC covers (Fig. 3.28).

The first serious attempts to build robust, low-cost bunker storages were made during the late 1970s (Andrews 1979; Wightley 1979; Champ & McCabe 1984; Yates & Sticka 1984) when large 30–50 m wide by 100–200 m long structures were built, primarily in New South Wales, Victoria and Western Australia. They consisted of earthen or portable (concrete or steel framed) walls up to 2 m high and were filled using specially designed conveyor/stacker equipment.

Bunker storages were initially built to accommodate the additional quantities of wheat received during

Fig. 3.26 Some examples of the store types used in Queensland, a storage complex with a steel shed (500 tonnes) at the rear, a grain dryer in front of it with four free-standing silos (1880 tonnes each) beside it, and two free-standing silos (in the foreground 4800 tonnes each) at Capella, Queensland, Australia.

seasons of high production. However, their designs have now been so well refined that they may be used to replace ageing permanent facilities where inloading and outloading rates are slow or existing structures cannot be sealed to permit fumigation. Grain losses due to water ingress or mould growth in modern-day bunkers are commonly less than 0.1%.

The structures are filled using purpose-built portable grain throwers known as Lobstars (Figs 3.29 and 3.30) or side delivery inclined belt conveyors which move down the length of the bunker as it is filled (Yates & Sticka 1984). The latter are of two types: those with fixed hoppers and belt conveyors delivering grain to the conveyor, or portable 'drive over' hoppers that move with the conveyor as the storage is filled. This equipment allows modern bunkers to be filled to a height of 8 m at rates of around 300 th⁻¹.

The bunker covers are laid progressively onto the grain surface as the bunker is filled (Fig. 3.31). The sheets are stitched together using conventional bag sewing machines (Fig. 3.32) and acrylic pastes are applied to the stitched joins to ensure that they are waterproof and gas-tight. After a bunker has been filled with wheat, the cover and bottom sheet are brought together over the wall and clamped with a baton and excess sheet secured by hook-shaped clamps (Fig. 3.33). The gas-tightness of the whole enclosing cover, along the length of the bun-

Fig. 3.30 Grain throwers can fill bunkers to a depth of 8 m at a rate of about 300 tonnes per hour; aeration ducting is visible in the foreground at Esperance, Western Australia.

Fig. 3.28 Bunker storage of wheat at Esperance, Western Australia.

Fig. 3.29 Portable grain throwers ready to load grain into empty bunkers at Esperance, Western Australia.

Fig. 3.31 Placing cover sheets over the wheat as the bunker is filled, at Esperance, Western Australia.

Fig. 3.32 Cover sheets being joined at the edges by sewing; the joints are then made gas-tight by the application of acrylic paste, at Esperance, Western Australia.

Fig. 3.33 After a bunker has been filled with wheat, the cover and bottom sheet are brought together over the wall and clamped with a batten (arrow) and excess sheet secured by hook-shaped clamps, at Bogan Gate, New South Wales (photograph courtesy of Dr H.J. Banks).

ker, is ensured by various means, depending on the type of bunker constructed.

Bunkers are outloaded using the same equipment used to fill them. The Lobstar is capable of loading grain into road vehicles at rates of up to 300 th⁻¹ when converted to outloading mode. Similar outloading rates can be achieved using front-end loaders to transfer grain from the bunker face to a hopper that feeds grain back onto the inclined conveyor.

The use of bunker storage in Australia has allowed bulk handling companies to concentrate wheat receival facilities into major centres. This has increased weighbridge and sample stand utilisation, significantly increased wheat receival rates during the busy harvest period and reduced pest control costs (Dick 1997; Satterley 2001). Development of techniques allowing bunkers to be fumigated effectively has been crucial to their success (Banks & Sticka 1981).

Sample stands and weighbridges

Country receival sites in Australia are normally equipped with a sample stand elevated to a height of around 2.5 m and equipped with vacuum probes so that load-by-load samples can be extracted from trucks (Fig. 3.34). Typically three vertical probes must be taken from each truck, totalling at least 3 L of grain. An adjoining grain laboratory (Fig. 3.35) determines the grade of each load; receival standards are applied to individual truckloads and are not averaged over a number of loads. Test equipment measures moisture content, protein content, test weight (density), screenings and impurities, defective (damaged) grain and weed seed counts. Several of these quality variables are measured using near infra-red whole grain analysis (Newman 1988). Wheat quality is again carefully monitored during outturn to the domestic market or during receival into export terminals before being loaded into each ship.

Fig. 3.34 A wheat sample being withdrawn from a truck on delivery to the storage complex at Esperance, Western Australia.

Fig. 3.35 The quality of grain samples taken from trucks being analysed in a small laboratory to determine the grade of each load at Esperance, Western Australia.

After sampling the trucks proceed to a weighbridge. These are of the electronic load cell type with a capacity of 60–150 t. This allows all road vehicles currently used to transport wheat in Australia to be weighed (Newman 1988; Dick 1997).

Grain drying facilities

In Australia, wheat is harvested at the height of summer under hot and dry conditions so there is little need for artificial or mechanical grain drying facilities (Kotzur 1998b). The hot dry finish normally experienced during the harvest has allowed a maximum receival moisture content of 12.5% to be established as an industry-wide standard (Desmarchelier & Ghaly 1993; Adamson 2002). There are a few areas where this standard cannot be achieved regularly. In these situations, grain drying facilities are often used, although there is rarely a need to reduce moisture contents by more than 1% or 2%. In certain places, raising the limit to 13.5% would have the advantage of reducing the possibility of rain damage by harvesting the ripe crop earlier; following experimental studies, this has been adopted in some temperate areas (Caddick & Shelton 1998; Adamson 2002).

Dust management

A feature of the Australian wheat storage industry is its very low incidence of grain dust explosions (Annis 1996; Viljoen in press; Wilson in press). This is because grain dust extraction systems are installed at all transfer points

in all terminal and sub-terminal facilities (Moore 1988; Newman 1988). The rising costs associated with disposal of grain dust have resulted in studies to determine alternative uses for this product, including stockfeed, garden mulch and agricultural fertiliser (Desmarchelier & Hogan 1978).

Objectives of storage

The primary purpose of developing a central storage system in Australia was to enable growers to deliver their wheat to a local receival site as the harvest progressed without the need to store significant quantities in on-farm storage structures (Golden 1998). This enhanced the industry's ability to store wheat safely, to segregate it according to marketing (Dines 1998; McMullen 1998, 2002, in press; Pheloung & Macbeth 2002) and other requirements, and to implement sound and effective stored grain pest control practices. These were, and remain, essential ingredients for an industry so heavily reliant on export for its long-term success.

While these objectives have not changed significantly, in the last 20 years there has been a significant increase in the use of on-farm storage. The drivers of this change have been the dramatic increase in the speed at which grain is harvested, the increase in the size of the farms themselves and deregulation of the marketing system which has in turn allowed growers to sell to a large number of buyers rather than to the one single monopoly marketer (Whitwell & Sydenham 1991; Hunter & Hooper 1992; Collins 1998a; Kotzur 1998a; Newman 1998; Anon. 2002c; Bridgeman 2002; Nugent 2002; Turner *et al.* 2002).

As the central storage system developed, so has the industry's ability to segregate wheat according to quality (Collins 1996b; Coetsee 1998; Gentilcore 1998; Tutt & Burton 1998; Satterley 2001; Anon. 2002a, c). One of the main features of the system today is its well-developed ability to test accurately all grains for a range of quality parameters, to segregate deliveries according to the marketers' receival standards and to deliver wheat of very specific quality to domestic customers or to export markets.

Major sources of quality decline

Storage insect pests

In Australia, wheat is harvested in hot, dry conditions when the relative humidity is generally low and grain moisture contents are low. Throughout the wheat belt stored product insects pose the greatest threat to wheat quality, with storage conditions almost perfect for insect attack.

The industry has implemented a range of successful pest control strategies to prevent insects from becoming a serious problem, but continual vigilance is necessary if this situation is to be maintained. While the central storage system is well placed to effectively implement insect pest control, the growth of on-farm storage capacity, the development of insect resistance to insecticides and fumigants, consumer concerns regarding pesticide residues and environmental concerns will present serious challenges to effective pest control in the future.

Currently the most serious pest of stored wheat stored in Australia is *Rhyzopertha dominica*, although *Tribolium castaneum*, *Tribolium confusum*, *Cryptolestes* spp and *Oryzaephilus surinamensis* are also common. Less common but still of some significance are the weevils *Sitophilus granarius* and *Sitophilus oryzae* (Collins 1998a; Champ 2002). However, psocids such as *Liposcelis* spp have recently become increasingly important (Beckett 1998; Rees 1998; Weller & Beckett 2002; Beckett & Moreton 2003; Nayak *et al.* 2003). Stored product moths are generally not significant pests in bulk stored wheat but are common in flour and provender mills.

Moisture migration

As the moisture content of wheat taken in to storage is generally at or below 12.5% (Caddick & Shelton 1998) moisture migration problems are rare (Griffiths 1964). However, where bunker construction has been haphazard, water ingress can result in significant losses. Fortunately, improvements in bunker design have minimised the incidence of such events. In addition, recent advances in bunker aeration design are expected to enhance the storability of wheat (and other) grains held in bunkers.

Commodity and pest management regimes

For more than 30 years, Australia has maintained a reputation for exporting clean, insect-free, high-quality grain (George 1998). However, this has not always been the case. Until the 1960s wheat frequently arrived at export terminals in Australia in an infested state and had to be disinfested. Complaints from overseas customers were numerous (Lamshed 1962; Bailey 1976; Whitwell & Sydenham 1991) and in 1965 alone 1m t of exported wheat was subject to claims due to insect infestation on outturn overseas (van S. Graver & Winks 1996). This was then, and remains today, unacceptable in an industry so valuable to Australia's economy (Adamson 2002). In response to these problems, the Commonwealth Government introduced its Export (Grain) Regulations in 1963. The regulations empower the Australian Quarantine Inspection Service to inspect all bulk wheat (and later, other grains) in the course of delivery from export terminal facilities to ships' holds to ensure that it is not infested with insect pests – a policy of 'nil tolerance' (Anon. 1976; Whitwell & Sydenham 1991; Coombs 1994). Application of these measures over the past 30 years has allowed the Australian wheat industry to gain, and hold, its reputation for exporting grain substantially free from insect infestation (Pheloung & Macbeth 2002).

Enforcement of nil tolerance was made possible by the introduction, in the early 1960s, of the residual insecticide malathion (Bailey 1976; Whitwell & Sydenham 1991). All wheat received into the central handling system was treated with malathion and the industry's ability to control stored grain pests improved dramatically. However, resistance soon developed and by the early 1970s alternatives had to be found (Collins & Wilson 1986). The industry responded to this crisis by developing a wide-ranging 'Integrated Plan for Pest Control' with short-term and long-term objectives (Bailey 1976; Simmonds 1989).

The short-term objective was to facilitate the continued use of malathion and other residual insecticides and to this end the AWB established a Working Party on Grain Protectants that consisted of representatives from bulk handling companies, the Stored Grain Research Laboratory, State Departments of Agriculture and other entities. This group sponsored research and field trials that identified the efficacy of a mixture of fenitrothion and bioresmethrin, which became the replacement for malathion in time for treatment of the 1976–77 harvest (Bengston *et al.* 1975; Ardley & Desmarchelier 1978). The Working Party has continued to provide a number of alternative protectants that are effective and acceptable to international markets (Desmarchelier 1975; Desmarchelier *et al.* 1981, 1987; Bengston *et al.* 1983; Snelson 1987; Bengston 1988). Subsequently, membership of the Working Party was extended to reflect the changing quality demands for Australian wheat in overseas and domestic markets (McMullen 1998, 2002, in press). Now renamed the National Working Party on Grain Protection, it continues to provide new and alternative grain protection techniques and technologies for the Australian grain storage and handling industry (Bengston & Strange 1998; Murray 1998, in press).

There has certainly been a need to provide alternatives. *Rhyzopertha dominica* had developed significant resistance to the bioresmethrin/fenitrothion mixture by the mid-1990s (Collins 1998b) and the registration of bioresmethrin was withdrawn in 2001. The replacement for control of *R. dominica* was the insect growth regulator methoprene, first applied to the 1992/93 harvest. However, in some localities *R. dominica* has now developed significant resistance to this and the more active form, s-methoprene, is now used in those places.

The long-term objectives were based on the premise that residual insecticides did not have a long-term future. The industry decided to invest in the adoption of storage facilities in which a wide range of pest and quality control methods (e.g. fumigants, controlled atmospheres, aeration) could be applied. The objective was the creation of a nationwide wheat storage system with the capacity to implement an effective programme of integrated commodity management with the flexibility to meet specific market requirements. In 1972 the Stored Grain Research Laboratory at Canberra was officially opened with the responsibility to undertake the research required to reduce the industry's reliance on residual insecticides (Adamson 2002; DiLeo 2002; Wilson 2002; Bridgeman in press). Its operating costs, at that time, were jointly funded by the AWB and Commonwealth Scientific and Industrial Research Organisation (CSIRO) (Bailey 1976; Whitwell & Sydenham 1991). The research programme included technologies for sealing grain storages, effective use of fumigants in sealed storages, development of controlled atmospheres to commercial adoption, development of bunker storage technology, aeration strategies to cool and protect grain, development of inert dusts for use as structural treatments, and the development up to commercial application of the SIROFLO® fumigation system (CSIRO 1970–1987, 1987–2001).

Fumigants

Coinciding with these developments were efforts by bulk handlers to seal their existing storages and to build new gas-tight storages to ensure effective fumigation with phosphine or methyl bromide (Banks 1981; Newman 1990, 1998; Tutt 2002). This trend has continued to the present, with more than 90% of Australia's wheat now being exported free of pesticide residues. Phosphine is now very widely employed to disinfest grain as use of methyl bromide is phased out (Botta *et al.* in press).

The development of the SIROFLO® system has been an important step in increasing the range of store types fumigated with phosphine since it can be used in stores that are not fully sealed – conferring the same advantage as the close loop fumigation (CLF) system in the USA. It allows bulk handling companies to fumigate these types of stores without the need to make costly structural modifications. The technique employs a pressurised distribution system that compensates for phosphine losses that would reduce fumigant concentrations during the exposure period (Fig. 3.36). It can be designed to suit a wide range of storage types – from silo complexes in export terminals to individual farm storage bins, and is now in widespread commercial use in Australia (Winks 1993; Winks & Russell 1994, 1997; APTC 1998, 2000, 2003).

Recently, due to the emergence of low-level resistance to phosphine in Australia, dosage rates for SIROFLO® treatments have had to be raised (APTC 2000, 2003). As a result the need has arisen to change the type of supply

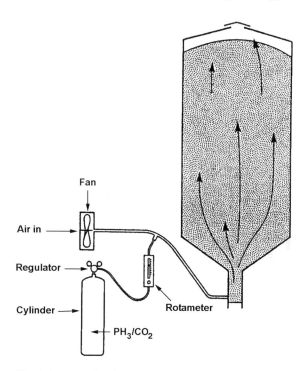

Fig. 3.36 The SIROFLO fumigation system is designed for use in open-top silos. Phosphine is released from a cylinder (or generator) into a constant stream of air blown through grain in a silo and diluted in the air stream to provide a constant concentration throughout the grain bulk. Long exposure periods of 21–28 days are required with this technique.

from cylinders with low concentrations of phosphine to pure phosphine, or by using solid aluminium phosphide to produce phosphine on-site. Both techniques are currently under development and in the future will eliminate the need to transport large numbers of cylinders to the fumigation site (APTC 2000, 2003).

Aeration

Aeration with ambient air is a common means of pest control in Australia, but warm climatic conditions prevent the use of this technique by itself and so it is combined with the application of insecticides and/or fumigants. However, new techniques using refrigerated aeration systems are being developed that may permit this in the future (APTC 1998).

Over the last decade there has been renewed interest in aeration in Australia, primarily because of the development of programmable microprocessor control and monitoring systems that allow best use of favourable ambient conditions to maintain grain quality. Thus fans can be triggered to operate whenever the temperature and relative humidity of the outside air is low enough to have a positive cooling effect on the stored commodity. These fan set points are then progressively lowered as the commodity cools, and the autumn/winter season approaches (APTC 1998, 2000, 2003).

Aeration systems of this type are progressively being introduced throughout Australia and have also been successfully incorporated into bunker storages.

Economics of operation of the system

The original central storage system in Australia was based on a single-use strategy for country storage facilities in any one season. Such a strategy means that the return on capital is small, is achieved over a very long time period and does not fit in with the current storage and handling business environment. Bunker storage construction costs are less than 5% of vertical storage and less than 15% of horizontal shed type storage. As a result, there is very little replacement permanent storage being constructed in Australia at the present time.

Future developments

Since privatisation Australian grain storage companies have reviewed their previous storage construction philosophies to ensure that future capital expenditure is more carefully targeted into areas where it will bring about greater efficiencies and will provide the least cost path between country receival site and export terminal or domestic end-user (Anon. 1997). This has resulted in:

- greater use of bunker storage (Gentilcore 1998; Satterley 2001);
- construction of permanent storage facilities only where multiple usage is possible in any one season (Coetsee 1988; Satterley 2001);
- increasing need for raising growers' technical awareness and ability (Cameron *et al.* in press; Hughes *et al.* in press);
- construction of much faster road receival facilities to ensure that as much wheat as possible can be delivered during the harvest period; and
- construction of new rail outloading facilities to permit rapid rail transport turnaround times (Satterley 2001).

There have been a substantial number of mergers and alliances formed between bulk handling companies. It is likely that further mergers will occur as companies attempt to grow to a size that uses resources more efficiently. Coupled to these developments is the expansion by grain storage organisations into other related areas of operation such as grain marketing, plant breeding, fertiliser and chemical supply, farm financing, flour milling, malt barley production and grain transport (Collins 1997; Watson 1996). Alliances and mergers are also being formed between these companies and the old state-based marketing organisations. It is likely that the performance and relevance of the remaining export monopoly marketing organisations will also be carefully examined in the near future. While it is difficult to predict the outcome of such enquiries, the most probable outcome is that regulators will create changes to encourage even more competition among industry participants in all parts of the supply chain from farm gate to export terminal.

The introduction of genetically modified crops and desire to develop quality-based niche markets is also likely to influence bulk handling in Australia in the future. However, the resulting requirements for detailed quality assurance and identity preservation programmes may lead to greater usage of on-farm storage and an expansion in container storage and transport rather than modification of the existing central storage system (Bridgeman 2002).

References

Adamson, D. (2002) The benefit of the CSIRO Stored Grain

Research Laboratory to the Australian grain industry. *Technical Paper No. 39*. CSIRO Division of Entomology, Canberra, Australia.

AFHB/ACIAR (1989) *Suggested Recommendations for the Fumigation of Grain in the ASEAN Region. Part 1 Principles and General Practice.* ASEAN Food Handling Bureau, Kuala Lumpur, Malaysia and Australian Centre for International Agricultural Research, Canberra, Australia.

Allen-Stevens, T. (2000) Being sure of what's in store. *Farmers' Weekly*, 22 July 2000, 12–13.

Anderson, K. & Noyes, R.T. (1989) *Grain Storage Costs in Oklahoma*. OSU FS No. 210. Oklahoma Co-operative Extension Service, Oklahoma State University, OK, USA.

Andrews, A. (1979) *Storing, handling and drying grain: a management guide for farms*. Queensland Department of Primary Industries, Brisbane, Australia.

Annis, P.C. (1996) The use of oil and water to reduce air-borne dust in grain handling facilities. In: *Proceedings of the 1996 Dust Explosion Summit*. IBC Conferences Pty Ltd, Sydney, Australia. http://www.ibcoz.com.au.

Anon. (1976) Inspection for pests, diseases and contaminants. *Wheat Industry Review Conference*, March 1976, Sydney, Australia.

Anon. (1986a) *Schering Grain Quality Guide*. Schering UK.

Anon. (1986b) Issues for the Royal Commission's inquiry. *Discussion Paper No. 1*, Royal Commission into Grain Storage, Handling and Transport, Commonwealth of Australia.

Anon. (1987a) Institutional framework for the Australian grain storage, handling and transport system. *Discussion Paper No. 2*, Royal Commission into Grain Storage, Handling and Transport, Commonwealth of Australia.

Anon. (1987b) Land transport. *Working Paper No. 6*, Royal Commission into Grain Storage, Handling and Transport, Commonwealth of Australia.

Anon. (1997) Flexible handling is the name of the game in volatile market. *Australian Bulk Handling Review*, **21**, 90–91.

Anon. (1998) Metro grain centre puts store in added value. *Australian Bulk Handling Review*, **3**, 9.

Anon. (2002a) CBH fast-tracks capital works for swifter grain harvesting. *Australian Bulk Handling Review*, **7**, 9.

Anon. (2002b) Borders ignored in business hunt. *Australian Bulk Handling Review*, **6**, 8–9.

Anon. (2002c) Grain transport and storage solutions outlined at summit. Making a play for niche markets. *Australian Bulk Handling Review*, **7**, 20, 22.

APTC (1998) *Proceedings of the Australian Postharvest Technical Conference*, Canberra, 26–29 May 1998, CSIRO Entomology, Canberra, Australia. http://www.sgrl.csiro.au/aptc1998/default.html.

APTC (2000) *Proceedings of the Australian Postharvest Technical Conference*, 1–4 August 2000, Adelaide, Australia. CSIRO Entomology, Canberra, Australia. http://www.sgrl.csiro.au/aptc2000/default.html.

APTC (2003) *Proceedings of the 2003 Australian Postharvest Technology Conference*, 26–27 June 2003, Canberra, Australia. CSIRO Entomology, Canberra, Australia. http://www.sgrl.csiro.au/aptcabstracts2003/default.html.

Ardley, J.H. & Desmarchelier, J.M. (1978) Field trials of bioresmethrin and fenitrothion combinations as potential grain protectants. *Journal of Stored Products Research*, **14**, 65–67.

Armitage, D.M. (1994) Some effects of grain cleaning on mites, insects and fungi. In: E. Highley, E.J. Wright, H.J. Banks & B.R. Champ (eds) *Proceedings of the 6th International Working Conference on Stored Product Protection*, 17–23 April 1994, Canberra, Australia, 2, 896–901. CABI, Wallingford, UK.

Armitage, D.M. & Burrell, N.J. (1978) The use of aeration spears for cooling infested grain. *Journal of Stored Products Research* **14**, 223–226.

Armitage, D.M. & Cook, D.A. (1999) *Limiting moisture uptake at the grain surface to prevent mite infestation*. Report 201, Home Grown Cereals Authority, London, UK.

Armitage, D.M. & George, C.L. (1986) The effect of three species of mites upon fungal growth on wheat. *Experimental and Applied Acarology*, **2**, 111–125.

Armitage, D.M. & Woods, J.L. (1999) The development of a storage strategy for malting barley. In: J. Zuxun, L. Quan, L. Yongsheng, T. Xianchang & G. Lianghua (eds) *Proceedings of 7th International Working Conference on Stored Product Protection*, 14–19 October 1998, Beijing, P.R. China, 2, 1358–1366. Sichuan Publishing House of Science and Technology, Chengdu, Sichuan Province, P.R. China.

Armitage, D.M., Wilkin, D.R. & Cogan, P.M. (1991) The cost and effectiveness of aeration in the British climate. In: F. Fleurat Lessard & P. Ducom (eds) *Proceedings of the 5th International Working Conference on Stored Product Protection*, September 1990, Bordeaux, France, III, 1925–1933.

Armitage, D.M., Cogan, P.M. & Wilkin, D.R. (1994) Integrated pest management in stored grain: combining surface insecticide treatments with aeration. *Journal of Stored Products Research*, **30**, 303–319.

Armitage, D.M., Cook, D.A. & Duckett, C. (1996) The use of an aspirated sieve to remove insects, mites and pesticides from grain. *Crop Protection*, **15**, 675–680.

Armitage, D.M., Kelly, M.P., Amos, K., Schaanning, S. & Spagnoli, W. (1999a) The white-marked spider beetle, *Ptinus fur* (L.) in stored grain – biology, seasonal occurrence and control using a surface insecticidal admixture. In: J. Zuxun, L. Quan, L. Yongsheng, T. Xianchang & G. Lianghua (eds) *Proceedings of 7th International Working Conference on Stored Product Protection*, 14–19 October 1998, Beijing, P.R. China, 1, 51–57. Sichuan Publishing House of Science and Technology, Chengdu, Sichuan Province, P.R. China.

Armitage, D.M., Wildey, K.B., Bruce, D., Kelly, M.P. & Mayer, A. (1999b) *The Grain Storage Guide*. Home Grown Cereals Authority, London, UK.

Armstrong, M.T. & Howe, R.W. (1963) The saw-toothed grain beetle, *Oryzaephilus surinamensis* in home-grown grain. *Journal of Agricultural Engineering Research*, **8**, 256–261.

Ayris, C. (1999) *A Heritage Ingrained. A History of Co-operative Bulk Handling Ltd. 1933–2000*. Co-operative Bulk Handling Ltd, West Perth WA, Australia.

Bailey, S.W. (1976) Grain pest control part 1. *Wheat Industry Review Conference*, March 1976, Sydney, Australia.

Baloch, U.K., Irshad, M. & Ahmed, M. (1994) Loss assessment and loss prevention in wheat and storage: technology development and transfer in Pakistan. In: E. Highley, E.J. Wright, H.J. Banks & B.R. Champ (eds) *Proceedings of* the 6th Working Conference on Stored Products Protection, Canberra, Australia, 2, 902–905, CABI, Wallingford, UK.

Banfield, E.C. (1949) Planning under the Research and Marketing Act of 1946: a study in the sociology of knowledge. *Journal of Farm Economics*, **31**, 48–75.

Banks, H.J. (1981) Phosphine fumigation of large temporary bulk grain storages. *Bulk Wheat*, **15**, 50–51.

Banks, H.J. (1998) Foreword. In: H.J. Banks, E.J. Wright & K.A. Damcevski (eds) *Stored Grain in Australia. Proceedings of the Australian Postharvest Technical Conference*, 26–29 May 1998, Canberra, Australia. CSIRO Entomology, Canberra, Australia.

Banks, H.J. & Sticka, R. (1981) Phosphine fumigation of earth-walled bulk grain storages: full scale trials using a surface technique. CSIRO Division of Entomology *Technical Paper No. 18*. CSIRO Entomology, Canberra, Australia.

Barker, N. (1992) Bunker storage technology in Australia: an historical perspective. *International Seminar on Silos and Storage Installations*, 26–28 October 1992, Tehran, Islamic Republic of Iran.

Bartlett, D., Armitage, D. & Harral, B. (2003) Optimising the performance of vertical aeration systems. In: P.F. Credland, D.M. Armitage, C.H. Bell, P.M. Cogan & E. Highley (eds) *Advances in Stored Product Protection, Proceedings of the 8th International Working Conference on Stored Product Protection*, 22–26 July 2002, York, UK, 946–955. CABI Publishing, Wallingford, UK.

Beckett, S.J. (1998) Treating psocids with heat: an alternative grain disinfestation treatment for a new pest. In: H.J. Banks, E.J. Wright & K.A. Damcevski (eds) *Stored Grain in Australia. Proceedings of the Australian Postharvest Technical Conference*, 26–29 May 1998, Canberra, Australia. CSIRO Entomology, Canberra, Australia.

Beckett, S.J. & Moreton, R. (2003) The mortality of three species of Psocoptera, *Liposcelis bostrychophila* Badonnel, *Liposcelis decolor* Pearman and *Liposcelis paeta* Pearman, at moderately elevated temperatures. *Journal of Stored Products Research*, **39**, 103–115.

Bengston, M. (1988) Protectant insecticides for stored wheat. *Stored Grain Australia*, **2**, 5–8.

Bengston, M. & Strange, A.C. (1998) Report of the national working party on grain protection. In: H.J. Banks, E.J. Wright & K.A. Damcevski (eds) *Stored Grain in Australia. Proceedings of the Australian Postharvest Technical Conference*, 26–29 May 1998, Canberra, Australia. CSIRO Entomology, Canberra, Australia.

Bengston, M., Cooper, L.M. & Grant-Taylor, F.J. (1975) A comparison of bioresmethrin, chlorpyrifos-methyl and pirimiphos-methyl as grain protectants against malathion resistant insects in wheat. *Queensland Journal of Agriculture and Animal Sciences*, **32**, 51–78.

Bengston, M., Davies, R.A.H., Desmarchelier, J.M., *et al.* (1983) Organophosphorothioates and synergised synthetic pyrethroids as grain protectants on bulk wheat. *Pesticide Science*, **14**, 373–384.

Blanco, C, Quiralte, R. & Delgado, J. (1997) Amphylaxis after ingestion of wheat flour contaminated with mites. *Journal of Allergy Clinical Immunology*, **99**, 308–313.

Botta, P., Cameron, J., Hughes, P., *et al.* (in press) National grain storage extension – phosphine awareness, use, changed practices and challenges. *Australian Postharvest Technology Conference*, 26–27 June 2003, Canberra, Australia.

Borlaug, N.E. (1989) Wheat Research and Development in Pakistan. Pakistan Agricultural Research Council/CIMMYT Collaborative Program. CIMMYT, Mexico DF, Mexico.

Bridgeman, B. (2002) Evolution of on-farm storage. In: E. Highley, H.J. Banks & E.J. Wright (eds) *Stored Grain in Australia 2000. Proceedings of the Australian Postharvest Technical Conference*, 1–4 August 2000, Adelaide, Australia. CSIRO Entomology, Canberra, Australia.

Bridgeman, B. (in press) Pest control considerations in a least cost supply chain. *Australian Postharvest Technology Conference*, 26–27 June 2003, Canberra, Australia.

Brooks, J.E. & Ijaz, A. (1988) *Reference manual vertebrate pest management in grain storage centres*. GoP/USAID/DWRC Vertebrate Pest Control Project, National Agricultural Research Council, Islamabad, Pakistan.

Brown, R.H. (1918) The farmer and federal grain inspection. *Yearbook of Agriculture*. US Department of Agriculture, Washington DC, USA.

Burton, R.H. (1998) Sealing permanent storages for fumigation using controlled atmospheres; the WA experience. In: H.J. Banks, E.J. Wright & K.A. Damcevski (eds) *Stored Grain in Australia. Proceedings of the Australian Postharvest Technical Conference*, 26–29 May 1998, Canberra, Australia. CSIRO Entomology, Canberra, Australia.

Butt, M.S., Faqir M Anjum, Amjad Ali & Atta-ur-Rehman (1997) Milling and baking properties of spring wheats. *Journal of Agricultural Research*, **35**, 406.

Caddick, L.P. & Shelton, S.P. (1998) Storability of wheat at 13% moisture content: a field study and review. In: H.J. Banks, E.J. Wright & K.A. Damcevski (eds) *Stored Grain in Australia. Proceedings of the Australian Postharvest Technical Conference*, 26–29 May 1998, Canberra, Australia. CSIRO Entomology, Canberra, Australia.

Callaghan, A.R. & Millington, A.J. (1956) *The Wheat Industry in Australia*. Angus & Robertson, Sydney, Australia.

Cameron, J., Hughes, P., Burrill, P., *et al.* (in press) National grain storage and extension – grower information needs and attitudes – a national scoping study. *Australian Postharvest Technology Conference*, 26–27 June, Canberra, Australia.

Chadwick, L. (1996) *Farm Management Handbook 1996/97*. SAC Publications.

Champ, B.R. (2002) Insect pests of stored products in Australia. In: A. Prakash, J. Rao, D.S. Jayas & J. Allotey (eds) *Insect Pests of Stored Products: a global scenario*. Applied Zoologists Research Association, Division of Entomology, Central Rice Research Institute, Cuttack, India.

Champ, B.R. & McCabe, J.B. (1984) Storage of grain in earth-covered bunkers. In: *Proceedings of the Third International Working Conference on Stored Product Entomology*, 23–28 October 1983, Kansas State University, Manhattan, KS.

Christensen, C.M. & Kaufmann, H.H. (1969) *Grain storage. The role of fungi in quality loss*. University of Minnesota Press, Minneapolis, MN, USA.

Coetsee, T. (1998) Metro grain centre moves handling into the next century. *Australian Bulk Handling Review*, **3**, 7–8.

Collins, D.A. & Cook, D.A. (1999) Periods of protection provided by different formulations of pirimiphos-methyl and etrimfos, when admixed with wheat, against four susceptible storage beetle pests. *Crop Protection*, **17**, 521–528.

Collins, P.J. (1998a) Insect pest trends in the farm system. In: H.J. Banks, E.J. Wright & K.A. Damcevski (eds) *Stored Grain in Australia. Proceedings of the Australian Postharvest Technical Conference*, 26–29 May 1998, Canberra, Australia. CSIRO Entomology, Canberra, Australia.

Collins, P.J. (1998b) Inheritance of resistance to pyrethroids in *Tribolium castaneum* Herbst. *Journal of Stored Products Research*, **34**, 395–401.

Collins, P. J. & Wilson, D. (1986) Insecticide resistance in the major coleopterous pests of stored grain in Southern Queensland. *Queensland Journal of Agricultural and Animal Science*, **43**, 107–114.

Collins, R. (1996a) Handling facilities gear up for the grain bonanza. *Australian Bulk Handling Review*, **1**, 7–8.

Collins, R. (1996b) Two models of efficiency. *Australian Bulk Handling Review*, **1**, 13.

Collins, R. (1997) Diversified handling levels out Vicgrain's seasonal blips. *Australian Bulk Handling Review*, **2**, 82–84.

Connell, M., Ford, J.R. & Nickson, P.J. (1992) Bunker grain stores, establishment and operations of a low capital cost grain storage system. Grain Elevators Board of Victoria. *International Seminar on Silos and Storage Installations*, 26–28 October 1992, Teheran, Islamic Republic of Iran.

Connell, P., Barrett, D., Berry, P & Hanna, N. (2002) Crops. *Australian Commodities: Forecasts and Issues*, **9**, 568–574.

Conyers, S.T., Bell, C.H., Llewellin, B.E. & Savvidou, N. (1996) Strategies for the use of modified atmospheres for the treatment of grain. Home Grown Cereals Authority Project *Report No. 125*. Home-Grown Cereals Authority, London.

Cook, D.A., Armitage, D.M. & Collins, D.A. (1999) Diatomaceous earths as alternatives to organophosphorus (OP) pesticide treatments on stored grain in the UK. *Postharvest News and Information* 10, 3, 39N–43N.

Cook, J.S. (1980) Low airflow fumigation method. U.S. Patent No. 4,200,657. U.S. Patent and Trademark Office, Washington DC, USA.

Coombs, R. (ed.) (1994) *Australian Grains*. Morescope Publishing Pty Ltd, Camberwell, Australia.

Cox, P.D. & Wilkin, D.R. (1998) A review of the options for biological control against invertebrate pests of stored grain in the UK. *IOBC/WPRS Bulletin*, **21**, 27–32.

CSIRO (1970–1987) Annual and biennial reports. CSIRO Division of Entomology, Canberra, Australia.

CSIRO (1987–2001) Annual reports and research reports. CSIRO Stored Grain Research Laboratory, Canberra, Australia.

Cuperus, G.W. (2000) *Insect Management for On-Farm Wheat Storage*. Co-operative Extension Service, Oklahoma State University, Stillwater OK, USA.

Day, D.L. & Nelson, G.L. (1962) Predicting performances of cross-flow systems for drying grain in storage in deep cylindrical bins. *ASAE Paper No. 62–925*. American Society of Agricultural Engineers, St Joseph MI, USA.

Day, D.L. & Nelson, G.L. (1964) Drying effects of cross-flow air circulation on wheat stored in deep cylindrical bins. *Technical Bulletin No. T-106*, Oklahoma Agricultural Experiment Station, Oklahoma State University, OK, USA.

Delmenico, R.J. (1992) Controlled atmosphere and fumigation in Western Australia: a decade of progress. In: S. Navarro & E. Donahaye (eds) *Proceedings of the International Conference on Controlled Atmosphere and Fumigation in Grain Storages*, June 1992, Winnipeg, Canada, Caspit Press Ltd., Jerusalem, Israel.

Denne, T. (1988) An expert system for stored grain pest management. PhD Thesis, Silwood Centre for Pest Management, Imperial College, University of London, UK.

Desmarchelier, J.M. (1975) The development of new grain protectants. *Bulk Wheat*, **9**, 36–37.

Desmarchelier, J.M. & Ghaly, T. (1993) Effects of raising the receival moisture content on the storability of Australian wheat. *Australian Journal of Experimental Agriculture*, **33**, 909–914.

Desmarchelier, J.M. & Hogan, J.P. (1978) Reduction of insecticide in grain dust by treatment with alkali. *Australian Journal of Experimental Agriculture and Animal Husbandry*, **18**, 453–456.

Desmarchelier, J M., Bengston, M., Henning, R., *et al.* (1981) Extensive pilot use of grain protectant combinations, fenitrothion plus bioresmethrin and pirimiphos-methyl plus bioresmethrin. *Pesticide Science*, **12**, 365–374.

Desmarchelier, J.M., Bengston, M., Davies, R., *et al.* (1987) Assessment of the grain protectants chlorpyrifos-methyl plus bioresmethrin, fenitrothion plus (1*R*)-phenothrin, methacrifos and pirimiphos-methyl plus carbaryl under practical conditions in Australia. *Pesticide Science*, **20**, 271–288.

Devereau, A.D. (2002) Physical factors in post-harvest quality. In: P. Golob, G. Farrell & J.E. Orchard (eds) *Crop Post-Harvest: Science and Technology. Volume 1 Principles and Practice.* Blackwell Science, Oxford, UK.

Dick, A. (1997) Flexible handling is the name of the game in volatile market. *Australian Bulk Handling Review*, **2**, 90–91.

DiLeo, J. (2002) The Australian grains industry and its relationship with the CSIRO Stored Grain Research Laboratory. In: *Proceedings of GEAPS Exchange 2002*, Vancouver, Canada. Grain Elevator and Processing Society, Omnipress Omnipro CD. http://www.geaps.com/proceedings/2002/.

Dines, J. (1998) Trends in market requirements and implications for storage; a flour miller's perspective. In: H.J. Banks, E.J. Wright & K.A. Damcevski (eds) *Stored Grain in Australia. Proceedings of the Australian Postharvest Technical Conference*, 26–29 May 1998, Canberra, Australia. CSIRO Entomology, Canberra, Australia.

FAO (1972) *Food composition tables for use in East Asia 1972*. Food and Agriculture Organization of UN, Food Policy and Nutrition Division, Rome, Italy.

Fleming, D.A. & Armitage, D.M. (2003) Lowering the moisture content of stored grain can gain extra time for cooling to prevent infestation. In: P.F. Credland, D.M. Armitage, C.H. Bell, P.M. Cogan & E. Highley (eds) *Advances in Stored Product Protection, Proceedings of the 8th International Working Conference on Stored Product Protection*, 22–26 July 2002, 696–701. York, UK. CABI Publishing, Wallingford, UK.

Fornari, H. (1973) *Bread Upon the Waters*. Aurora Publishers, Nashville TN, USA.

Fox, E.A., Garthwaite, D.G., Thomas, M.R. & Bankes, J.A. (2001) *Pesticide usage survey report 180. Farm grain stores in Great Britain 1998/1999.* DEFRA (PB6170), London, UK.

Garthwaite, D.G., Thomas, M.R. & Bankes, J.A. (2001) *Pesticide usage survey report 180. Commercial grain stores in Great Britain 1998/1999.* DEFRA (PB6171), London, UK.

Gentilcore, J. (1998) Grain handling gains from new strategy in South Australia. *Australian Bulk Handling Review*, **3**, 26–29.

George, L. (1998) The consumer's perspective: trends in market requirements and implications for storage. In: H.J. Banks, E.J. Wright & K.A. Damcevski (eds) *Stored Grain in Australia. Proceedings of the Australian Postharvest Technical Conference*, 26–29 May 1998, Canberra, Australia. CSIRO Entomology, Canberra, Australia.

Gibbs, P.A. (1994) Development of a programmable aeration controller. In: E. Highley, E.J. Wright, H.J. Banks & B.R. Champ (eds) *Proceedings of the 6th International Working Conference on Stored Product Protection*, 17–23 April 1994, Canberra, Australia, 1, 286–289. CABI, Wallingford, UK.

Ghluam, J. (1992) Botanicals and pheromones in IPM. In: U.K. Baloch (ed.) *Integrated Pest Management in Food Grains*, FAO/Pakistan Agricultural Research Council, Islamabad, Pakistan.

Golden, M. (1998) Changing needs of export versus domestic markets: the NSW experience. In: H.J. Banks, E.J. Wright & K.A. Damcevski (eds) *Stored Grain in Australia. Proceedings of the Australian Postharvest Technical Conference*, 26–29 May 1998, Canberra, Australia. CSIRO Entomology, Canberra, Australia.

Golob, P. (1997) Current status and future perspectives for inert dusts for control of stored product insects. *Journal of Stored Products Research*, **33**, 69–79.

Green, E.J.U. (1986) Current use of controlled atmospheres and fumigants in grain storages of Western Australia. In: B. de Mesa (ed.) *Grain Protection in Postharvest Systems. Proceedings of the 9th ASEAN Technical Seminar on Grain Postharvest Technology*, 26–29 August 1986, Singapore.

Griffiths, H.J. (1964) *Bulk storage of grain: a summary of factors governing control of deterioration.* CSIRO Division of Mechanical Engineering, Melbourne, Australia.

Harner, J.P. & Hagstrum, D.W. (1990) Utilizing high airflow rates for aerating wheat. *Applied Engineering in Agriculture*, **6**, 315–321.

Henderson, S. (1987) A mean moisture content–relative humidity relationship for nine varieties of wheat. *Journal of Stored Products Research*, **23**, 143–147.

Hetmanski, M.T. (1997) *Production of ochratoxin A on wheat and barley.* Rep FD96/70. CSL Food Science Laboratory, Norwich, UK.

Hinchy, M. (1989) On-site costs of segregating wheat in New South Wales country storage. *Agriculture and Resources Quarterly*, **1**, 312–316.

Hill, L.D. (1990) *Grain Grades and Standards: historical issues shaping the future.* University of Illinois Press, Chicago and Urbana IL, USA.

Hodges, R.J. (2002) Pests of durable crops – insects and arachnids. In: P. Golob, G. Farrell & J.E. Orchard (eds) *Crop Postharvest: Science and Technology. Volume 1 Principles and Practice.* Blackwell Science, Oxford, UK.

Holman, L.E. (1960) *Aeration of Grain in Commercial Storages*, Marketing Research Report No. 178 USDA ARS, Washington DC, USA.

Howe, R.W. (1962) A study of the heating of stored grain caused by insects. *Annals of Applied Biology*, **50**, 137–158.

Hughes, A.M. (1976) The mites of stored food and houses. *Technical Bulletin 9*, HMSO, London, UK.

Hughes, P., Burrill, P., Botta, P., *et al.* (in press) National grain storage extension: satisfying grower needs regarding aeration. *Australian Postharvest Technology Conference*, 26–27 June 2003, Canberra, Australia.

Hunter, R.D. & Hooper, S. P. (1992) On-farm grain storage in Australia. *Agriculture and Resources Quarterly*, **4**, 242–256.

Jacob, T.A. (1996) The effect of constant temperature and humidity on the development, longevity and productivity of *Ahasverus advena* (Watl.) (Coleoptera: Silvanidae). *Journal of Stored Products Research*, **32**, 115–121.

Jayas, D.S. & Mann, D. (1994) Presentation of airflow resistance data of seed bulks. Applied Engineering in Agriculture. *American Society of Agricultural Engineering*, **10**(1), 79–83.

Jayas, D.S. & Muir, W.E. (1991) Airflow-pressure drop data for modelling fluid flow in anisotropic bulks. *Transactions of the ASAE*, **34**(1), 251–254.

Jeyamkondan, S., Noyes, R.T. & Phillips, T.W. (2003a) *Effective phosphine fumigation in concrete silos.* ASAE Paper No. 036151. Presented at the International Meeting, American Society of Agricultural Engineers, 26–30 July 2003, Las Vegas NV, USA.

Jeyamkondan, S., Ho, Y. & Noyes, R.T. (2003b) *A software program for grain blending at commercial elevators.* ASAE Paper No. 036002. Presented at the International Meeting, American Society of Agricultural Engineers, 26–30 July 2003, Las Vegas NV, USA.

Knight, J.D., Armitage, D.M. & Wilkin, D.R. (1997) *Integrated Grain Store Manager* (Problem solving and quality management software). Obtainable from J.D. Knight, School of Environment, Earth Sciences and Engineering, Imperial College, Centre for Environment Technology, Silwood Park, Ascot, Berkshire, SL5 7PY, UK.

Koehnke, M. (1986) *Kernels and Chaff: a history of wheat market development.* Marx Koehnke, Lincoln NB, USA.

Kotzur, A. (1998a) The future of on-farm storage: a silo manufacturer's perspective. In: H.J. Banks, E.J. Wright & K.A. Damcevski (eds) *Stored Grain in Australia. Proceedings of the Australian Postharvest Technical Conference*, 26–29 May 1998, Canberra, Australia. CSIRO Entomology, Canberra, Australia.

Kotzur, A. (1998b) Ambient air in-store grain drying: recent Australian experience. In: H.J. Banks, E.J. Wright & K.A. Damcevski (eds) *Stored Grain in Australia.* Proceedings of the Australian Postharvest Technical Conference, 26–29 May 1998, Canberra, Australia. CSIRO Entomology, Canberra, Australia.

Krischik, V. & Burkholder, W. (1995) Stored product insects and biological control agents. *Stored Product Management.* Circular E-912. Co-operative Extension Service, Oklahoma State University, Stillwater OK, USA.

Lamshed, M. (1962) Grain is better in bulk. The story of South Australian Co-operative Bulk Handling Limited.

Larson, H.M. (1926) *The Wheat Market and the Farmer in Minnesota, 1858–1900*, Columbia University Press, Columbia NY, USA.

Lasseran, J. (1977) Dry aeration or delayed aeration cooling. *Perspectives Agricoles*, **6**, 59–66.

Lee, G.A. (1937) The historical significance of the Chicago grain elevator system. *Agricultural History*, **2**, 16–32.

Maxon, R.C., Acasio, U.A. & Shamsher, H.K. (1994) Manual on handling, management and marketing of cereal grains. *USAID Report No. 15*. Food and Feed Grains Institute, Kansas State University, KS, USA.

Maier, D. & Navarro, S. (2001) Chilling of grain by refrigerated air. In: S. Navarro & R.T. Noyes (eds) *The Mechanics and Physics of Modern Grain Aeration.* CRC Press, Boca Raton FL, USA.

Maier, D.E., Mason, L.J. & Woloshuk, C.P. (1994) Maximise grain quality and profits using S.L.A.M. *Bulletin No. ID-207.* Grain Quality Task Force, Co-operative Extension Service, Purdue University, Purdue IL, USA.

Manis, J.M. (1992) Sampling, inspecting and grading. In: D.B. Sauer (ed.) *Storage of Cereal Grains and Their Products.* 4th edition. American Association of Cereal Chemists, St. Paul MN, USA.

McLean, K.A. (1989) *Drying and storing combinable crops.* Farming Press, Ipswich, UK.

McLean, K.A. (1993) The use of auger-stirring in bulk grain dryers, implications for quality. Cereal Quality III. *Aspects of Applied Biology*, **36**, 457–463.

McMullen, G. (1998) Trends in export markets. In: H.J. Banks, E.J. Wright & K.A. Damcevski (eds) *Stored Grain in Australia. Proceedings of the Australian Postharvest Technical Conference*, 26–29 May 1998, Canberra, Australia. CSIRO Entomology, Canberra, Australia.

McMullen, G. (2002) Trends in domestic and international markets. In: E. Highley, H.J. Banks & E.J. Wright (eds) *Stored Grain in Australia 2000. Proceedings of the Australian Postharvest Technical Conference*, 1–4 August 2000, Adelaide, Australia. CSIRO Entomology, Canberra, Australia.

McMullen, G. (in press) Prospects for marketing in 2010. *Australian Postharvest Technology Conference*, 26–27 June 2003, Canberra, Australia.

Mehl, J.M. (1954) The futures market. *Yearbook of Agriculture.* US Department of Agriculture, Washington DC.

Meyer, A.N. & Belmain, S.R. (2002) Rodent control. In: P. Golob, G. Farrell & J.E. Orchard (eds) *Crop Post-Harvest: Science and Technology. Volume 1 Principles and Practice.* Blackwell Science, Oxford, UK.

Moore, J.V. (1980) Building bigger silos faster. *Bulk Wheat*, **14**, 47–48.

Moore, J.V. (1988) Port handling facilities for grain. In: B. R. Champ & E. Highley (eds) *Bulk Handling and Storage of Grain in the Humid Tropics.* ACIAR Proceedings No 22. ACIAR, Canberra, Australia.

Murray, W. (1998) Grains industry initiatives on regulation. In: H.J. Banks, E.J. Wright & K.A. Damcevski (eds) *Stored Grain in Australia. Proceedings of the Australian Postharvest Technical Conference*, Canberra, 26–29 May 1998,

CSIRO Entomology, Canberra, Australia.

Murray, W. (in press) Grains industry interaction with regulatory organisations. *Australian Postharvest Technology Conference*, 26–27 June 2003, Canberra, Australia.

Nash, M.J. (1978) *Crop Conservation and Storage in Cool, Temperate Climates.* Pergamon, Oxford, UK.

Navarro, S. & Noyes, R.T. (eds) (2001) *The Mechanics and Physics of Modern Grain Aeration.* CRC Press, Boca Raton FL, USA.

Nayak, M.K., Collins, P.J. & Kopittke, R.A. (2003) Residual toxicities and persistence of organophosphorus insecticides mixed with carbaryl as structural treatments against three liposcelidid psocid species (Psocoptera: Liposcelidiidae) infesting stored grain. *Journal of Stored Products Research*, **39**, 343–353.

Nellist, M.E. (1979) Safe temperatures for drying grain. *NIAE Report 29.* A report to the Home Grown Cereals Authority, London, UK.

Nellist, M.E. (1988) Near ambient grain drying. *Agricultural Engineer*, **3**, 93–101.

Nellist, M.E. & Bartlett, D.I. (1988) A comparison of fan and heater control policies for near-ambient drying. *Report 54.* Institute of Engineering Research, AFRC, London.

Nellist, M.E. & Bruce, D.M. (1987) Drying and cereal quality. *Aspects of Applied Biology*, **15**, 439–456.

Newman, C.J.E. (1988) Storage and ancillary equipment for bulk handling of grain. In: B.R. Champ & E. Highley (eds) *Bulk Handling and Storage of Grain in the Humid Tropics.* ACIAR Proceedings No 22. ACIAR, Canberra, Australia.

Newman, C.J.E. (1990) Specification and design of enclosure for gas treatment. In: B.R. Champ, E. Highley & H.J. Banks (eds) *Fumigation and Controlled Atmosphere Storage of Grain.* ACIAR Proceedings No 25. ACIAR, Canberra, Australia.

Newman, C.J.E. (1996) Design parameters for aeration and in-store drying. In: B.R. Champ, E. Highley & G.I. Johnson (eds) *Grain Drying in Asia.* ACIAR Proceedings No. 71. ACIAR, Canberra, Australia.

Newman, C.R. (1998) A model for improved fumigant use on farms in Australia. In: H.J. Banks, E.J. Wright & K.A. Damcevski (eds) *Stored Grain in Australia. Proceedings of the Australian Postharvest Technical Conference*, 26–29 May 1998, Canberra, Australia. CSIRO Entomology, Canberra, Australia.

Northholt, M.D. & Bullerman, L.B. (1982) Prevention of mold growth and toxin formation through control of environmental conditions. *Journal of Food Protection*, **45**, 519–526.

Noyes, R.T., Clary, B.L. & Cuperus, G.W. (1991) Maintaining quality of stored grain by aeration. OSU FS No. 1100. Oklahoma Co-operative Extension Service, Oklahoma State University, Stillwater OK, USA.

Noyes, R.T., Weinzierl, R., Cuperus, G.W. & Maier, D.E. (1995a) Stored grain management techniques. *Stored Product Management.* Circular E-912. Co-operative Extension Service, Oklahoma State University, Stillwater OK, USA.

Noyes, R.T., Kenkel, P. & Tate, G. (1995b) Closed loop fumigation systems. *Stored Product Management.* Circular E-912. Co-operative Extension Service, Oklahoma State University, Stillwater OK, USA.

Noyes, R.T., Kenkel, P., Criswell, J.T. & Cuperus, G.W. (1997)

Manifolding and sealing installation methods for manifolded phosphine recirculation systems in multiple concrete silos in US elevators. In: E.J. Donahaye, S. Navarro & A. Varnava (eds) *Proceedings of International Conference on Controlled Atmospheres and Fumigation in Stored Products*, April 1996, Nicosia, Cyprus, 359–369. Printco Ltd.

Noyes, R.T., Navarro, S. & Armitage, D.M. (2001) Supplemental aeration systems. In: S. Navarro & R.T. Noyes (eds) *The Mechanics and Physics of Modern Grain Aeration*. CRC Press, Boca Raton FL, USA.

Noyes, R.T., Jeyamkondan, S., Anderson, K.B., Ho, Y. & Adam, B. (2003) *Development of a grain segregation (electronic elevator bin board) and blending computer software model*. Invention Disclosure Application Report for Office of Intellectual Property, Oklahoma State University, Stillwater OK, USA.

Nugent, T. (2002) *Stalk to Store: the on-farm grain handling and storage manual*. 2nd edition. Kondinin Group, Cloverdale, Australia.

O'Driscoll, K. (2000) SACBH revamp in line with new growth trend. *Australian Bulk Handling Review*, **5**, 117.

Pheloung, P. & Macbeth, F. (2002) Export inspection; adding value to Australia's grain. In: E. Highley, H.J. Banks & E.J. Wright (eds) *Stored Grain in Australia 2000. Proceedings of the Australian Postharvest Technical Conference*, 1–4 August 2000, Adelaide, Australia. CSIRO Entomology, Canberra, Australia.

Phillips, T., Bolin, P. & Criswell, J. (2003) Letter to Oklahoma grain elevator managers advising them of the immediate availability of Storcide® (Product of Gustafson, Inc., Dallas, TX) for direct application to grain as a residual pesticide. Department of Entomology & Plant Pathology, Division of Agricultural Sciences and Natural Resources (DASNR), Oklahoma State University, Stillwater OK, USA.

Pixton, S.W. & Warburton, S. (1971) Moisture content/relative humidity equilibrium of some cereal grains at different temperatures. *Journal of Stored Products Research*, **6**, 283–293.

Prattley, C. (1995) *Australian Food. The Complete Reference to the Australian Food Industry*. Morescope Publishing Pty Ltd., Camberwell, Australia.

Prickett, A.J. (1988) English farm grain stores (1987) Part 1. Storage practice and pest incidence. *Report 23*. ADAS Central Science Laboratory, London, UK.

Prickett, A.J. & Muggleton, J. (1991) Commercial grain stores 1988/89, England and Wales. Pest incidence and storage practice. Home Grown Cereals Authority, London.

Post, R.E. & Murphy, E.J. (1954) Wheat, a food grain. *Yearbook of Agriculture*. US Department of Agriculture, Washington DC, USA.

Rees, D. (1998) Psocids as pests of Australian grain storages. In: H.J. Banks, E.J. Wright & K.A. Damcevski (eds) *Stored Grain in Australia. Proceedings of the Australian Postharvest Technical Conference*, 26–29 May 1998, Canberra, Australia. CSIRO Entomology, Canberra, Australia.

Reeves, G. & Cottingham, I. (1994) Grain industry legislation. In: B. Coombs (ed.) *Australian Grains: a complete reference book on the grain industry*. 63–72. Morescope Publishing Pty Ltd, Camberwell, Australia.

Ripp, B.E., Banks, H.J., Bond, E.J., Calverley, D.J., Jay, E.G. & Navarro, S. (1984) Controlled atmosphere and fumigation in grain storages. *Proceedings of an International Symposium*, 11–22 April 1983, Perth, Western Australia. Elsevier, Amsterdam.

Satterley, J. (2001) Strategic site makes grain receival facility a winner. *Australian Bulk Handling Review*, **6**, 44–45.

Schmidt, J.L. (1955) Wheat storage research. *Technical Bulletin No. 1113*. US Department of Agriculture, Washington DC, USA.

Scott, O., Hayes, J. & Adlington, D. (1998) Upgrading grain terminal in northern New South Wales. *Australian Bulk Handling Review*, **3**, 11–15.

Scudamore, K.A., Patel, S. & Breeze, V. (1999) Surveillance of stored grain from the 1997 harvest in the United Kingdom for ochratoxin A. *Food Additives and Contaminants*, **16**, 281–290.

Shah, F.H & Abdul, H. (1992) Micro-organism and mycotoxins in storage. In: U.K. Baloch (ed.) *Integrated Pest Management in Food Grains*, FOA/Pakistan Agricultural Research Council, Islamabad, Pakistan.

Shuman, D., Epsky, N.D. & Crompton R.D. (2003) Commercialisation of a species-identifying automated stored-product insect monitoring system. In: P.F. Credland, D.M. Armitage, C.H. Bell, P.M. Cogan & E. Highley (eds) *Advances in Stored Product Protection, Proceedings of the 8th International Working Conference on Stored Product Protection*, 22–26 July 2002, York, UK, 110–114. CABI Publishing, Wallingford, UK.

Simmonds, D.H. (1989) *Wheat and Wheat Quality in Australia*. Australian Wheat Board Melbourne and CSIRO Melbourne, Australia.

Small, M.A. (1981) Wheat and rail – the partnership. *Bulk Wheat*, **15**, 18–20.

Snelson, J.T. (1987) *Grain Protectants*. ACIAR Monograph No 3. ACIAR, Canberra, Australia.

Stathers, T.E. (2002) Pest management – Inert dusts. In: P. Golob, G. Farrell & J.E. Orchard (eds) *Crop Post-Harvest: Science and Technology. Volume 1 Principles and Practice*. Blackwell Science, Oxford, UK.

Tariq, M., Ahmad, M.S. & Javed, M.A. (1994) Phosphine fumigation in hexagonal bins. In: R.C. Maxon, U.A. Acasio, H.K. Shamsher (eds) *Manual on Handling, Management and Marketing of Cereal Grains*. USAID Report No. 15. Food and Feed Grains Institute, Kansas State University, KS.

Thearle, J. (2000) Counting the cost of grain storage. *Crop Protection Magazine*, **24**, 4–10.

Tontz, R.L. (1955) Legal parity: implementation of the policy of equality for agriculture, 1929–1954. *Agricultural History*, **29**, 174–181.

Turner, S., Connell, P., Hooper, S. & O'Donnell, V. (2002) On-farm grain storage in Australia. In: E. Highley, H.J. Banks & E.J. Wright (eds) *Stored Grain in Australia 2000. Proceedings of the Australian Postharvest Technical Conference*, 1–4 August 2000, Adelaide, Australia. CSIRO Entomology, Canberra, Australia.

Tutt, C. (2002) Conversion of CBH storage to sealed structures and fumigation. In: *Proceedings of GEAPS Exchange 2002, Vancouver*, Canada. Grain Elevator and Processing Society, Omnipress Omnipro CD. http://www.geaps.com/proceedings/2002/.

Tutt, C. & Burton, J. (1998) Segregation and value adding in WA: the Metro Grain Centre. In: H.J. Banks, E.J. Wright & K.A. Damcevski (eds) *Stored Grain in Australia. Proceedings of the Australian Postharvest Technical Conference, 26–29 May 1998*, Canberra, Australia. CSIRO Entomology, Canberra, Australia.

USDA (1940) *Agricultural Marketing Service: Organization and Functions*. Agricultural Marketing Service, US Department of Agriculture, Washington DC, USA.

USDA (1979) *Chronological Landmarks in American History*. USDA Agriculture Information Bulletin No. 425. US Department of Agriculture, Washington DC, USA.

USDA (1988) *The Official United States Standards for Grain*, Federal Grain Inspection Service, U.S. Department of Agriculture, Washington DC, USA.

USDA (1999) *Inspecting Grain: practical procedures for grain handlers*. Federal Grain Inspection Service, Grain Inspection, Packers and Stockyards Administration, US Department of Agriculture, Washington DC, USA.

van S. Graver, J. & Winks, R.G. (1994) A brief history of the entomological problems of wheat storage in Australia. In: E. Highley, E.J. Wright, H.J. Banks & B.R. Champ (eds) *Stored Product Protection. Proceedings of the 6th International Working Conference on Stored Product Protection, 17–23 April, 1994*, Canberra, Australia. CAB International, Wallingford, UK.

Viljoen, J. (in press) Grain dust suppression by spraying water on grain. *Australian Postharvest Technology Conference, 26–27 June 2003*, Canberra, Australia.

Wareing, P.W. (2002) Pest of durable crops – moulds. In: P. Golob, G. Farrell & J.E. Orchard (eds) *Crop Post-Harvest: Science and Technology. Volume 1 Principles and Practice.* Blackwell Publishing, Oxford, UK.

Watson, S. (1996) Principles of grain marketing: some lessons from Australian experience. *ACIAR Technical Report No 38*. ACIAR, Canberra, Australia.

Weller, G.L. & Beckett, S.J. (2002) Can SIROFLO label rates control psocid infestation? In: E. Highley, H.J. Banks & E.J. Wright (eds) *Stored Grain in Australia 2000. Proceedings of the Australian Postharvest Technical Conference, 1–4 August 2000*, Adelaide, Australia. CSIRO Entomology, Canberra, Australia.

Whitaker, A.P. (1929) The Spanish contribution to American agriculture. *Agricultural History*, **3**, 1–14.

Whitwell, G. & Sydenham, D. (1991) *A Shared Harvest: the Australian Wheat Industry, 1939–1989.* Macmillan Education Australia, Melbourne, Australia.

Wightley, A.C. (1979) GEB experiments with temporary storage. *Bulk Wheat*, **13**, 29–31.

Wildey, K.B., Cox, P.D., Wakefield, M., Price, N.R., Moore, D. & Bell, B. (2002) The use of entomopathogenic fungi for stored product pest control – the 'Mycopest' project. In: C.

Adler, S. Navarro, M. Scholler & L. Stengard-Hansen (eds) *Proceedings of the IOBC/WPRS Working Group* 'Integrated Protection in Stored Products', **25**, 15–20.

Wilson, P. (2002) Technology transfer: from research to commercial application. In: *Proceedings of GEAPS Exchange 2002*, Vancouver, Canada. Grain Elevator and Processing Society, Omnipress Omnipro CD. http://www.geaps.com/proceedings/2002/.

Wilson, P. (in press) Environmental issues covering export terminal operations. *Australian Postharvest Technology Conference, 26–27 June 2003*, Canberra, Australia.

Winks, R.G. (1993) The development of SIROFLO® in Australia. In: S. Navarro & E. Donahaye (eds) *Proceedings of an International Conference on Controlled Atmosphere and Fumigation in Grain Storages, June 1992*, Winnipeg, Canada. Caspit Press Ltd., Jerusalem, Israel.

Winks, R. G. & Russell, G. (1994) Effectiveness of SIROFLO® in vertical storages. In: E. Highley, E.J. Wright, H.J. Banks & B.R. Champ (eds) *Stored Product Protection. Proceedings of the 6th International Working Conference on Stored-product Protection*, Canberra, April 1994. CAB International, Wallingford, UK.

Winks, R.G. & Russell, G.F. (1997) Active fumigation systems. In: E.J. Donahaye, S. Navarro & A. Varnava (eds) *Proceedings of the International Conference on Controlled Atmosphere and Fumigation in Stored Products, 21–26 April 1996*, Nicosia, Cyprus.

Winterbottom, D.C. (1922) *Weevil in wheat and storage of grain in bags. A record of Australian experience during the war period (1915 to 1919).* Government Printer, North Terrace, Adelaide, Australia.

Wisniewiski, H.M., Sigurdarson, S., Rubenstein, R., Kaskcak, R.J. & Carp, R.I. (1996) Mites as vectors for scrapie. *The Lancet*, **347**, 1114.

Wood, G., Mann, P.J., Lewis, D.F., Reid, W.J. & Moss, M.O. (1990) Studies on a toxic metabolite from the mould *Wallemia*. *Food Additives and Contaminants*, **7**, 69–77.

Yates, C.J. & Sticka, R. (1984) Development and future trends in bunker storage. In: B.E. Ripp, H.J. Banks, E.J. Bond, D.J Calverley, E.G. Jay & S. Navarro (eds) Controlled atmosphere and fumigation in grain storages. *Proceedings of an International Symposium, 11–22 April 1983*, Perth, Western Australia. Elsevier, Amsterdam.

Zdarkova, E. (1996) The effect of mites on germination of seed. *Ochrana Rostlin*, **32**, 175–179.

Zekulich, M. (1997) *The Grain Journey: the history of the grain pool of WA*. The Grain Pool of Western Australia, Perth, Australia.

Zettler, J.L. & Cuperus, G.W. (1990) Pesticide resistance in *Tribolium castaneum* (Coleoptera: Tenebrionidae) and *Rhyzopertha dominica* (Coleoptera: Bostrichidae) in wheat. *Journal of Economic Entomology*, **83**, 1677–1681.

Chapter 4
Malting Barley: Europe

D. M. Armitage, E. D. Baxter, J. Knight, D. R. Wilkin and J. L. Woods

Barley (*Hordeum vulgare*) is an ancient cereal crop (Plate 4), probably first grown in the fertile crescent of the Middle East about 40 000 BP. Barley has a great tolerance of climatic extremes but does not grow well on wet or acidic soils. In 1997, Europe produced about 99m t of barley. A major use of the cereal is by maltsters, who prepare malt for use in brewing, whisky making and the manufacture of various foodstuffs. Barley may also be used in animal feed. When barley is used for malt, the grain is soaked in water, under controlled conditions, allowing it to germinate or sprout. It is then dried or roasted in a kiln, cleaned and stored. According to the Brewers of Europe the total production of malt in the EU in 1999 was 6.6m t (equivalent to about 7.8m t of malting barley). Around 39% of this was exported. The largest producers were Germany (2m t), the UK (1.6m t) and France (1.3m t) (CBMC 2000).

In Europe before 1945 barley was commonly imported from the USA and stored in sacks. Subsequently, home-produced barleys maintained in cool, aerated bulk storage have been more typical. At harvest barley grains will not germinate, although they are normally fully viable. This condition, known as dormancy, has evolved naturally in many seeds. Grain is said to be dormant when it is completely viable but will not germinate when supplied with adequate moisture and air. Dormancy is genetically controlled, so some barley varieties tend to be more prone to dormancy than others. However, it is also affected by climatic conditions, and is more prevalent after cool, cloudy summers than after sunny ones. If a variety has very little natural dormancy it can sometimes germinate on the ear (pre-germination), particularly if conditions at harvest are very wet. Dormancy declines naturally during the weeks after harvest, and has normally disappeared by about November, although in some years it may persist for longer.

Since the malting process depends so much on predictable germination of the barley, storage practices to break dormancy and to preserve germination are crucial. Germination is preserved by careful high-temperature drying, for instance by ensuring dryer temperatures do not exceed 65°C if the grain moisture content is 20% or more. In earlier times, dormancy was managed by allowing it to pass naturally and so barley was held for prolonged periods. Modern practices are different and today dormancy is often managed by temperature-controlled storage regimes, facilitated by models of dormancy break and germination decline. The main threat to stored malting barley is that poor storage will allow germination to proceed sufficiently that it has to be downgraded to animal feed, with consequent loss of financial premium. Rapid germination tests used with predictive models should assist early diagnosis of quality so that this can be prevented.

Malt is made by soaking the barley in water (steeping), allowing the grains to sprout (germinate) and then drying them (kilning). During this process the cell walls and much of the stored protein is digested, transforming the tough barley grain into readily friable malt. At the same time amylolytic enzymes are synthesised. These will convert the barley starch into fermentable sugars during the brewing process. By the end of the steeping stage the grain should have sprouted and the cream-coloured root initials, the chit, should have emerged (Fig. 4.1). During the germination period the grain produces a number of enzymes which break down the cell walls and some of the protein in a process known as modification. Kilning halts modification before significant breakdown of starch has occurred. Germination is thus the vital step in the production of malt (Fig. 4.1): malting barley must be fully viable and able to germinate before it can be malted.

Fig. 4.1 Steps in the process of preparing malt, from left to right: barley grain, sprouting barley, chitted barley and malted barley (photographs by David Crossley).

Historical perspectives

Before 1945, deficiencies in output and quality of UK-grown malting barleys were made up by imports from California during the autumn, while native barleys were still maturing in the stack. After World War II the shortage of ships and exchange problems increased the reliance on native barleys (Bishop 1947). Combine harvesting reduced damage that sometimes occurred before the war when the crop was cut by binder and left in stooks, where it was vulnerable to weather damage. However, the concentration of barley intake at harvest increased the reliance on good drying and storage techniques and facilities. Storage in sacks was replaced by storage in bulk, which in the UK was mostly at maltsters' own sites. Initially this was in bins of 30–50 t but more recently floor storage has become common and ventilation systems to cool the grain began to be introduced after about 1960.

On the European mainland, unlike their UK counterparts, maltsters have traditionally relied on farmers or commercial stores to hold malting barley during the period between harvests. In addition, Belgian, Dutch and Danish maltsters have a long history of importing barley from France. Hence the French have been the leaders in barley storage systems. European cereal production developed in much the same way as described for the UK in that the combine harvester replaced the binder in the 1940s and 1950s. This move to bulk grain prompted the establishment of many large, co-operative stores in France and some of them developed specialist skills in the storage of malting barley. One major difference between mainland Europe and the UK is that within Europe great emphasis is placed on the use of rail or inland waterways to move grain, whereas in the UK road transport is dominant.

Traditionally, maltsters in countries prone to dormancy would keep sufficient stocks of barley from the preceding year's harvest to last until Christmas, by which time all dormancy in that crop should have passed. However, this is an expensive practice and more recently warm/cool storage treatments have been explored which can break dormancy more quickly, thus reducing the need for a carry-over stock. One of the side benefits of the developments in drying, storage and ventilation in the 1950s and 1960s was the facility to break dormancy by storing barley warm as it comes from the dryer and then cooling it quickly to preserve viability.

Physical facilities

In the UK most malting barley is stored for only a short time on farms as maltsters are not confident of farmers' ability to maintain quality. The majority of maltsters buy their annual barley requirements at harvest time and then store their own grain or use specialist, contract stores. A survey of 14 UK maltings' managers, covering nearly 1m t, indicated that over 60% stored some grain for a year or more; about 59% was held at the maltings, 27% at commercial sites and only 14% on farms. Approximately 46% was held in floor stores, about 9% in internal bins or silos, 28% in external bins or silos (Fig. 4.2) and 17% in large silos of 500–1000 t.

In Denmark maltsters only store sufficient grain for their immediate needs and so most is stored on-farm or at commercial sites. Generally barley is stored on-floor and dried using on-floor drying systems.

In France there is almost no on-farm storage of malting barley, which goes directly from the farm to farmer-owned co-operative stores or to commercial silos. In practice, a large part of the storage is controlled by a small number of large malting companies. Storage companies owned or linked to the malting industry buy the grain from the farmer or co-operatives and arrange for deliveries to the maltings, who only hold stocks sufficient for two to three weeks of production. France supplies malting barley by barge or rail to Belgium, which grows little of its own, and by ship to Portugal, which grows only 5–10% of the crop that is processed there. In northern Spain barley is only held for short periods on farms before collection by local co-operatives where it is mainly kept in floor stores with facilities for aeration but not for drying or cleaning. Maltsters hold the grain for less than a month in silo storage without aeration facilities, but most grain is cleaned before processing.

Fig. 4.2 One thousand-tonne aerated 'jumbo' bins used to store malting barley (UK).

Objectives of storage

In drying and subsequent storage a balance needs to be reached between maintaining viability, breaking dormancy and avoiding water sensitivity – a state in which the grain may 'drown' during steeping if exposed to too much water. These factors may be measured by different germination tests.

Viability (germinative capacity) determines how many grains are alive. Ideally 99–100% should be in this state but, in Europe, barley is generally accepted as long as at least 95% of grains are alive. This is determined by soaking grains in hydrogen peroxide solution at about 20°C for three days, then recording the proportion that shows signs of root or shoot growth. An alternative method is the Vitascope test, in which grains are cut longitudinally and soaked in a tetrazolium solution in a vacuum at 55°C. Red staining of the root and shoot initials in the germ indicates that seeds are viable. This test is used for a rapid assessment of viability at barley intake (Anon. 1997).

Another measure, germinative energy, indicates whether or not the grain is dormant and will germinate during the malting process. This is checked by allowing grains to germinate on moist filter paper. In the UK the minimum requirement for malting would be at least 95%

germination within three days, but in some countries a five-day test is used. An additional test, in which excess water is present, may also be used. This tests the water sensitivity of the barley. Grain that is water sensitive but not dormant can still be malted, but different steeping regimes would be employed.

If grain becomes damaged during handling, germination may be impaired. Detachment of the husk (skinning) is the most common damage but, more importantly, some grains may be broken and the portion lacking the germ will not germinate. In addition, starch and protein will leach out from the damaged grain during steeping and encourage microbial growth. Many maltsters set a maximum limit of 2% for broken grains. It is acknowledged that emptying of large floor stores using buckets on tractors leads to increased levels of damage.

Nitrogen content is also important for malting barley. A low nitrogen content is preferable as this means a higher proportion of carbohydrate and thus a higher potential to produce fermentable sugars and consequently alcohol. Conversely, high nitrogen is associated with less alcohol and is more likely to give a cloudy beer. It is therefore important during storage to segregate batches of barley according to nitrogen content.

Contract targets for germination and other qualities in malting barley throughout Europe are shown in Table 4.1. However, it should be noted that the definition of, and the methods of assessing, the same quality may vary. For example, in the UK barley viability is specified, whereas in other countries only germination energy will be measured as it is a reasonable indicator of viability when dormancy is low, which is not the case in northern Europe. Moisture is measured by the ISO method (ISO 1985), drying ground grain in a ventilated oven at 130°C for 2 h, against which standard moisture meters are then calibrated.

Major sources of quality decline

According to reports from UK maltsters there are two common causes of rejection of grain from farm or commercial stores. These are insect infestation or a failure of the stock to correspond to the pre-delivery sample. This sample is taken in the store before the barley is emptied and delivered to the maltsters and is the basis of the sale price. Unsatisfactory germination, excess nitrogen and mould contamination were other reasons given for rejection and the frequency of these problems varied from season to season. However, the total level of rejection for all reasons was small, ranging between 1.5 and 5.5% of the grain delivered.

Table 4.1   A comparison of European contract requirements for malting barley qualities.

	UK	Belgium	France	Denmark	Germany	Spain
Germination (%)	98	95	95	> 95	> 95	97
Protein (%)	9.7–11.6	10–12	9–12 target 10.5	9.8–11.0	8–11.75 (2R) 8–11.5 (6R)	9–11.5
Kernel size	> 90% 2.5 mm < 8% 2.2–2.5 mm < 2% 2.2 mm	> 90% 2.5 mm	> 90% 2.5 mm	> 90% 2.5 mm	> 85% 2.5 mm	65% > 2.5 mm
Broken grains (%)	< 2	< 2	< 2	< 1	< 2	< 3
Impurities (%)	< 2	< 3	< 0.5	< 0.5		< 8
Moisture (%)	< 15	< 15	< 14.5	< 14.5	< 14	< 12
Varietal purity (%)	> 97	> 93	> 93	> 99	> 95	> 90
Insects	0 live or dead	0 live	0 live	—	0	—

2R = 2 row barley. 6R = 6 row barley varieties. 'Germination' in Denmark and Germany means germinative energy, whereas in the UK it means viability.

Maintaining viability and breaking dormancy are major issues for malting barley storage that differentiate it from other commodities. Avoiding insect infestation is also important but this is common to other stored foods. These dynamic factors can be affected by drying and storage procedures.

Barley drying

Drying can be done in two ways: using large volumes of near ambient air and drying slowly or using a short exposure to much hotter air. Ambient on-floor drying techniques using airflows of about 180 m^3 h^{-1} t^{-1} should dry the grain in ten days and may be considered a gentler approach than hot-air drying, but if recommendations are not followed and drying is prolonged, germination decline can be rapid (Table 4.2). If using a hot-air dryer, the general guideline is to use a maximum air-in temperature of 65°C when barley is at 20% m.c., reducing by 1°C for every 1% increase in moisture content.

Predicting germination changes

The change in the germination characteristics of malting barley during storage can be considered as a combination of two processes:

(1) A break of dormancy where the percentage of the viable seeds that can germinate under given conditions, g_d, is increasing.
(2) A loss of viability (germinative capacity) where the number of seeds that can ultimately germinate, in the absence of dormancy, is declining.

This is illustrated in Fig. 4.3. The combination of the two effects results in the characteristic germination history curve, which predicts the overall germinative energy, g, as a product of the percentage viable grains (g_v) and the percentage of the viable grains to have broken dormancy (g_d).

The changes in g_v and g_d with time can be predicted using probit analysis. This assumes that the lengths of time to loss of viability and to break of dormancy are nor-

Table 4.2   Time in days for germination of barley to fall to 95% when stored at different temperatures. Source: Burrell (1966).

Moisture content %	°C							
	0	5	10	15	20	25	30	35
14	*	*	*	350	245	140	70	35
16	*	*	350	147	105	42	28	10.5
18	*	350	161	70	42	17.5	10.5	< 7
20	*	175	42	32	16	7	< 7	< 7
22	259	49	17.5	10.5	7	< 7	< 7	< 7
24	168	14	7	< 7	< 7	< 7	< 7	< 7

*Over 1 year.

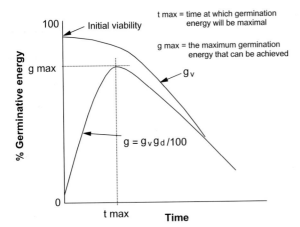

Fig. 4.3 Germination history of a barley in storage, g_v = germinative capacity (viability) and g_d = % of viable grains that are non-dormant

mally distributed. The values of g_v and g_d are calculated from the probability function:

$$p = \frac{1}{\sqrt{2\pi}} \int_{-\alpha}^{X} e^{-X^2/2} dX$$

where $X = (t - t_{50})/\sigma$, t is time, t_{50} is the half-life of dormant or decaying populations and σ is the standard deviation of the normal distribution of viability loss or dormancy break.

The value of σ for viability loss is given by Ellis and Roberts (1980) as:

$$\log_{10}\sigma_v = 9.983 - 5.896 \log_{10}M - 0.04\,T - 0.000428\,T^2$$

where M is the moisture content and the value of σ for break of dormancy is taken from work on the variety Triumph (Briggs & Woods 1993; Woods *et al.* 1994) as:

$$\log_{10}\sigma_d = 1.91 - 0.0352\,T$$

The standard deviations, σ_v and σ_d, have the unit of time and define the rates of the two processes. It should be noted that σ_v is a strong function of moisture content and temperature, while σ_d is a function of temperature only. The model reflects the experimental finding that moisture content does not significantly affect rate of break of dormancy.

Break of dormancy

In order to apply these equations to predict germination changes during storage, the following assumptions are made:

(1) The minimum required germinative energy is 95%.

(2) A typical worst-case germinative energy at intake is 10%. However, in barleys collected with the objective of acquiring dormant samples in the 1990 and 1991 harvest, 4 of 33 had germinative energies of only 6%, 8%, 6% and 8.5%.

(3) The initial viability is 98%. This viability refers to a value based on an ageing test (Woods *et al.* 1994) and is not derived from values in the hydrogen peroxide or staining test. The value is based on five barleys tested at 38°C, one of which was replicated at four moisture contents. All values were above 98%.

(4) The mean moisture content in storage is 12%. Due to variations in dryer performance, 11% and 13% were also examined.

(5) All germination values refer to the germination test being undertaken with grain placed on filter paper wetted with 4 ml of water and left for three days (Anon. 1997), which is the recognised test for germinative energy in the UK.

Based on these assumptions and the double probit analysis, the effect of constant storage temperature and moisture content on the storage time to raise germinative energy from 10% to 95% is illustrated in Table 4.3. This also shows that at lower temperatures and longer times it is possible to achieve higher germinative energies due to the sensitivity of viability loss to temperature.

Loss of viability during cooling and storage

Cooling can be considered as a series of 'fronts' passing through the grain. Data on completion times for cooling fronts of 15°, 10° and 5°C (Wilkin *et al.* 1990) can be used to predict worst case viability loss during warm storage (30°C, where it would take 24.2 days for germination energies to go from 10% to 95%) followed by ambient cooling. Predictions for grain at 12% or 13% m.c. are presented in Table 4.4 for the case of barley aerated by fan at a rate of 6.8 m³ h⁻¹ t⁻¹ with a starting date of 1 August. There is an initial period of blowing (8.7 days) after which it was assumed that the grain would have at least been cooled to 20°C. Cooling the grain like this, by continuous aeration, immediately after warm, dormancy-breaking storage, would be good practice commercially as it reduces temperature immediately. The predicted loss in viability is considerable, and noticeably higher at 13% compared to 12% m.c., although it is less than if there had been no cooling. Current work suggests that assuming an initial viability of 98% was

Table 4.3 Time in days to break dormancy at a range of storage temperatures and moisture contents.

°C	Moisture content = 11%			Moisture content = 12%			Moisture content = 13%		
	t_{10-95}	g_{max}	t_{max}	t_{10-95}	g_{max}	t_{max}	t_{10-95}	g_{max}	t_{max}
10	115	97.6	176	116	97.4	168	118	96.9	161
15	77	97.6	116	78	97.2	110	79	96.7	105
20	52	97.4	76	52	97.0	72	54	96.3	68
25	35	97.2	49	35	96.7	46	37	95.7	44
30	23	96.9	32	24	96.1	30	—	94.6	28
35	16	96.3	20	19	95.0	19	—	92.5	17
40	12	95.3	13	—	92.8	12	—	88.1	11
45	—	93.0	8	—	88.3	7	—	79.3	6

t_{10-95} time to increase germinative energy from 10% to 95%; g_{max} maximum germinative energy that can be achieved at a given temperature; t_{max} time to achieve g_{max}.

Table 4.4 Predicted dormancy and viability changes during cooling (6.8 m³/h/tonnes, start date 1 August) for grain at 12% or 13% moisture content.

Stage	Duration (days)	T (°C)	g (%)	g_v (%)	g_d (%)
12% moisture content					
Warm storage at 30°	24.2 (t_{10-95})	30	95.0	96.7	98.1
up to arrival 20° front	8.7	30	96.0	96.0	99.9
up to arrival 15° front	15.3	20	95.7	95.7	100.0
up to arrival 10° front	40.0	15	95.3	95.3	100.0
up to arrival 5° front	104.0	10	94.5	94.5	100.0
13% moisture content					
Warm storage at 30°	24.2 (t_{10-95})	30	93.9	95.5	98.2
up to arrival 20° front	8.7	30	94.1	94.2	99.9
up to arrival 15° front	15.3	20	93.5	93.5	100.0
up to arrival 10° front	40.0	15	92.5	92.5	100.0
up to arrival 5° front	104.0	10	90.7	90.7	100.0

t_{10-95} time to increase germinative energy from 10% to 95%; g (%) germinative energy; g_v (%) proportion of viable grains , g_d (%) proportion of grain past dormancy.

conservative and that 99.5% is more typical, giving a safer storage than predicted above.

Insect development during dormancy break and subsequent cooling

A survey of UK maltings in the 1960s indicated that over 50% had suffered from infestation (Tebb 1967). The most serious pests were grain weevil (*Sitophilus granarius*), Khapra beetle (*Trogoderma granarium*) and the saw-toothed grain beetle (*Oryzaephilus surinamensis*) which occurred on 32%, 20% and 15% of premises, respectively. Of these, *T. granarium* was found mainly in maltings while the others were bulk grain pests and also occurred on farms. By 1973 it was noted that *Ptinus tectus* and the moths *Endrosis sarcitrella* and *Hofmannophila pseudospretella* had become equally important as pests of maltings, possibly due to the fact that, although storage moisture was closely controlled, temperature was neglected (Hunter *et al.* 1973). Since about 1975, modern kiln designs have eliminated the food residues that allowed insects to be insulated from the heat and breed successfully. *Trogoderma granarium* has largely disappeared, perhaps as a consequence of this. The design of modern maltings is thought likely to have eliminated most infestation from the germination floor.

The need to break dormancy by delaying cooling may give extra time for the development of insect infestations. This has been considered theoretically and experimentally by simulating insect development during stages of dormancy break and the subsequent cooling steps (Armitage & Cook 1997; Armitage & Woods 1997). Calculations based on speed of insect development and fecundity have shown that lower temperatures gave greater safety margins between insect development time and the break of dormancy. Saw-toothed grain beetles were the quickest developing insects at high temperatures with little margin for error between dormancy break and the time taken for insect development. Most fan control strategies for cooling allowed for possible insect development after dormancy break and during cooling. Rust red grain beetles, *Cryptolestes ferrugineus*, were the greatest threat at 35°C and saw-toothed beetles at 25–30°C, with few insects developing at 20°C, where grain weevils were predominant. Laboratory tests indicated that at an initial temperature of 40°C, only small numbers of *T. granarium* could develop, but at 35°C large populations occurred, as was the case with *O. surinamensis*, *Rhyzopertha dominica* and *S. granarius* at 30°C. At 25°C, the small numbers of *O. surinamensis* were eclipsed by large increases in *S. granarius*, which also increased vigorously at 20°C. To discourage insect development it would therefore seem best to break dormancy very quickly over 40°C or slowly below 20°C.

Moulds and mycotoxins

Maltsters prefer that malting barleys are dried to about 12% moisture for storage, which is too low for growth of *Penicillium verrucosum* and subsequent formation of ochratoxin A (OA). Barleys are examined at intake and any showing signs of mould infection (staining, weathering, odour, decreased germination) are rejected. Since the OA problem was first raised, there have been annual surveys of UK malting barleys. Only in 1994 and 1997 was OA detected in barley, in this case at a rate of 0.1–0.5 µg kg^{-1} in 11 (7%) samples. Malt itself was also tested, and between 1994 and 1997, 17–20% of malts had OA between 0.1 and 1 µg kg^{-1}, although in 1997 no OA was detected in malt. The EU limit for OA contamination is 5 µg kg^{-1}.

The levels of OA in barley are retained on malt but about one third to one half of any OA in malt is lost in the brewing process, largely degraded during mashing (thought to be via hydrolysis of the amide bond). Consequently, the levels in beer are very low – mostly less than 0.02 µg L^{-1}. This is just above the limit of detection (Baxter *et al.* 2001). However, after the wet harvest of 1996, OA was found in German malting barley and in malt (Table 4.5) (Gareis 1999) and 70–80% of beers were contaminated with between 0.01 and 0.29 (mean 0.03) µg L^{-1} (Bresch 2000).

Commodity and pest management regimes

Responses from 14 barley store managers at two malting companies in the UK, covering nearly 1m t of malting barley, indicated that about 75% of sites relied on cooling and only 11% used pesticides. In the UK pesticides for malting barley have to be approved by the Brewers' and Licensed Retailers' Association and the Scottish Whisky Association. Stocks may be fumigated with phosphine or admixed with etrimfos, pirimiphos-methyl or chlorpyriphos-methyl insecticides. However, if barley is treated with chlorpyriphos-methyl then it must be stored for at least eight weeks before malting.

Very often, grain is loaded into silos still warm from the dryer and at a moisture content of 12%. About 60% of the grain went into store at 30°C or above. The aim may be to leave the barley warm for a period to allow dormancy break, which is sometimes a UK concern with certain varieties, particularly in Scotland. Some years ago there

Table 4.5 Concentration of ochratoxin A and ochratoxin B (µg kg^{-1}) in barley and malt from 16 German malting plants. Source: Gareis (1999).

	Barley		Malt
	Freshly harvested	After 5 months storage	
Ochratoxin A			
% samples	7.10	10.10	11.70
median conc.	0.035	0.10	0.26
Ochratoxin B			
% samples	7.80	4.00	5.30
median conc.	0.036	0.14	0.04

were fears that swift or prolonged cooling could lead to 'secondary dormancy' but recent investigations have found no evidence for this phenomenon (Woods & McCallum 2000). Secondary dormancy is usually ascribed to excessive development of fungi and bacteria on the inner skin of the barley that restricts oxygen supply to the germ. It can therefore only occur at moisture contents that permit the growth of storage micro-organisms. This is unlikely to happen in well-managed dry storage. An apparent drop in germinative energy is bound to be observed occasionally, given the statistical error in sampling and testing. This offers an interesting explanation for this phenomenon. A more likely problem is that using cool barley extends the steeping period as it will cool the water.

Danish commercial practice generally involves drying the grain using a maximum air temperature of 60°C, to preserve viability, followed by cooling to temperatures as low as 0°C. Stores are cleaned and sprayed with malathion annually and phosphine is used against infestations. Stores are regularly monitored for temperature and also for pests using traps.

In northern Spain, silos at maltings have no aeration facilities so pest control depends on phosphine fumigation or contact pesticides, mainly pirimiphos-methyl. Farmers hold barley in short-term storage in the field before it is collected by local co-operatives. Most co-operative facilities are floor stores with aeration that can reduce temperatures to 10–12°C by November. This strategy replaces the older practice of treating with contact insecticides, often followed by a fumigation later in the year.

In Belgium, most barley is imported, much of it from France. In this case the grain is normally treated with dichlorvos as it is loaded onto barges for transport. The short residual life of this pesticide is thought ideal.

In Germany only phosphine, carbon dioxide, nitrogen, pirimiphos-methyl or diatomaceous earth are permitted to treat bulk grain and in practice most barley is not treated with a contact insecticide. A much wider range of pesticides can be used for treatment of empty stores including dichlorvos, alone and in mixture with piperonyl butoxide and pyrethrins, a mixture of piperonyl butoxide and pyrethrins, phoxim or pirimiphos-methyl.

The most widely used insecticides in French stores are pirimiphos-methyl and chlorpyrifos-methyl and in northern France, where temperature are generally lower, deltamethrin is also in frequent use. Bioresmethrin has never been used on a large scale, mainly due to high cost, while malathion is seldom used because it has a very short half-life. At the export terminals a mixture of dichlorvos and malathion is commonly applied although malathion is now often replaced by chlorpyrifos-methyl which gives better controls of grain mites. Currently, dichlorvos and chlorpyrifos-methyl are both under threat because of the high cost of European re-registration.

The fate of pesticide residues

Studies on the fate of pirimiphos-methyl residues during malting showed that 8–10% were washed off during steeping, there was little breakdown until after four to five days of germination and 40% of residues were destroyed by kilning. Less than 20% of the applied primiphos-methyl survived into the malt and less than 1% survived in the roots although these had in any case been separated from the malt.

Residues in beer are likely to be below the limit of detection because (1) levels in malt are low – in most commercial samples they are not detectable and in those few in which there is some contamination, concentrations are less than 0.5 mg kg^{-1}; (2) there is a 1–10 dilution in going from malt to beer; and (3) the partition coefficient of pirimiphos-methyl is high (log P_{ow} = 4.2) relative to other pesticides (fenitrothion 3.3, malathion 2.9), which means that primiphos-methyl is not very soluble in aqueous solutions and adheres to solid material. Thus most of the pirimiphos-methyl in the malt is removed with the spent grains. Of that which survives this part of the brewing process, more is lost with the solid precipitate (trub) which forms when the wort is boiled and some is also lost with the spent yeast (Miyake *et al.* 1999, 2002).

Economics of operation of the system

In the UK the costs of barley storage are similar to those for wheat (see Chapter 3). A number of factors need to be considered when deciding on the best strategy for the storage of malting barley. At one level is the capital cost of constructing stores either at the maltings, at a commercial site or on the farm. There may be economies of scale to be gained with stores at the maltings and commercial stores but this obviously involves an overall larger capital outlay. The most cost-effective strategy depends upon the capital costs of installing or replacing a store, the benefits it is likely to bring and the expected lifespan of the structure. The uncertainties of barley prices and the costs of production in the future obviously make this investment difficult to assess.

The economics of storing grain depend on a number of factors, most importantly the value of the grain itself, with malting barley selling at a premium compared to feed barley. There is a cost associated with maintaining the quality of the malting barley and if this exceeds the extra value of malting over feed barley then it is no longer sensible to carry out storage operations, assuming the barley can be sold as feed. Furthermore, the change in price of barley with time is also an important consideration; at or around harvest the value of the crop is lower than in the following year. If the increase in value over time is less than the cost of storing the grain then it makes sense to sell at harvest and not incur these costs. Of course, if storage facilities already exist, some of the cost has already been met and it may be worth storing the grain. The forecasting of future prices is imprecise and so there will always be an element of uncertainty.

The main operations that incur cost during the intake and storage of grain are the preparation of the store, which includes cleaning and/or washing, and possible surface treatment with pesticide. The major cost of this operation is labour. If there is a standing labour force then this expense can be met from existing overhead costs; if, however, labour needs to be hired then this would add to costs. Incoming grain may need to be cleaned to meet contract requirements and this will be an expense, firstly, because the process uses energy and labour, and secondly, as material is removed the weight of malting barley is reduced and unwanted or lower-value products are produced. The reduction in the quantity of malting barley needs to be added to the costs of this operation.

In most years, malting barley will require drying before it is put into store or when it is first in the store. Drying can be achieved using a variety of machines each having an associated capital and running cost. High-temperature drying can be used to reduce moisture content quickly but grain requires cooling afterwards. This is a relatively expensive operation although fuel costs are generally small relative to capital investment and the labour costs for operation and maintenance. Drying grain in bulk using ambient air is generally cheaper but much slower at reducing the moisture content and may not be able to cope with very wet grain. These systems are difficult to manage due to the unpredictability of the weather. Stocks are not susceptible to germination heat damage but the surface region, which in poor weather is the last to dry, is vulnerable to moulds. Any drying operation will reduce the weight of grain in store as the water is removed. If the grain is being dried a great deal then moisture shrinkage will become a large component of the drying cost, since there is no price adjustment for grain at less than the required moisture content.

Once the grain is in the store, lowering the temperature by blowing ambient air through the grain can prevent or retard infestation by insects and mites. The cost associated with this will be the energy required to run the fans. Savings can be made through the use of timers to restrict fans to off-peak electricity and by differential thermostats, which optimise running time (Armitage & Woods 1997). An alternative to this is to use pesticides to control infestations.

A summary of the factors to be considered when deciding on whether it is economically justifiable to clean, dry, cool or treat grain is given in Fig. 4.4. For a comprehensive description of the energy consumption and other potential costs of storage operations see McLean (1989).

Future developments

It is clear that the preference on the part of the maltsters in, notably, Denmark and Germany is to encourage farmers to store more malting barley on-farm and thus to save the costs of storage and improve cash flow. Elsewhere, maltsters seem to lack confidence in farmers' ability to protect grain viability adequately, since this quality is absolutely essential to the success of the malting process.

An increased reliance on storage on-farm may however be facilitated by the outcome of a current project in the EU FAIR programme 'Building a decision support system for management and control of quality of malting barley'. This collates current information on breaking dormancy, germination decline, pest populations and control practices and presents predictions and advice on storage strategies. These strategies have been validated on a farm scale in several countries. One important outcome of the project was a possible 24 h method of germination determination using image analysis, which should assist initial diagnosis of the suitability of barley for malting. The resulting software will also provide a quality report for that batch of grain. This is regarded internationally as an essential part of the requirement for increased traceability. Due to the need to avoid sprouting in the field it is not possible to breed out dormancy. Thus dormancy currently exists throughout the industry but need not be a problem if it is well managed, for instance by carry-over stocks, by warm storage treatments or by making up shortfalls by the import of non-dormant barley. The warm storage approach is economically attractive compared with the other management options.

As part of the EU FAIR project, discussions with maltsters throughout Europe revealed that their main concern

Costs of barley storage operations

Drying loss (DL)
Loss in weight (tonnes) × value of barley (cost/tonne)

Cleaning losses (CL)
Loss in weight (tonnes) × value of barley (cost/tonne)

Pesticide costs (PC)
Pesticide cost per tonne × tonnage of barley + labour costs

Cost of drying
Energy cost (cost/tonne/%m.c.) × tonnage of barley × % change in m.c. required + labour costs

Cost of cooling
Energy cost (cost/tonne/$^{\circ}$C) × tonnage of barley × temperature change ($^{\circ}$C) + labour costs

Value of barley stock

Value of 'cleaning products' (CP)
Weight (tonnes) × value of product (price/tonne)

Value of barley sold for malting (VM)
Price of malting barley per tonne × tonnage

Value of barley sold for feed (VF)
Price of feed barley per tonne × tonnage

To decide whether barley should be store for malt or for feed, first calculate the total costs (TC) of expected storage operations as follows -
$$TC = DL + CL + PC + CD + CC - CP$$
If it is cost-effective to keep barley for malting (VM – TC > VF) then the cost of storage operations can be justified. If it is not cost effective to keep barley for malting (VM – TC < VF) then storage operations are only required if the barley will become unsuitable for feed.

Fig. 4.4 Calculation of costs used to decide whether barley grain should be used for malting or for animal feed.

was the need to replace pesticides with equally effective physical control strategies. Knowledge about cooling with ambient air varied. In France and Denmark there was no concern about cooling grain to nearly zero, except for the cost of heating the steeping water. In the UK the effects of rapid cooling on dormancy are regarded with suspicion by some operators. This is not the opinion of the Maltsters Association of Great Britain, which represents most of the maltsters in the UK, and recent work on the FAIR project, together with that of Woods and McCallum (2000), confirms that cooling does not induce dormancy.

Specific storage strategies for malting barley may be required that incorporate the possibility of breaking dormancy and preserving germination without encour-

aging infestation. If cooling is to be the main weapon of malting barley preservation, then automatic control of ventilation needs to be more widely applied with the possibility of developing microprocessor control units with specific strategies for malting barley. An often-expressed concern about aeration, especially of a dry commodity like malting barley, is that cooling with ambient air will dampen the grain. Although this does not usually occur, because the r.h. of the cool air drops as it is warmed by the grain and by heat from the fan (a rise of 0.5–2°C), occasionally a dampening front can be pulled down from the surface by suction. In addition, evaporative cooling is less evident in dry grain, which consequently appears to be more difficult to cool.

References

Anon. (1997) *Institute of Brewing Methods of Analysis.* Sections 1.1 to 1.13. Institute of Brewing, London, UK.

Armitage, D.M. & Cook, D.A. (1997) Laboratory experiments to compare the development of populations of five species of insect during dormancy break and subsequent cool storage of malting barley. *Journal of the Institute of Brewing,* **103**, 245–249.

Armitage, D.M. & Woods, J.L. (eds) (1997) The development of an integrated storage strategy for malting barley. *Project Report 138.* Home Grown Cereals Authority, London, UK.

Bishop, L.R. (1947) Post-war malting barley problems. *Chemistry and Industry,* **66**, 779–783.

Baxter, E.D., Slaiding, I.R. & Kelly, B. (2001) The behaviour of ochratoxin A in brewing. *Journal of the American Society of Brewing Chemists,* **59**(3), 98–100.

Bresch, H. (2000) Ochratoxin A in coffee, tea and beer. *Archiv fur Lebensmittelhygiene,* **51**, 89–94.

Briggs, D.E. & Woods, J.L. (1993) Dormancy in malting barley: studies on drying, storage, biochemistry and physiology. *Project Report No. 84.* Home Grown Cereals Authority, London, UK.

Burrell, N.J. (1966) Refrigerated damp grain storage. *Pest Infestation Research,* 1965, 17–19.

CBMC (2000) Confédération des Brasseurs du Marché Commun, Brussels, Belgium. http://www.cbmc.org/ukpages/stats/st13ma.htm.

Ellis, R.H. & Roberts, E.H. (1980) The influence of temperature and moisture content on seed viability in barley (*Hordeum distichum* L.). *Annals of Botany,* **45**, 31–37.

Gareis, M. (1999) Contamination of German malting barley and of malt produced from it with the mycotoxins ochratoxin A and B. *Archiv fur Lebensmittelhygiene,* **50**, 83–87.

Hunter, F.A., Tulloch, J.B.M. & Lambourne, M.G. (1973) Insects and mites of maltings in the east Midlands of England. *Journal of Stored Products Research,* **9**, 119–141.

ISO (1985) *ISO 712: Cereals and cereal products: determination of moisture content (basic reference method).* International Organization for Standardization, Geneva, Switzerland.

McLean, K.A. (1989) *Drying and Storing Combinable Crops.* Farming Press Books, Ipswich, UK.

Miyake, Y., Koji, K., Matsuki, H., Tajima, R., Ono, M. & Mine, T. (1999) Fate of agrochemical residues, associated with malt and hops, during brewing. *American Society of Brewing Chemists,* **57**, 46–54.

Miyake, Y., Hashimoto, K., Matsuki, H., Ono, M. & Tajima, R. (2002) Fate of insecticide and fungicide residues on barley during storage and malting. *American Society of Brewing Chemists,* **60**, 110–115.

Tebb, G. (1967) A survey of infestation in maltings and breweries. *Journal of the Institute of Brewing,* **74**, 207–219.

Wilkin, D.R., Armitage, D.M., Cogan, P.M. & Thomas, K.P. (1990) Integrated pest control strategy for stored grain. *Project Report No. 24.* Home Grown Cereals Authority, London, UK.

Woods, J.L., Favier, J.F. & Briggs, D.E. (1994) Predicting the germinative energy of dormant malting barley during storage. *Journal of the Institute of Brewing,* **100**, 257–269.

Woods, J.L. & McCallum, D.J. (2000) Dormancy in malting barley: quantifying the effect of storage temperature on different varieties and the response to sudden cooling. *Journal of the Institute of Brewing,* **106**, 251–258.

Chapter 5
Sorghum

R. A. Boxall, M. Gebre-Tsadik, K. Jayaraj, C. P. Ramam, B. B. Patternaik and R. J. Hodges

Sorghum (*Sorghum bicolor*) is a small grain (Plate 5) fourth in importance among the cereals after wheat, rice and maize. It is grown for various purposes in warm subtropical areas and, owing to its relative drought tolerance, is a cereal staple and fodder crop in many tropical dryland regions. It is known by a variety of names: guinea corn (West Africa), *mtama* (East Africa), *durra* (Sudan), *jowar* (India) and milo (US). It appears to have originated in Africa and was probably domesticated in Ethiopia by selection from wild sorghum between 5000 and 7000 years ago (Kent 1984; Peacock & Wilson 1984; Narayana 1989). The crop was certainly being cultivated in Egypt in 4200 BP, as recorded in wall paintings of that period, and is likely to have been one of the first plants domesticated for use as human food and fodder (Vinall *et al.* 1936). Thence it spread throughout Africa, and via Arabia to India and the Middle East and to China (Burkhill 1952; Wayne-Smith & Frederiksen 2002).

This chapter focuses on sorghum handling and storage by farmers in South India and a centralised food security reserve in Ethiopia, starting with a special consideration of pit storage. This is one of the earliest types of grain storage and a method that is still relatively commonly used to conserve sorghum in parts of Ethiopia, Sudan, Somalia and India (Gilman & Boxall 1974). Capacities of pit stores range from less than 1 t to more than 100 t and both subsistence farmers and commercial farmers use this method. The movement of soil moisture into pit stores is common and mould growth on the stored grain can sometimes be extensive. Improvements to pits stores have focused on development of linings to prevent water penetration, and polythene liners have been found to be cost effective (Boxall 1974a; Lynch *et al.* 1986). Concrete linings incorporating a moisture barrier are currently being promoted in Ethiopia.

Farmers in dryland areas of South India, like many other arid zones, grow sorghum mostly as a subsistence crop and the majority of grain remains on the farm for local consumption. The grain is stored in a variety of structures and protected from insects by repeated sun-drying and winnowing during the course of the storage season. Neem leaves may be admixed with the grain and, if grain is stored in sacks, insecticide may be applied to sack surfaces. The Government of India has a centralised food security strategy based around food grain distribution to the poor through the Public Distribution System. However, this is based on wheat and rice; sorghum is not included in this system although there is increasing interest in developing decentralised food security by encouraging village groups to pool their sorghum in grain banks (Jayaraj *et al.* 2003).

The Ethiopian Emergency Food Security Reserve (EFSR), established in the mid-1980s, currently holds around 450 000 t of bagged grain. Wheat is the predominant commodity while (unlike in India) sorghum contributes between 5% and 10% and may be stored for up to two years. These reserves are accommodated in purpose-built warehouses at five strategic locations and serve as a bridging stock for emergency and development projects. They initially consisted entirely of imported food grains but now include both imported and locally purchased supplies. Other relief and development agencies, with warehouse facilities similar to those of the EFSR, hold larger stocks of sorghum but for shorter periods. The diverse range of suppliers to the EFSR of both imported and local sorghum presents grain of widely varying quality at intake. This poses a considerable risk of quantitative and qualitative loss. However, a high standard of storage management and quality control ensures that overall losses (due to pests, spillage and grain drying) are kept to less than 2%.

Underground storage

Historical perspectives

Underground storage of grain is an ancient practice still in use in some dryland areas where sorghum is a staple food grain. It is believed to have been the main method of long-term grain storage from pre-Neolithic times up to the early nineteenth century, especially in Spain and Morocco in the West, across to India and China in the East, and including southeastern Europe, the Mediterranean area and the Middle East (Sigaut 1980; Sterling *et al.* 1983). Tribal movement may, in some instances, have been responsible for the introduction of pit storage into new areas, particularly in Africa, but it is apparent that the method developed independently in many countries.

Physical facilities

Underground pit stores require very little in the way of materials and can therefore be used in a wide variety of situations. A number of different designs with capacities ranging from less than 1 t to more than 100 t can be found. In Sudan for example, pits (which may be large and hold several tonnes) are little more than shallow, circular hollows in the ground in which the sorghum is piled. The mound of grain above ground level is covered first with a layer of straw and then a layer of soil. Similar pits are found in eastern and southern India though this method of storage is becoming less common, partly due to the rise in the water table. In northern Nigeria, pits are rectangular or cylindrical with straight sides; again the grain bulk may extend above ground level and may be covered with a layer of straw or matting before finally being covered with soil. The dome shape of the completed storage structure facilitates water runoff during the rainy season.

Pits in which the grain bulk is completely below ground are found in Yemen, Ethiopia, parts of southern Africa and the state of Andhra Pradesh in India. The pits, with capacities usually in the range 0.5–5 t typically have a narrow mouth just below ground level, opening via a short neck to a round or conical chamber. The mouth of the pit is covered by a stone or strips of wood, sealed in place by mud or cowdung, and the space above is filled to ground level with soil. Although these pits are often sited in the open and are not easily detected, for added security they may be located inside the dwelling house, cattle shed or cattle enclosure.

The use of linings to the floor and walls of pits is variable and may depend on the soil type. Some pits are unlined, but more usually some effort is made to line them to limit moisture damage to stored grain. None of the methods entirely exclude moisture, although, quite obviously, the drier the site the more likely they are to succeed. There are two categories of lining: first, the use of plant materials such as grass, straw, chaff and stalks of sorghum, which, unless applied in considerable thickness probably does little more than keep the grain from direct contact with the soil. The second category includes the use of clay, animal dung or termite mound soil plastering, which may be hardened by lighting a fire in the pit. This type of lining may restrict water entry but will not prevent it. Pits in which part of the grain bulk extends above ground level may be opened relatively quickly, since it is a simple matter to strip away the shallow covering of soil and the underlying straw or matting. Pits with the opening buried some way below ground level may take 30–45 minutes to open. Farmers are aware of the risks of increased levels of carbon dioxide in pits and may wait for a further 30 minutes before entering to remove grain.

Objectives of storage

Underground pit storage is often considered to be inexpensive and a method adopted by the poorer, subsistence farmer. This is true in some parts of the world; however, large-scale commercial farmers also use the method. Underground pit storage predominates in drier areas where rainfall is often erratic and there may be a need to store food grain for several seasons to guard against famine. The storage periods are typically between six and nine months. However, successful long-term storage of up to ten years has been reported in Sudan. In dry areas, wood and grass suitable for the usual traditional types of grain store are often in short supply, and so pit storage is the obvious choice for most producers. Some grain traders in Sudan and Somalia use pits with capacities of around 100 t (Gilman & Boxall 1974).

It is sometimes considered advantageous in rural communities if the amount of reserve grain kept by farmers is not known to others – a situation easily achieved in pit storage. Pits are often difficult to locate and, even if their whereabouts are known, pilfering is not an easy matter. There is no risk from fire and the method also provides some protection against rodents.

Major sources of quality decline

It might be expected that the principles of airtight storage would apply to the use of underground pits, and that damage and loss due to insects and some harmful storage micro-organisms would be minimised. Certainly sorghum stored in pits is protected to some extent from insect attack because of reduced oxygen levels; however, infestations by storage beetles including *Rhyzopertha dominica* and *Sitophilus* spp sometimes cause extensive damage. Infestation may be confined to surface layers of grain in the first instance but progressive emptying of a pit exposes fresh grain to insect attack. There may be a carry-over of infestation from one year to the next if pits are not thoroughly cleaned before reuse. Insects may be introduced in new grain from the fields at harvest time, and in very dry areas (for example in parts of Sudan) insects may enter pits after they have been closed, particularly if the soil covering is thin (Darling 1959).

The movement of moisture into pit stores can be comparatively rapid and extensive, although actual losses to mould damage are not often excessive. Some mould damage in the vicinity of the pit sides (even when lined) and on the surface of the grain bulk is to be expected. Mould damage will reduce the nutritional value of the grain but more serious are the health risks associated with the consumption of mould-damaged grain that may contain mycotoxins (Vol. 1 p. 128, Wareing 2002). In a study of the mycoflora of sorghum stored in pits in Ethiopia more than 50 species of fungi were identified, many of which were storage or soil species. Over half of the species identified belonged to the mycotoxin-producing genera *Aspergillus*, *Penicillium* and *Fusarium*, which pose a significant risk to the health of rural consumers (Niles 1976).

Sorghum stored in pits in the highland areas of Yemen is said to remain dry and to suffer only very minor mould damage, perhaps because the pits are sited on well-drained land as well as being lined with sorghum stalks and leaves or plastered with mud (Greig 1970).

Sorghum grains may change in texture, appearance and flavour due to insect infestation and more so because of microbiological activity. Grain stored in pits is often not suitable for seed as viability can be severely reduced. However, unthreshed sorghum heads are sometimes stored in pits on top of the grain bulk and may be used as a source of seed grain. It is possible that grain stored in this way will be more susceptible to insect and mould attack because of the large air spaces.

Commodity and pest management regimes

Successful pit storage depends on restricting the supply of air and moisture. Reduction of damage by insects will be achieved when concentrations of oxygen are low and carbon dioxide high. When oxygen is at 2% or lower a complete kill of all stages of insects infesting grain is possible (Oxley & Wickenden 1963; Burrell 1980; Annis 1987). It is doubtful whether pit stores are ever completely airtight but they have been shown to be reasonably airtight, and result in a significant reduction in damage caused by insect infestation. Measurements of oxygen concentrations in the headspace of underground pits in Ethiopia showed that around 2% oxygen was achieved within five weeks of closing the pits (Boxall 1974a). If the pits remained undisturbed insect infestation was either eliminated or severely restricted.

The restriction of insect activity is also closely associated with soil moisture, which moves into the grain from the surrounding soil. It is probable that the growth of moulds brought about by the increase in moisture utilises the oxygen in the atmosphere of the pit and contributes to an increase in carbon dioxide, which may result in death by asphyxiation. Insect respiration and to a lesser extent the respiration of moist grain itself also contribute. Mould growth will be inhibited as the oxygen supply is depleted.

Considerable care is taken in the preparation of pits for storage: grain residues are cleaned out and often a fire will be burnt inside the pit for a day or two as a means of disinfesting the pit or drying (curing) newly applied plaster coatings on the walls and floor. Cleaning and drying of the grain before storage is also important. In Yemen insect infestation was insignificant or even non-existent in farmers' pits that were cleaned thoroughly before filling with clean, dry, newly harvested grain (Greig 1970).

Farmers in Ethiopia have occasionally treated grain stored in pits with insecticides, but the use of traditional methods of controlling or repelling insects is more common. These include the admixture of wood ash with grain and the use of various unspecified dried and pounded plant materials or extracts in water applied directly to the grain. This practice is well known for the protection of grain stocks in other situations in other parts of the world (Vol. 1 p. 280, Belmain 2002).

Economics of operation of the system

The time taken to prepare a pit depends on the soil type,

the capacity and the labour available. In Ethiopia, for example, it is estimated that a pit with a capacity of 1–2 t will take 10–30 man-days to complete. This will include the application of a mud/cowdung lining fired to harden it. Pits are expected to last anything between 20 and 30 years, with very little maintenance, and some pits have been reported to be in use for between 50 and 100 years. The mud/cowdung lining will eventually have to be replaced, but minor repairs before each new storage season will usually suffice. Construction costs are minimal since family labour is used and excavation of pits is usually undertaken during periods of low agricultural activity.

Future developments

The major problem associated with underground storage of grain is the ingress of water and the associated development of mould. Improvements to pit stores have therefore focused on development of linings to prevent or restrict water penetration. Three pit linings, namely matting and straw, polythene, and concrete, tested in Ethiopia were found to be effective in restricting moisture ingress and reducing mould damage to stored sorghum over a period of four months. At the end of the storage period the mean moisture content of the sorghum in lined pits ranged from 13.2% to 13.9% compared with 18.1% in unlined pits (Boxall 1974b).

In Morocco, underground storage of grain (mainly wheat) is widespread and studies have demonstrated the value of polythene sheet linings as replacements for the traditional straw linings. A study was made of the storage of wheat in improved and traditional underground pits of 1.5 t capacity over a period of 16 months. The moisture content of grain within polythene-lined pits remained virtually unchanged at around 12.5%. In the traditional straw-lined pits the moisture content increased to at least 18%. Losses in polythene-lined pits were significantly lower at 3% (dry weight) compared to 19% (dry weight) in the traditional pits. The cost of the lining was equivalent to approximately 100 kg of grain and therefore considered an economically viable improvement (Bartali 1994). However, as in the earlier study in Ethiopia (Boxall 1974b), it was found that the polythene sheets are easily damaged during handling or while unloading a pit. If damaged sheets are reused in subsequent seasons, water may enter through holes in the sheet, causing localised moulding of the grain. The problem could be overcome by using polythene sacks. The floor of the pit should first be lined in the usual way but using a small polythene sheet or overlapping empty polythene sacks. A number of sacks should then be loosely filled with grain (to facilitate handling) and placed around the sides of the pit. The centre space can then be filled with loose grain and will remain protected from moisture ingress through the walls of the pit by the bulkhead of partly filled sacks. Additional layers of sacks can be added as necessary and the central space filled as before (Boxall 1974b).

Work in Ethiopia in the early 1970s, on the feasibility of lining underground pits with concrete, was based on earlier investigation of protective coatings for mud/straw plaster walls of dwelling houses. This had shown that a layer of clay reinforced with cement reduced damage caused by rainwater. A similar coating when applied to storage pits had the effect of hardening the walls, thus preventing the mixing of soil and grain, especially when the grain was being removed. However, although the lining was more durable than the traditional mud/cowdung plastering, there was still a considerable maintenance cost. Consideration was therefore given to developing a more permanent lining. An improved lining consisting of two layers of a cement/sand/aggregate mix, each 5 cm deep, with an embedded reinforcing layer of chicken wire, performed well but it was found advantageous to include a waterproof barrier. This was achieved by applying a coating of bitumen between the two layers of rendering, or by applying a bitumen emulsion to the final layer of rendering (Boxall 1973). A sloping concrete collar was built around the mouth of the pit with the objective of directing rainwater from the opening (Figs 5.1 and 5.2).

Later studies confirmed the effectiveness of concrete linings. The moisture content of sorghum stored for nine months in 12 concrete-lined pits was found to range from 13.8% to 16.8%. With a few exceptions, the moisture content generally did not rise significantly above the initial level of 13.5%. In unlined control pits the moisture content after nine months ranged from 15.9% to 20%. Weight losses, primarily due to insect infestation, in lined and unlined pits averaged 4.4% and 14.5%, respectively (Lemessa 1990).

More recently there has been renewed interest in this method of improving pit stores in Ethiopia and the technique is being actively promoted by the Agricultural Extension Service (AES). However, the latest design incorporates a polythene sheet as a moisture barrier rather than a layer of bitumen. Cost estimates for lining a typical pit of around 2 t capacity are given in Table 5.1. Reports by the AES indicate that there has been good

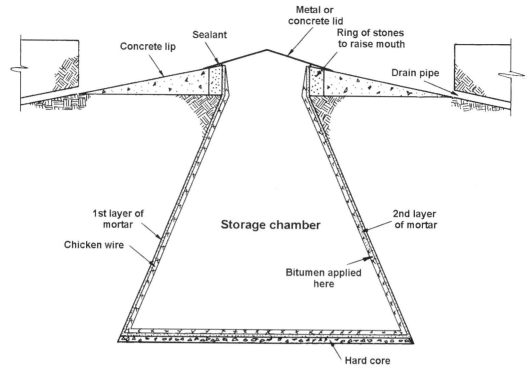

Fig. 5.1 Cross-section of improved storage pit (Boxall 1973).

Fig. 5.2 Improved storage pit during construction showing sloping concrete collar.

Table 5.1 Construction costs for concrete-lined pit store, 1999.

Item	Cost (Et birr)
6×50 kg cement	270.00
6 m^2 chicken wire	24.00
8 m^2 plastic/polythene sheet	32.00
Sand, aggregate (collected locally)	Nil
Total	326.00
	= \$US39.75

uptake of the design and that the costs are acceptable for the medium to large-scale producer.

In Sudan, a slightly different approach is being taken to address the problem of moisture ingress in pit stores. Traditionally, pits up to 1.5 m deep are filled with sorghum until a dome shape above ground level is formed by gravity at the angle of repose. A layer of chaff is then spread on top of the sorghum, and this is then capped with a layer of soil 25–50 cm deep and extending about 30 cm from the rim of the pit. However, it has been found that both subsistence and commercial farmers have benefited from using shallower (50 cm deep) but wider pits and adding a lining of chaff or, in the case of larger-scale commercial farmers, a lining of polythene. These innovations keep the sorghum drier, first by reducing the amount of water seepage through cracks, and second by minimising diffusion of water vapour from the wetter soil into the grain. The approach also leads to an increase in grain temperatures and consequently lower levels of insect infestation, hence adding to cost effectiveness. Moreover, the quality of the sorghum at the end of the storage period shows a substantial improvement over that of traditionally stored grain. Further improvements can result from increasing the thickness of the soil cap to at least 50 cm and extending it to about 1 m from the rim of a filled pit. This provides additional protection by covering over cracks in the cap that would otherwise allow seepage of rainwater into the pits (Bakhiet *et al.* 2001; Abdalla *et al.* 2002).

Whether or not it is justifiable to spend time and effort on the improvement of traditional underground pit stores must be decided by individual circumstances. The extent of losses in pits and in alternative, traditional and improved above-ground structures must be taken into account as well as the associated costs of pest control. The availability of materials for suitable alternative structures will also be an important factor. In Ethiopia, where underground storage is well established and often dominant and where losses in both above-ground and traditional underground stores can be significant, the use of polythene liners and waterproofed concrete liners are regarded as cost-effective methods of minimising storage losses. Similarly, in Sudan simple modifications to pit design and the use of chaff or polythene linings are proving to be popular with both small-scale and commercial farmers.

South India

Historical perspectives

Sorghum, known as *jowar* in India, is the most important food and fodder crop of dryland agriculture. Its cultivation is concentrated in peninsular and central India. The stored grain is used primarily as human food in various forms, such as *roti* or *chapatti* or *bhakri* (unleavened bread) or is cooked, either whole or broken like rice, as *sankati* (porridge). It is grown as a *kharif* or winter crop (July–November), harvested at the end of the wet monsoon period, and a *rabi* or summer crop, harvested during the dry season (October–February). Only a very small area of this crop is grown under irrigation, usually during summer (January–April). Natural selection and domestication over thousands of years have resulted in the development of numerous varieties highly local in their adaptation. Pure line selection practised among the principal local strains has produced improved varieties.

Development of commercial hybrids started in 1962. A choice of hybrids and varieties is now available for the different sorghum growing states and situations (Anon. 1984). The cultivation of high-yielding varieties (HYVs) has increased production and productivity but has also brought storage problems as these sorghums are softer and so more susceptible to insect infestation. Untimely rains, especially just before harvest, result in blackening of the *kharif* crop, especially as these have panicles that stand upright rather than hang down, which results in them retaining water long enough for mould damage to occur.

The majority of present-day crop storage methods have been known and used for thousands of years. A knowledge of how to store food, as distinct from how to obtain a supply of food, is responsible for the development of civilisation (Nash 1985). In India, the concept of storage started with underground pits. As the grain was used as a negotiable instrument for procuring other essential commodities, the concept was to conceal rather than to conserve. When money replaced grain in trading, storage above ground developed. The use of underground pits, which had once been prevalent in both *rabi* and *kharif* areas, is still practised in parts of South India, Maharashtra and Orissa for storage of sorghum and paddy rice but elsewhere has been largely abandoned.

India today has a centralised strategy of food and nutritional security, the main pillar of which is the distribution of food grain by the Public Distribution System (PDS). The PDS purchases and stores rice and wheat from surplus areas for distribution at subsidised rates to poor people through fair-price shops. This system does not operate with sorghum although this is the staple for a large proportion of the inhabitants of the dryland areas. Sorghum is not normally procured by the PDS. This is said to be due to its short shelf-life; however there are no data to support this contention and the perception may reflect the difference in condition of the grain at the time of procurement.

Physical facilities

The types of traditional grain storage structures used in different parts of India depend on the available local materials, local skills, cost of the structure and ease of use. Production season also plays a role in the choice of store type since, compared to *rabi*, yields in the *kharif* season are higher, so larger storage structures are required. It is generally observed that the *rabi* farmers have a strong convergence of practice, as all of them grow improved

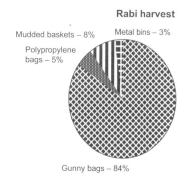

Rabi harvest

Mudded baskets – 8%   Metal bins – 3%

Polypropylene
bags – 5%

Gunny bags – 84%

Kharif harvest

Others – 9%

Metal bins
– 3%

Gunny bags
– 32%

Mudded
baskets – 31%

Polypropylene
bags – 25%

Fig. 5.3 Store types used by
farmers harvesting a *rabi* or *kharif*
crop of sorghum in South India in
June 1997 (Hodges *et al.* 2000).

varieties and most use only gunny bags (Fig. 5.3). In
contrast, farmers harvesting sorghum in the *kharif* season usually grow a wider range of varieties and use a
more diverse selection of storage structures; including
gunny bags, polypropylene bags, bamboo baskets, and
so on (Hodges *et al.* 2000). Some of the storage structures that are used for storage of sorghum are described
below.

Underground storage structures

These are mostly dug-out pits with a capacity of 2–5 t.
The walls of the pit are lined with stones, paddy straw
ropes, loose straw or other plant material. After filling,
straw is placed on the top of the grain before closing
with a plaster of mud and cowdung mixture. Pits are
used mostly for paddy and sorghum but are slowly
being discontinued due to groundwater seepage and
the time-consuming process of unloading which makes
withdrawals inconvenient. However, as the structure is
hermetic, considerable concentrations of carbon dioxide
accumulate during storage, which has the advantage of
reducing insect attack on the grain.

Earthen storage structures

Earthern storage structures (see Fig. 5.4) are either large
mud rings with tapering tops or pots. Both are closed
with a mud pan or other small pot. Higher-capacity
structures are constructed using a series of mud rings.
They are made of fired or unfired clay and are kept in
the house. If burnt clay is used it gives protection against
rats. Both types are popular for sorghum and other food
grains. The capacity range is 0.2–0.5 t.

Bamboo or reed storage bins

Storage bins (Fig. 5.5) are either indoor or outdoor struc-

Fig. 5.4 Indian farmer storing *kharif* sorghum in mud
pots.

Fig. 5.5 Indian farmer with sorghum storage bins, made
of split bamboo and sealed with a mixture of mud, paddy
straw and cowdung.

tures. The main materials for construction are bamboo
splits and locally available reed or cane. Bamboo splits
are used for indoor structures while those constructed

from reeds may be placed outdoors. Both are made relatively airtight by plastering with a mixture of mud, chopped paddy straw and cowdung. The paddy straw is included to reinforce the plastering but the resulting finish is not as fine as when only mud and cowdung is used. The indoor bins are used mainly for storage of sorghum and have capacities of 0.75–1 t. The outdoor structures have a capacity of 2–5 t and a thatched roof of palm leaves or hay.

Masonry storage structures

For the storage of sorghum or other food grains a brick compartment (Fig. 5.6) may be constructed in a house using mud, lime or cement mortar and completed with a cement rendering. The floor is of cement or stone slabs joined with lime or cement mortar. The roof may be of clay tiles, wood or cement depending upon the roof adopted for the house. The capacity of the structure varies from about 3–25 t although a portion of the store may be used for other purposes if it is not all needed for grain storage.

Bag storage

Sorghum is commonly stored in jute bags (gunny bags) (Fig. 5.7) and in some parts of the country polypropylene bags are popular. Both bag types are usually 100 kg capacity and typically farmers store up to 1 t in bags.

Fig. 5.7 Storage of *rabi* sorghum in jute bags in a farmer's house. Grain is being sampled using a bag sampler.

Seed storage

For seed storage, small baskets (10–25 kg capacity) (Fig. 5.8) made of bamboo splits are used. These baskets

Fig. 5.6 Farmer with a large stock of *kharif* sorghum stored in bulk in a compartment built in the house. Grain is being sampled using a multi-compartment grain spear.

Fig. 5.8 Storage of seed in a split bamboo basket fully sealed with mud and cowdung.

are plastered with mud and dung paste. After thoroughly cleaning and drying the grain, the basket is filled with sorghum and covered with a layer (2–3 cm thick) of ash or fine sand before finally being plastered with a mud and dung paste. The basket is made reasonably airtight and provides a physical barrier, thus reducing the risk of insect infestation.

Market systems and quality standards

It has been estimated that of the sorghum grain harvested about 30% enters the market. Official figures suggest that of this about 10% passes through regulated market channels and the remainder through unregulated channels. The market for sorghum is diverse and includes human food, poultry and dairy feed, industrial alcohol production, starch production and brewing. Not all is for local consumption as some is exported.

Quality standards

The Government of India stipulates quality standards when it or its agents procure sorghum grain from farmers (Table 5.2). Only in the state of Maharastra is sorghum procured by the government for the purpose of price support. The standards are not enforced for non-governmental transactions.

Major sources of quality decline

Sorghum is stored in a variety of storage structures for varying periods. Factors that determine the period for which sorghum can be stored without significant quality decline include moisture content and temperature of the grain, the abundance of fungal spores in and on grains, amount of foreign matter and presence of insects and mites. Each of these factors is closely related and

dependent on the others. Moisture content of the grain is of paramount importance and excess moisture will lead to serious losses. As the moisture content rises to values above 10%, insect infestation may occur and generate heat and moisture that supports fungal growth. Bulk density, flow characteristics and germination properties are also affected. Storage fungi cause discoloration of the kernel, reduction in germination, heating, mustiness and the production of mycotoxins.

In a study of selected villages in the states of Karnataka, Maharashtra and Andhra Pradesh it was found that *rabi* sorghum, harvested in the dry season, enters storage with better quality characteristics than its *kharif*-produced counterpart. The latter may suffer mould damage, visible as blackening, before storage if the monsoon rains are prolonged. For grain that is already in store, the monsoon period is associated with a major decline in quality for both crops, in particular a rise in mould damage and the start of insect attack. Both these factors are somewhat greater for *kharif* sorghum, presumably due to the rather lower quality of *kharif* grain at the onset of the monsoon. However, an intrinsically greater susceptibility of *kharif* varieties to mould and insect attack cannot be ruled out. Both grain types are of rather poor quality by the end of the storage season, whether it be October in the case of the *kharif* crop or January in the case of the *rabi* (Hodges *et al.* 2000).

Although farmers of *kharif* sorghum used a wider range of storage methods, this appeared not to be a factor affecting the grain quality, since the quality of *kharif* grain in gunny bags (the storage technique of most *rabi* farmers) differed little from the grain stored by other methods. It appears that choice of storage method is dependent on a wider set of factors than storage efficiency alone. Differences between *rabi* and *kharif* can be explained to some extent by the amount of grain to be stored, particularly the larger *kharif* crop. Ease of use,

Std No.	Quality variable	Maximum limits (%)
1	Foreign matter*	1.5
2	Other foods grains	3.0
3	Shrivelled and immature grains	4.0
4	Damaged grains	1.5
5	Slightly damaged and discoloured	1.0
6	Weeviled grains	1.0
7	Moisture	14.0

Table 5.2 Government of India quality standards for the *kharif* sorghum harvest in marketing season 2000–01.

*Within the overall limit of 1.5% for 'foreign matter' the inorganic matter shall not exceed 0.5% and poisonous seeds too shall not exceed 0.5%, of which *Datura* and akra seeds (*Vicia* spp) are not to exceed 0.025% and 0.2%, respectively.

availability of appropriate artisanal skills and costs are among the issues mentioned by farmers. Regional preferences also undoubtedly play a role.

There are qualitative and quantitative losses due to insects, mites, rodents and fungal attack. According to a study conducted by the College of Home Science, Andhra Pradesh Agricultural University (now the Acharya N.G. Ranga Agricultural University) total loss due to insects, taking into consideration quantitative and nutritive losses, was in the range 10–15%. However, rather lower losses have also been suggested, with the *kharif* crop at the end of the storage season (October) losing on average only 1.7%, while at the same time the *rabi* crop appeared to have lost only about 0.9% (Hodges *et al.* 2000). The two crops differed with respect to the predominant insect pest, with *Rhyzopertha dominica* and *Sitophilus oryzae* equally common on *rabi* but *R. dominica* predominant on *kharif*. Although the two species differ in their abilities to tolerate dry conditions (Haines 1991), *S. oryzae* is seriously limited on grain with moisture contents below 11% but it seems that this is not a major consideration in this case as this species was more common on the drier, *rabi* crop. The predominance of *R. dominica* on the *kharif* crop is presumably a reflection of the susceptibility of the different grain varieties to the two species. In connection with this, it is interesting to note that Reddy and Nusrath (1988), studying insect infestation and mycotoxin production in *kharif* sorghum varieties, list *R. dominica* as a major pest of this grain and make no mention of any *Sitophilus* spp. Furthermore, high-yielding *kharif* varieties appear to be more susceptibile to *R. dominica* as the percentage damage caused by this species was many times greater than that caused by *S. oryzae* (Kishore *et al.* 1977).

In South India, mycotoxin contamination generally remains below levels that would represent a human health hazard (Bhat & Rukmini 1978; Sashidhar *et al.* 1992; Hodges *et al.* 2000). In one survey, it was found that there was no notable increase in prevalence of mycotoxins on stored grain after the monsoon season even though mould attack rose significantly during this period (Hodges *et al.* 2000). Fumonisin B1 was almost exclusively restricted to the *kharif* crop where it was found in all samples of stored grain, even before the monsoon. It seems likely that this mycotoxin is associated with pre-harvest mould damage. The picture given here, of relatively slight mycotoxin contamination, should not be taken to imply that there are no potential problems with mycotoxicosis as other researchers have reported significant contamination of *kharif* sorghum by aflatoxin

(Mall *et al.* 1986), *Alternaria* toxin (Anasari & Shrivastava 1990) and fumonisin (Bhat *et al.* 1997). In the case of fumonisin, a disease outbreak was reported in a few villages on the Deccan plain in households where rain-damaged mouldy grain was being consumed (Bhat *et al.* 1997). In the storage system investigated by Hodges *et al.* (2000), mycotoxin contamination rates may sometimes be higher in those years where weather conditions are less favourable or otherwise due to poor storage by individual farmers. Pre-harvest grain blackening of *kharif* grain therefore remains an issue of concern for marketing (due to the mould growth) and for reasons of health, due to mycotoxin contamination.

Overall, the grain storage practices of farmers in South India do not appear to be a constraint to the production and consumption of sorghum. Mycotoxin contamination and grain losses due to insect attack appear to remain low, although towards the end of their respective storage seasons the *kharif* and *rabi* crops show considerable quality decline. This seems not to be a significant problem as this decline was limited to only a small portion of the remaining stock. However, if farmers wish to retain stocks between seasons, to market grain strategically, then their current practices are likely to be inadequate. Farmers have developed strategies to keep these quality changes within acceptable limits and the changes that occur apparently do not influence farmers' choice of crop or variety. This reflects to a certain extent the fact that, for many households, the ability to store grain for periods of more than six months is constrained by production resources rather than post-harvest practice. To be more specific, those farmers who might be significantly affected by serious grain quality deterioration towards the end of the harvest are those without grain at this time.

Commodity and pest management systems

Prior to storage, the freshly threshed commodity is sun-dried for 15 days. Dried grain is cleaned to remove foreign matter and may be mixed with *Azadarichta indica* (neem) leaves and then loaded in store. Storage structures are placed on dunnage such as wooden planks, stones, polythene sheets and bamboo mats. Generally, insect infestation becomes apparent after four to five months in *kharif* grain and after six to seven months in *rabi* grain. Frequent sun-drying and cleaning by hand-picking and winnowing is used to control insect infestation. Farmers generally do not use synthetic pesticides.

Since 1970, the Government of India has supported a Save Grain Campaign (SGC) to educate farmers about good storage practice. The SGC extension workers demonstrate various insect and rodent control methods in villages. For prophylactic surface treatments against insects, a 0.5% emulsion of malathion is sprayed at the rate of 3 L per 100 m^2. Fumigation treatments are rarely if ever undertaken in farm storage since most stores are not sufficiently airtight for effective treatment and are in any case placed too close to dwellings for its safe use.

SGC also promotes a range of storage structures to suit the needs of farmers. Reinforced cement concrete ring bins, reinforced brick bins and metal bins have become popular outdoor storage structures. Indoor stores, called *pucca kothi,* created by building two extra walls in the corner of a house and providing a top-loading hatch and bottom-emptying port, have also been successful.

Rodents damage both the grain and the storage structure in the house. Non-chemical methods of controlling rodents include the use of different types of traps, environmental manipulation such as rodent proofing of the structures and maintaining high standards of sanitation and hygiene, and use of predators such as cats and dogs. However, rodents can be controlled effectively with poisons. In domestic situations the anticoagulant bromadiolone is recommended and is widely used in rat control operations. Rat burrows may be fumigated with phosphine. Farmers prefer to use poison baiting rather than trapping as this gives much quicker results.

Economics of storage

There have been few studies to compare the costs of storage at farm level using traditional and improved stores in India and none that focuses on sorghum in particular. However, the results of a long-term study of losses of farm-stored paddy rice demonstrated that the use of improved traditional basket stores and metal grain bins resulted in substantially lower levels of loss. Moreover, the social cost–benefit ratios for improved traditional stores and metal bins were 1.51 and 1.14, respectively (Boxall *et al.*1979). It was recognised that because of the physical characteristics of paddy rice, the levels of loss are likely to be lower than for other grains such as sorghum, which will be more susceptible to insect infestation. Consequently higher social cost–benefit ratios might be expected with this commodity.

Future developments

In Indian villages, many households have to rely for their food security on the government sponsored Public Distribution System that operates through fair-price shops. These offer relatively poor-quality rice or wheat at low prices to those people entitled to such provision. However, even in the medium term, the cost of public sector grain management to ensure food security is unsustainable. For this reason, central and state authorities have encouraged research into policy and practical guidelines for village-level food security. The advantages of decentralised grain storage are that it offers the poor access to food and it links poor farmers to the grain market, thus enhancing livelihoods.

One approach being tested is to develop grain banks with self-help groups in dryland areas of Andhra Pradesh where the cereal staple is sorghum. The self-help groups are formed with the help of an NGO that recognises that any initiative in community storage should be developed through a process in which demand is established and the way of working elaborated within the group. This is especially important in view of the past history of grain banks in other parts of the world. For example, the formation of cereal banks introduced in the Sahel in the 1970s were popular at first but by the 1990s, of the 4000 or so banks that had been constructed, very few were still functional (Günther & Mück 1995). It is not clear whether the causes of failure are inherent in the system or can be resolved by better management systems. Nevertheless, India has had successful community grain storage in various parts of the country but most of these appear to have fallen into disuse as better communications lead to greater reliance on markets.

Besides the formation of effective groups, another important element in the establishment of communal grain management is that the stock can be retained with minimal quality decline during the period that the group require it for their own food security. The grain will either be consumed or, if in excess, sold on the local market when prices have risen sufficiently. In the case under study, the sorghum crop is rain-fed and cultivated during the *kharif* season and would normally remain in store until the start of the next monsoon in June. This implies a maximum storage period of about nine months (Jayaraj *et al.* 2003). The method of storage adopted by the groups in question is a small concrete silo (13 t capacity) preferably with at least two compartments, light, well-fitting top hatch covers, well-sealed discharge pipes and walls painted inside and out with epoxy paint to prevent moisture ingress. The grain management strategy involves solarising grain by exposing a thin layer of grain sealed under a plastic sheet to six hours of sunshine. Normally farmers sun-dry their produce to a moisture content of 12% or less before bag storage. In the case of grain destined for silo storage, a period of solarisation follows sun-drying. By sealing the grain under a plastic sheet, high temperatures are achieved that kill any infesting insects. Once the grain has cooled it is placed in the silo. This is then well sealed to prevent the entry of insects or moisture. If these procedures are followed then the grain remains in good condition until discharge. Periodic inspection is required to ensure that grain continues in good condition, and between harvests silo compartments must be cleaned thoroughly. This commodity management approach is appropriate to local conditions as it:

- builds on existing practice – solarisation is an extension of grain drying which farmers know very well;
- requires minimal inputs, as plastic sheeting is easily and cheaply obtained locally; and
- can be implemented without any external support, in contrast to special treatments such as fumigation or the application of pesticide.

Ethiopia

Historical perspectives

Ethiopia is a food-deficit country, vulnerable to severe droughts, flooding and outbreaks of pests and diseases of crops and livestock. All seriously affect national food supply and are recognised as the main causes of widespread famines in the past. Famine results in dislocation of family life, widespread suffering and loss of life; people give up whatever meagre land holdings they have and migrate to other areas in search of food. Under these circumstances the government has to call for international assistance, including the provision of food aid. Mobilisation of external food aid takes time, during which many lives might be lost. However, the suffering of famine-affected people can be eased, especially during the early stages of an emergency, if they have access to strategically positioned food reserves.

The concept of an Emergency Food Security Reserve (EFSR) in Ethiopia was initiated in 1982. Following the famine of 1984/85, government, donors and others involved in the distribution of food aid recognised the value of establishing such a reserve to provide readily available grain stocks for use during future food emergencies.

Since its inception the size of the reserve steadily increased through the late 1980s and early 1990s from about 180 000 t to its current level of more than 450 000 t. Initially the reserve consisted entirely of grain imported as food aid. Wheat was the most common cereal supplied (approximately 88%) with maize and sorghum accounting for most of the balance. Since 1996, donor agencies have been providing food aid assistance both in kind and in cash, and as a consequence, significant local purchases of grain have been made. Of the locally produced grains, priority is given to the purchase of sorghum and maize since these crops are the main food staples of drought-prone areas and can be preserved for longer periods by the EFSR than by farmers or small-scale grain traders. Studies of post-harvest losses in Ethiopia have shown that farmers lose more than 10% of their sorghum and that they frequently dispose of much of their grain very soon after harvest in order to avoid damage by insects and mould.

The proportion of sorghum in the reserve is variable, but usually between 5% and 10% depending on availability in the local market and donor contributions. It is held in jute or polypropylene bags, mostly of 50 kg capacity. Other relief and development agencies with similar storage facilities to those of the EFSR hold larger proportions of locally produced sorghum but for much shorter periods, to service their ongoing programmes.

Physical facilities

The EFSR currently owns or rents warehouses with a total capacity in excess of 400 000 t at five strategic locations. Warehouses with capacities in the range 5000–10 000 t are purpose-built, steel portal frame structures, with steel cladding or concrete block walls (Fig. 5.9).

Fig. 5.9 Open-eaved bag stores used by the Ethiopian Food Security Reserve for the storage of bagged wheat and sorghum.

Warehouse sites are paved with adequate space to allow ready access by lorries.

Objectives of storage

The EFSR was established to meet emergency food needs for famine-affected groups in case of widespread crop failure, pending arrival of grain from abroad. Besides its role as a strategic reserve, the EFSR also acts as a buffer stock to help manage the food aid pipeline. Grain destined for the reserve may remain in store for up to two years. Hence, the success of the system can be judged on the ability of the reserve (1) to provide beneficiaries with grain of acceptable quality (i.e. with minimum qualitative deterioration) and (2) to meet demands for a calldown of stocks in times of emergency.

In Ethiopia, commercially traded sorghum is usually expected to contain no more than 12% of 'impurities' (foreign matter and defective grains). However, this is deemed to be too high for sorghum destined for the EFSR and consequently a quality specification (based on the Ethiopian national standard for sorghum grades 1 and 2) has been set to meet the special long-term needs of the reserve (Table 5.3).

The quality of food grains is checked both on entering and issuing from the reserve and, although there are minor deviations from time to time, broadly speaking grain at issue conforms to specification.

The reserve has essentially operated by providing a bridging stock for relief agencies that need food grains immediately for their ongoing projects, but which promise to repay that same quantity of grain once they have completed the import or local purchase several months later. Hence, the presence of the reserve has facilitated a prompt response to need.

Major sources of quality decline

All stored grain is vulnerable to damage and loss caused by insects, rodents and moulds, and grain in long-term storage requires special care and attention if such losses are to be minimised. The EFSR warehouses are well constructed and suitable for long-term storage and a system of grain storage management and quality control has been implemented to ensure that, as far as possible, grain can be stored safely for long periods.

As the EFSR has evolved, an increasing number of food aid donors and NGOs have made use of the bridging facility. They may route grain imports to the EFSR initially and draw down their programme requirements from reserve stocks. Alternatively they may fulfil their pledges to the reserve by supplying locally procured surpluses. Such a diverse range of supplies to the reserve implies a more complex and risk-prone demand on quality maintenance operations within the storage system. The potential for both qualitative and quantitative loss is increased and consequently a high degree of management sophistication is required.

Imported sorghum poses a particular risk. Consignments have sometimes contained high proportions of foreign matter and broken grains, which are conducive to insect pest development, and occasionally grain has arrived infested by storage insect pests such as *Sitophilus* spp which thrive under Ethiopian storage conditions. It would appear that less attention is paid to the quality of this grain than other cereals, probably because most sorghum produced outside Africa is usually destined for fairly rapid processing for the livestock sector.

Although the EFSR has drawn up its own quality specifications for food grains, these do not necessarily apply to food aid donations. Donor countries usually have their own specification for food aid grain. This is not to say that donors ignore their responsibility to supply food grains of a quality matched exactly to specific requirements of recipient countries. Rather, they may respond to annual appeals by the World Food Programme for example, to supply grain to the global food aid basket and sometimes at short notice in response to emergen-

Quality factor	Maximum % permitted	
	Grade 1	Grade 2
Moisture content	13.0	13.0
Impurities		
Damaged, shrunken and weevilled grain	4.5	7.5
Ergot, smut	0	0
Foreign matter (weed seeds, dust, dirt, chaff, straw, etc.)	2.5	3.0
Total impurities	7.0	10.5

Table 5.3 Ethiopian Emergency Food Security Reserve grading standards for sorghum.

cies. Under these circumstances attention to quality may come second to the quantity needed.

Insect pest problems have also been encountered in locally purchased sorghum, although local varieties appear to be less susceptible to attack. Storage pests flying out from infested farm stores may infest sorghum in the field just before harvest and such low levels of infestation may be missed or ignored, first by farmers intent on selling their grain and then by traders with deadlines for procurement targets to be met. Stricter quality control is possible with local grains since consignments that do not conform to specification can be refused.

Although the design of EFSR warehouses aims to limit access by rodents, some inevitably gain entry and damage the stored grain. However, it is generally accepted that loss to rodents does not exceed 0.5% annually. Routine fumigation of grain in long-term storage, though primarily designed for control of insects, is also effective in controlling rodents within stacks.

Mould damage to grain is rarely a problem. Occasionally, locally procured sorghum with high moisture content may be offered to the reserve but such consignments are automatically turned away, although suppliers are permitted to redeliver grain after re-drying.

Commodity and pest management regimes

The dedicated Quality Control and Maintenance Team of the EFSR is responsible for ensuring a high standard of grain quality and storage pest control. Instructions for warehouse maintenance, cleaning, inspection and monitoring of stocks and pest control are set out in a *Procedures manual.*

A programme of planned maintenance is carried out. The warehouses are inspected regularly and structural repairs completed as quickly as possible. Special attention is paid to identifying and dealing with possible entry points for rodents and birds or gaps where pests or rainwater may enter. Vegetation around the store that may harbour rodents is regularly cut back.

The importance of store hygiene is recognised as an important and essential step in pest prevention and so a team of cleaners is employed at each site. They follow a schedule to ensure that all parts of the store are routinely cleaned to remove dust and residues from floors, walls, ledges, girders and stack surfaces. Walls and floors of the warehouse are sprayed with a residual insecticide (pirimiphos-methyl) before grain stocks are received.

No dunnage is used, but before receiving grain into store, a waterproof sheet of heavy gauge polythene is

laid on the floor to protect bag stacks against possible moisture uptake from the floor.

Incoming grain is checked to ensure that it conforms to specification and is in a fit state to store. First, a sample drawn from accessible bags on the truck is inspected and the moisture content checked. If the grain quality appears satisfactory, the truck is allowed to unload. Consignments in which insect infestation is discovered may be rejected or may be unloaded in a separate warehouse to await fumigation in order to reduce the risk of cross-infestation of stocks.

Consignments accepted on the basis of the preliminary inspection will be stacked together (Fig. 5.10) but are subjected to more intensive and representative sampling during unloading. Samples are analysed for quality and moisture content and a cumulative quality profile for each stack is calculated. This is achieved by relating the moisture content and quality values from samples to the weight of the specific consignments from which they are drawn, rather than taking a simple average of all sample results. Similar checks of quality and moisture content are made at the time the grain is issued from store and the quality profile of the remaining grain adjusted. These quality and moisture records related to specific consignments of grain entering and leaving the warehouse help to identify and explain any losses, especially those due to natural drying of grain over extended storage periods.

Once a warehouse has been filled all the grain is fumigated under gas-proof sheets as a matter of routine. Stocks are inspected at least quarterly and usually more frequently with the objective of detecting early signs of infestation. Grain stocks are fumigated under gas-tight

Fig. 5.10 Stacks of jute bags holding sorghum being built on waterproof sheets in the Ethiopian Food Security Reserve.

sheets using the fumigant phosphine and the warehouse fabric is sprayed with insecticide as necessary.

This highly efficient quality monitoring and pest management regime ensures that storage losses are maintained well below a commercially significant threshold.

Economics of operation of the system

The operation of a food security reserve carries a relatively high cost (Table 5.4). It is estimated that the total invested in the EFSR by donors and the Ethiopian Government has been around $US15m in warehouses, $1.5m in offices, equipment and vehicles and about $25m in grain stocks. At stocking levels and turnover for 1999, the fixed investment in the reserve is estimated to be around $400 t^{-1}.

Most operating, administrative and recurrent costs of the reserve are fixed and in the short term tend not to vary with either the stock held or the turnover of stock. Amortisation of warehouses and equipment amounts to around $2.2m a year and other fixed recurrent costs are around $0.5m. When total recurrent costs are expressed as a cost per unit of reserve stock held (averaging 130 000 t in July 1999) they come to about $21 t^{-1}. However, if costs were calculated on the total capacity of the warehouses the costs per tonne would more than halve. Costs that vary directly with the quantity stored (pest control, materials and labour) are generally low at less than $1.0 t^{-1}.

To these costs must be added:

(1) warehouse maintenance costs, estimated at around 1% of capital costs or $0.2m per annum;
(2) the notional or opportunity cost of interest on grain stocks at 12% per tonne of grain stored, $30 per annum; and
(3) the cost of weight losses in storage (pests, drying, spillage) at an upper limit of around 2% a year (2000 t overall) equivalent to $0.56m (or $6 t^{-1} of grain stored).

Future developments

No major changes in the status of the EFSR are anticipated in the immediate future. However, the reserve is monitored regularly to ensure an adequate stockholding, given that the lead times for repayment of loans have been sometimes longer than previously expected. The level of locally procured grains including sorghum is expected to increase as donors switch to provision of cash rather than donations in kind. It is expected that the reserve will be maintained for at least the next ten years; however, during this period the rationale for maintaining the reserve and the options for changing its structure will have to be addressed.

References

Abdalla, A.T., Stigter, C.J., Bakhiet, N.G., Gough, M.C., Mohamed, H.A, Mohammed, A.E & Ahmed, M.A. (2002) Traditional underground grain storage in clay soils in Sudan improved by recent innovations. *Tropicultura*, **20**(4), 170–175.

Anasari, A.A. & Shrivastava, A.K. (1990) Natural occurrence of *Alternaria* mycotoxins in sorghum and ragi from North Bihar, India. *Food Additives and Contaminants*, **7**(6), 815–820.

Annis, P.C. (1987) Towards rational controlled atmosphere dosage schedules: a review of current knowledge. In: E. Donahaye & S. Navarro (eds) *Proceedings of the 4th International Working Conference on Stored-Product Protection*, 21–26 September 1986, Tel Aviv, Israel.

Anon. (1984) *Handbook of Agriculture*. Indian Council of Agricultural Research, New Delhi, India.

Bakhiet, N.G., Stigter, K. & Ahmed el-Tayeb Abdalla (2001) Underground storage of sorghum as a banking alternative. *LEISA – Magazine on Low External Input and Sustainable Agriculture*, **17**(1), 13.

Bartali, El H. (1994) Amélioration du système de stockage souterrain des céréales. Quoted in: *Synthesis. African Experiences in the Improvement of Post-Harvest Techniques*, 4–8 July 1994. Agricultural Engineering Service, Agricultural Support Systems Division, Food and Agriculture Organization of the United Nations, Rome, Italy.

Belmain, S.R. (2002) Pest Management – Botanicals. In: P. Golob, G. Farrell & J.E. Orchard (eds) *Crop Post-Harvest: Science and Technology. Volume 1 Principles and Practice.* Blackwell Science, Oxford, UK.

Bhat, R.V. & Rukmini, B. (1978) Mycotoxins in sorghum: toxicogenic fungi during storage and natural occurrence of T2 toxin. *Proceedings of the International Workshop on Sorghum Diseases*, 1–15 December 1978, Hyderabad. ICRISAT, Patancheru, Andra Pradesh, India.

Bhat, R.V., Shetty, P.H., Amruth, R.P. & Sudershan, R.V. (1997) A foodborne disease outbreak due to the consumption of mouldy sorghum and maize containing fumonisin mycotoxins. *Journal of Toxicology & Clinical Toxicology*, **35**(3), 249–255.

Boxall, R.A. (1973) Ferrocement-lined underground grain silos in Ethiopia. In: *Ferrocement: Applications in Developing Countries.* Report of the National Academy of Sciences, Washington DC.

Boxall, R.A. (1974a) Underground storage of grain in Harar province, Ethiopia. *Tropical Stored Products Information*, **28**, 39–48.

Boxall, R.A. (1974b) Improvement of traditional grain storage pits in Harar province, Ethiopia; a preliminary investigation. *International Pest Control*, **16**(5), 4–7.

Boxall, R.A., Greeley, M. & Tyagi, D.S. (1979) The prevention of farm level food grain storage losses in India: a social cost

Table 5.4 Summary of costs of grain storage by the Ethiopian Emergency Food Security Reserve (1998/99 figures). Source: WAAS International: Draft report on EFSRA costs June 1999 (unpublished).

Grain costs	Unit	$	
CIF cost	tonne	212.50	
Inland transport	tonne	46.50	
Port handling and storage	tonne	18.00	
Unloading and loading	tonne	—	
Total			

Fixed costs		$	$
Warehouse capital cost:	*Tonnes*	*Cap. cost*	*Annual cost*
Nazareth and Sheshemane	20 000	1 401 577	187 641
Mekele	45 000	3 153 548	422 193
Kombolcha	60 000	4 204 731	562 924
Shinile	25 000	1 751 971	234 552
Annuity Nzth, Shesh, Komb			1 407 311
Woreta and Wolaiyta Sodo	79 500	3 872 750	518 479
Total warehouse building costs	229 500	14 384 577	1 925 790
Office buildings		195 061	26 115
Pest control equipment		248 707	70 470
Analysis equipment		65 845	11 654
Warehouse equipment		108 853	19 265
Bag conveyors		80 533	13 827
Vehicles		399 064	110 704
Office furniture and equipment		229 220	40 568
Total building and equipment costs		15 711 860	2 218 393
Approx unit cost/average stock	130 000	121	
Approx unit cost including grain	130 000	398	
Recurrent fixed costs:			
Equipment and maintenance			16 975
Grain/buildings insurance			73 408
Warehouse salaries			93 079
Utilities			10 764
Warehouse rent 70 000 tonnes			114 750
Head office salaries			97 674
Office supplies			25 534
Office and vehicle maintenance			36 157
Utilities (office)			9 584
Office rent			8 923
Training			37 125
Total recurrent fixed costs			523 973
Total fixed costs			*2 742 366*

Variable costs			$/tonne
Fumigation			0.31
Insecticide			0.19
Labour – spraying		174	0.04
Total cost per tonne			0.54

Total cost	Tonnes	$	$/tonne
at	100 000	2 796 366	27.96
	120 000	2 807 166	23.39
	150 000	2 823 366	18.82
	300 000	2 904 366	9.68

benefit analysis. *Tropical Stored Products Information,* **37,** 11–17.

Burkhill, I.H. (1952) Habits of man and origins of the cultivated plants of the world. *Proceedings of the Linnean Society, London,* **164,** 12–42.

Burrell, N.J. (1980) Effects of airtight storage on insect pest of stored products. In: J. Shejbal (ed.) *Controlled Atmosphere Storage of Grains.* Elsevier Scientific Publishing, Amsterdam, The Netherlands.

Darling, H.S. (1959) The insect pests of stored grain in the Sudan with special reference to the factors affecting the incidence of infestations. PhD Thesis, University of London, UK.

Gilman, G.A. & Boxall, R.A. (1974) The storage of food grain in traditional underground pits. *Tropical Stored Products Information,* **28,** 19–38.

Greig, D.J. (1970) Informal technical report on grain storage. Highlands Farm Development Project, Yemen. AGS SF/YEM9. Food and Agriculture Organization of the United Nations, Rome, Italy.

Günther, D. & Mück, O. (1995) Les banques de céréals ont-elles banqueroute? Perspectives et limites d'un modele villageois de securite alimentaire au Sahel. GTZ, Eschborn, Germany.

Haines, C.P. (ed.) (1991) *Insects and Arachnids of Tropical Stored Products: their biology and identification.* 2nd edition. Natural Resources Institute, Chatham, UK.

Hodges, R.J., Hall, A.J., Jayaraj, K., Jaiswal, P., Potdar, N. & Yogand, B. (2000) Quality changes in farm-stored sorghum grain grown in the wet or dry season in southern India: a technical and social study. In: G.I. Johnson, Le Van To, Nguyen Duy Duc & M.C. Webb (eds) *Proceedings of the 19th ASEAN/1st APEC Seminar on Postharvest Technology,* 9–12 November 1999, Ho Chi Minh City, Vietnam. ACIAR Proceedings **100.** Australian Centre for International Agricultural Research, Canberra, Australia.

Jayaraj, K., Reddy, T., Adolph, B. & Hodges, R.J. (2003) Management of community grain stocks in dryland areas of Andhra Pradesh, India. In: P.F. Credland, D.M. Armitage, C.H. Bell, P.M. Cogan and E. Highley (eds) *Proceedings of the 8ʰ International Working Conference on Stored Products Protection,* York, UK, 22–26 July 2002. CABI Publishing, Wallingford, UK.

Kent, N.L. (1984) *Technology of Cereals.* 3rd edition. Pergamon Press, Oxford, UK.

Kishore, P., Jotwani, M.G. & Sharma, G.C. (1977) Relative susceptibility of released high-yielding varieties and hybrids of sorghum to insect attack in storage. *Entomologists' Newsletter,* **7,** 14–15.

Lemessa, F. (1990) Unterresuchbgen zur Verminderung von Veluste bei Lagerung von Sorghum in Pits in Osoromia, Aethiopiaen. Dissertation, Technical University of Berlin, Germany.

Lynch, B., Reicher, M., Solomon, B. & Kuhne, W. (1986) *Examination and improvement of underground pits in Alemaya Wereda.* Report of the Alemaya University of Agriculture, Ethiopia.

Mall, O.P., Pateria, H.M. & Chauhan, S.K. (1986) Mycoflora and aflatoxin in wet harvested sorghum. *Indian Phytopathology,* **39**(3), 409–413.

Narayana, D. (1989) Jowar. In: J. Raghotham Reddy (ed.) *Agriculture in Andhra Pradesh* (Volume II). Society of Scientists for Advancement of Agriculture, Hyderabad, Andhra Pradesh, India.

Nash, M.J. (1985) *Crop Conservation and Storage in Cool Temperate Climates.* 2nd edition. Pergamon Press, Oxford, UK.

Niles, E.V. (1976) The mycoflora of sorghum stored in underground pits in Ethiopia. *Tropical Science,* **18**(2), 115–124.

Oxley, T.A. & Wickenden, G. (1963) The effect of restricted air supply on some insects which infest grain. *Annals of Applied Biology,* **51,** 313–324.

Peacock, J.M. & Wilson, G.L. (1984) Jowar. In: P.R. Goldsworthy & M.M. Fisher (eds) *The Physiology of Tropical Field Crops.* John Wiley & Sons, Brisbane, Australia.

Reddy, B.N. & Nusrath, M. (1988) Relationship between the incidence of storage pests and production of mycotoxin in jowar. *National Academy Science Letters,* **11**(10), 307–308.

Sashidhar, R.B., Ramakrishna, N. & Bhat, R.V. (1992) Moulds and mycotoxins in sorghum stored in traditional containers in India. *Journal of Stored Products Research,* **28**(4), 257–260.

Sigaut, F. (1980) Significance of underground storage in traditional systems of grain production. In: J. Shejbal (ed.) *Controlled Atmosphere Storage of Grains.* Elsevier Scientific Publishing, Amsterdam, The Netherlands.

Sterling, R.L., Meixel, G.D., Dunkel, F. & Fairhurst, C. (1983) Underground storage of food. *Underground Space,* **7,** 257–262.

Vinall, H.N., Stephens, J.C. & Martin, J.H. (1936) *Identification, history and distribution of common sorghum varieties.* USDA Technical Bulletin **506.**

Waas International. Draft report on EFSRA costs June 1999 (unpublished).

Wareing, P.W. (2002) Pests of durable crops – moulds. In: P. Golob, G. Farrell & J.E. Orchard (eds) *Crop Post-Harvest: Science and Technology. Volume 1 Principles and Practice.* Blackwell Science, Oxford, UK.

Wayne-Smith, C. & Frederiksen, R. (eds) (2002) *Sorghum: origin, history, technology and production.* John Wiley & Sons, Inc. New York.

Chapter 6
Common Beans: Latin America

C. Cardona

The common or dry bean (*Phaseolus vulgaris*) (Plate 6) originated in South America or Mexico and is now grown widely in tropical and subtropical regions around the world. Latin America is the largest bean producer in the world. A high proportion of beans in the region are consumed as a staple on the farm or traded in local markets. As a consequence, storage periods tend to be short (usually six months or much less). Bean stores in Latin America are normally small-scale, on-farm facilities of rather simple construction, or large-capacity sophisticated stores. Smallholder farmers usually keep beans for consumption or for use as seed. In a few cases they market beans strategically to take best advantage of prices. Government or privately owned warehouses usually store beans on behalf of marketing boards or for food security.

For proper storage the seed moisture content must be less than 15% (preferably 13%), and beans must be stored in conditions of less than 75% relative humidity. Failure to do so may result in discoloration, hard seed coat and hard-to-cook defects that seriously affect bean quality. Bruchid beetles, and to a lesser extent moulds, are the main causes of bean deterioration in storage. Large-scale operators usually reduce insect infestations by using insecticides and fumigants, whereas smallholders normally do not apply pest control. Most farmers try to sell or consume the produce as soon as possible in order to avoid problems with bruchid beetles. There is a need for an effective extension campaign aimed at reaching farmers with improved technologies, based on an understanding of the social and economic factors that prevent adoption of new approaches.

Historical perspectives

The common or dry bean (*Phaseolus vulgaris*) is the most important food legume for about 300 million people, most of them in the developing world. It is produced primarily in tropical low-income countries of Latin America, Africa and Asia which together account for over three-quarters of the annual world production of 18.8m t. Beans provide an inexpensive food for poor consumers, and rank fourth in Latin America as a source of protein. The crop is particularly important in the diets of women and children (CIAT 2001).

Apart from their contribution to human nutrition, beans have a large economic importance with global production valued at about $11bn annually (CIAT 2001). Latin America is the leading bean producer in the world. It contributes more than 28% of the total world production with an annual output of about 5.4m t (Pachico 1993; FAOStat 2000). In Latin America the area of beans planted has expanded by only 2% over the last decade whereas bean production has increased by 25% as a result of increasing yields. The annual rate of increase in production (2.7%) exceeds the rate of population growth. The result has been an increase in per capita consumption (CIAT 2001). A high proportion of beans in Latin America is consumed as a staple on the farm or traded in local markets. Income generation from bean production is becoming more important in Brazil, Mexico, Central America and the Andean zone. In the Southern Cone (Argentina and Chile) beans are produced mainly for export (Pachico 1993).

In general beans have a long storage life and can be easily stored on-farm; however, storage periods tend to be short – usually six months or less. This is because beans are a food staple in most of Latin America and as growing seasons overlap supplies tend to be more or less continuous. For these reasons storage facilities are generally not sophisticated and have remained unchanged for at least 30 years.

Physical facilities

Bean stores in Latin America are normally either small-scale, on-farm facilities of rather simple construction, or large-capacity regional stores of sophisticated construction. Farmers may store beans unthreshed (Fig. 6.1) until the opportunity for threshing arises, after which the beans may be stored in baskets, pots, drums, bins or sacks (Fig. 6.2), most often in their own households or in makeshift storage huts (Giraldo *et al.* 2000). Large, usually government-owned, sophisticated regional stores are common throughout Latin America (Figs 6.3 and 6.4). These warehouses hold beans in bags, have steel portal frame structures, brick or concrete block walls and are, quite often, provided with sophisticated temperature and humidity control devices and fumigation facilities. Cereals, pulses, oil seeds and other commodities are stored in these facilities for periods of one year or more with the purpose of regulating markets.

Fig. 6.3 Typical warehouse in South America, with steel portal frame and brick walls, used for storing beans in bags.

Fig. 6.4 Inside a bean warehouse showing bag stacks on wooden palettes and bag stacking equipment (left).

Fig. 6.1 Unthreshed beans stored in a South American farmhouse, awaiting threshing.

Fig. 6.2 Threshed beans stored in sacks in a South American farmhouse.

Objectives of storage

Smallholder farmers keep beans for consumption and for seed; most sell any surplus seed as soon as possible because they need cash and because they want to avoid damage by bruchid beetles (van Schoonhoven 1976). In a few cases farmers market beans strategically to take best advantage of prices. Government or privately owned warehouses usually store beans on behalf of marketing boards or to ensure national food security (Giraldo *et al.* 2000).

Good storage conditions are vital for maximising the survival of high-quality beans for long periods and for minimising storage losses induced by seed-borne fungal saprophytes and pathogens, such as *Aspergillus* or *Penicillium* spp (Schwartz & Morales 1989). Bean moisture content must be less than 15% (preferably 13%) and seed

Plate 1 Rice grain (*Oryzae sativa*) as seen at various stages of processing: paddy rice (left), brown rice (middle) and white rice (right). The *indica* grain illustrated is not all of the same variety. White milled rice consists of the grain endosperm after removal of husk, germ and bran.

Plate 2 Maize cobs (*Zea mais*) stored on the farm with the sheathing leaves removed (dehusked); many farmers may store the cobs undehusked.

Plate 3 Wheat grains (*Triticum* spp).

Plate 4 Barley grains (*Hordeum vulgare*).

Plate 5 Freshly harvested sorghum grain (*Sorghum bicolor*) in good condition (left), infested, discoloured sorghum after eight months of inadequate farm storage (right).

Plate 6 Common or dry beans (*Phaseolus vulgaris*).

Plate 7 Cowpeas or blackeyed beans (*Vigna unguiculata*).

Plate 8 Bulk canola seed (*Brassica napus* or *B. rapa*) in storage. The seed is small (usually about 1.5 mm diameter) and depending on variety 1000 seeds may weigh only 2.5–5.0 g.

Plate 9 Peanut pods and kernels (*Arachis hyogaea*).

Plate 10 Dried coconut kernel (*Cocos nucifera*) – copra.

Plate 11 Green robusta coffee beans (*Coffea canephora*).

Plate 12 Cocoa beans (*Theobroma cacao*).

Plate 13 A selection of dried fruit and nuts produced in the US. Top: walnuts; then clockwise dates, raisins, pecans, almonds, figs, hazlenuts and, in the centre, pistachio.

Plate 14 Dried smoked fish on sale in an African market.

must be stored in conditions of less than 75% relative humidity (López & Christensen 1962). As with many other commodities, temperatures lower than 10°C will extend the viability of seed beans (López & Crispín-Medina 1971).

High temperatures and humidities in storage result in beans of bad flavour and with defective cooked-grain texture (Giraldo *et al.* 2000), and are associated with two important phenomena: 'hard seed coat' and 'hard-to-cook'. Hard seed coat is a condition of seed dormancy where the seed coat becomes impermeable to water (Shellie-Dessert & Bliss 1991). This in turn prolongs cooking time and germination period. Although the development of a hard seed coat is controlled partly genetically (Rolston 1978), it is generally accepted that the incidence of this phenomenon is subject to environmental conditions and increases when beans are stored at high temperatures and low relative humidities. Hard seed coats can be softened using various treatments such as an acid soak, scarification, radiation and blanching (Rolston 1978).

In contrast to hard seed coat, the hard-to-cook defect is irreversible and develops during storage under high temperatures (above 21°C) and relative humidities. Seeds with the hard-to-cook defect absorb water but do not soften sufficiently during cooking (Jackson & Varriano-Marston 1981). Storage under high relative humidity is one of the key factors in initiating the hard-to-cook defect (Jones & Boulter 1983). Beans below 10% moisture content at 25°C did not show significant hard-to-cook defect after two years, whereas six months' storage above 13% moisture content at 25°C resulted in reduced cookability (Morris & Wood 1956). In addition, high moisture beans also darken more rapidly during storage; this leads to the beans being rejected by consumers in many parts of Latin America.

Resistance to mechanical damage during harvest and canning may be important in some countries like Argentina and Chile, which are major exporters of beans to the developed world (Gargiulo & Pachico 1986).

Major sources of quality decline

Apart from the hard coat and hard-to-cook defects, insects and moulds, and to a lesser extent rodents, are the main problems affecting stored beans in Latin America. A total of 28 insect species have been reported on stored beans (van Schoonhoven 1976) although most are of minor importance or only accidentally found on beans. By far the most important pests of stored beans are the

Fig. 6.5 Bean, *Phaseolus vulgaris*, damaged during storage; the mottled beans show fungal damage and all beans show the emergence holes of the bruchid beetles, *Acanthoscelides obtectus* or *Zabrotes subfasciatus*.

Mexican bean weevil, *Zabrotes subfasciatus*, and the bean weevil, *Acanthoscelides obtectus* (Cardona 1989; Vol. 1 p. 99, Hodges 2002). The larvae develop hidden within the beans and the adults emerge from the beans leaving large holes (Fig. 6.5). Literature on the economic importance of bruchids is scarce. It has been estimated that in Mexico and Central America storage losses may be as high as 35% although it is not known what proportion of this loss is attributable to insects (McGuire & Crandall 1967). It is said that bruchids are responsible for storage losses of 13% in Brazil, 7.4% in Colombia and 12% in Nicaragua (van Schoonhoven 1976), whereas in eastern Colombia losses were reported to be only 4.6% (Arias 1995) since storage periods were short, averaging only 44 days.

Several species of *Aspergillus* fungi are listed as sources of aflatoxin contamination of beans (Schwartz & Morales 1989). However, the number of stores holding aflatoxin-contaminated beans is usually low (< 1.5%) (Pinto *et al.* 1986; Soares & Rodriguez-Amaya 1989; Cárcamo *et al.* 1993; Muhlemann *et al.* 1997). The fungi *Penicillium* spp and *Botryodiplodia theobromae* have also been recorded as contaminants of stored beans in Latin America (López & Crispín-Medina 1971; Ellis *et al.* 1977).

Commodity and pest management regimes

Post-harvest operations for beans in Latin America follow a chain from on-farm to off-farm (Fig. 6.6). It is becoming more usual for large bean crops to be har-

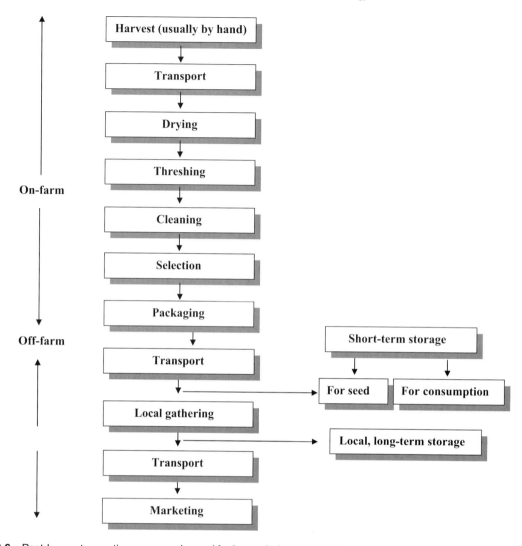

Fig. 6.6 Post-harvest operations commonly used for beans in Latin America.

vested mechanically but most smallholder crops are still harvested by hand. Smallholders usually dry their beans on patios, roads or wooden platforms, or sometimes in ceiling spaces. Co-operatives, large producers and government institutions use grain dryers. Mechanical threshing is common but most farmers still place dry plants on platforms with slatted floors and shell by beating the material with poles and collecting the beans underneath (O'Kelly & Forster 1983). In Ecuador, Honduras, El Salvador and other countries, it is common to place dry plants on roads so that the beans are run over by vehicles. Beans are usually winnowed by hand to remove unwanted material.

Smallholder farmers usually do not apply insecticides, although some may fumigate their beans in metal drums using carbon disulphide or phosphine. Most will try to sell or consume the produce as soon as possible in order to avoid bruchid problems. Some farmers use vegetable oils (van Schoonhoven 1978; Hall & Harman 1991) or mix the seeds with ash, sand, lime, silica or other inert materials (van Huis 1991; Vol. 1 p. 271, Stathers 2002).

Large-scale operators recognise the importance of store hygiene as a means of preventing insect problems. Usually beans are inspected at commodity intake and then submitted to periodic inspections to detect bruchids or other insects. Infestation problems are usually reduced successfully through a programme of fumigation with phosphine under gas-proof sheeting and fabric spraying with malathion or phoxim (Giraldo *et al.* 2000).

Other approaches to the control of storage pests are possible. For example, high levels of genetic resistance to the Mexican bean weevil (*Z. subfasciatus*) have been identified (Cardona *et al.* 1990) and incorporated in commercial bean lines (Cardona & Kornegay 1999). However, this resistance has not reached farmers in Latin America in the form of weevil-resistant cultivars because the cultivars continue to lack resistance to attack by pre-harvest fungi and viruses. Breeding for resistance to these pathogens is ongoing. The lack of host-plant resistance is unfortunate because savings in insecticide costs to smallholder farmers could reach hundreds of millions of US dollars annually.

Future developments

The demand for beans will continue to increase in Latin America, but the crop does not receive as much research attention as coffee, cotton, cut flowers, soyabeans and other major export commodities. Thus in 2001, the Consultative Group on International Agricultural Research (CGIAR) spent approximately $11.1m on bean research, representing about 3.8% of the total CGIAR commodity investment (CGIAR 2001). The relatively limited resources allocated may be due to the fact that beans are largely a subsistence crop. Nevertheless, major storage problems (seed discoloration, hard coat seed, hard-to-cook defect, bruchids and pathogens) need to be addressed in order to avoid storage losses. Much would be gained if available technological solutions to storage problems could be extended to farmers in the region and if more appropriate storage facilities could be made available to farmers. This will require an understanding of social and economic factors that prevent adoption of new technologies, and seeking a means of extending the technologies to millions of small, resource-poor farmers. In areas where the Mexican bean weevil is prevalent, deployment of existing bruchid-resistant bean varieties would also help in reducing storage losses in tropical Latin America. The lines developed so far have the seed size and seed colour required by growers and consumers, but unfortunately still lack resistance to the major bean pre-harvest diseases.

References

Arias, J.H. (1995) Evaluación de pérdidas postcosecha de frijol y papa en el Carmen de Viboral. *Actualidades Corpoica (Colombia)*, **9**, 9–14.

Cárcamo, R., Pinel, L. & de Malo, V. *et al.* (1993) Caracterización del sistema postcosecha en frijol común a nivel del pequeño agricultor y del intermediario, en el Municipio de Moroceli, Honduras. In: Memorias XXXIX Reunión Anual del PCCMCA, 175–177. Guatemala City, Guatemala.

Cardona, C. (1989) Insects and other invertebrate bean pests in Latin America. In: H.F. Schwartz & M.A. Pastor-Corrales (eds) *Bean Production Problems in the Tropics*. 2nd edition. 505–570. Centro Internacional de Agricultura Tropical, Cali, Colombia.

Cardona, C. & Kornegay, J. (1999) Bean germplasm resources for insect resistance. In: S.L. Clements & S.S. Quisenberry (eds) *Global Plant Genetic Resources for Insect Resistant Crops*. 85–99. CRC Press, Boca Raton FL.

Cardona, C., Kornegay, J., Posso, C.E., Morales, F. & Ramirez, H. (1990) Comparative value of four arcelin variants in the development of dry bean lines resistant to the Mexican bean weevil. *Entomologia Experimentalis et Applicata*, **56**, 187–206.

CGIAR (2001) CGIAR Research; Areas of Research; Beans (*Phaseolus vulgaris*). Consultative Group on International Agricultural Research, Washington DC, USA. http://www.cgiar.org/research/res_beans.html.

CIAT (2001) Improved beans for Africa and Latin America. Centro Internacional de Agricultura Tropical, Cali, Colombia. http://www.ciat.cgiar.org/about_ciat/beans.htm.

Ellis, M.A., Gálvez, G.E. & Sinclair, J.B. (1977) Efecto del tratamiento de semillas de frijol (*Phaseolus vulgaris*) de buena y mala calidad sobre la germinación en condiciones de campo. *Turrialba*, **27**, 37–39.

FAOStat (2000) Production Statistics. United Nations Food and Agriculture Organization, Rome, Italy. http://apps.fao.org.

Gargiulo, C.A. & Pachico, D. (1986) Análisis descriptivo del sector porotero del noroeste Argentino. *Publicación Miscelánea No. 80*. Estación Experimental Agro Industrial Obispo Colombres, Tucumán, Argentina.

Giraldo, G., Méndez, M. & Franco, J. (2000) *Manual para el Manejo Pre y Poscosecha de Semilla Producida de Manera Artesanal bajo el Esquema de Pequeñas Empresas de Semillas*. Centro Internacional de Agricultura Tropical, Cali, Colombia.

Hall, J.S. & Harman, G.E. (1991) Efficacy of oil treatments of legume seeds for control of *Aspergillus* and *Zabrotes*. *Crop Protection*, **10**, 315–319.

Hodges, R.J. (2002) Pests of durable crops – insects and arachnids. In: P. Golob, G. Farrell & J.E. Orchard (eds) *Crop Postharvest: Science and Technology. Volume 1 Principles and Practice*. Blackwell Science, Oxford, UK.

Jackson, M.G. & Varriano-Marston, E. (1981) Hard-to-cook phenomenon in beans: effects of accelerated storage on water absorption and cooking time. *Journal of Food Science*, **46**, 799–803.

Jones, P.M.B. & Boulter, D. (1983) The cause of reduced cooking rate in *Phaseolus vulgaris* following adverse storage conditions. *Journal of Food Science*, **48**, 623–626.

López, L.C. & Christensen, C.M. (1962) Efectos del ataque de hongos en el frijol almacenado. *Agricultura Técnica en México*, **2**, 33–37.

López, L.C. & Crispín-Medina, A. (1971) Resistencia varietal del grano de frijol almacenado al ataque por hongos. *Agricultura Técnica en México*, **3**, 67–69.

McGuire, J.U. & Crandall, B.S. (1967) *Survey of Insect Pests and Plant Diseases of Selected Food Crops of Mexico, Central America and Panama.* 123–157. International Agricultural Development Service, Agricultural Research Service, United States Department of Agriculture and Agency for International Development, Washington DC, USA.

Morris, H. J. & Wood, E. (1956) Influence of moisture content on keeping quality of dry beans. *Food Technology*, **10**, 225–229.

Muhlemann, M., Luthy, J. & Hubner, P. (1997) Mycotoxin contamination of food in Ecuador. *Mitteilungen aus dem Gebiete der Lebensmitteluntersuchung und Hygiene*, **88**, 474–496.

O'Kelly, E. & Forster, R.H. (1983) *Processing and Storage of Food Grains by Rural Families.* FAO Agricultural Services Bulletin No. 53.

Pachico, D. (1993) The demand for bean technology. In: G. Henry (ed.) Trends in CIAT Commodities 1993. *Working Document No. 128.* Centro Internacional de Agricultura Tropical, Cali, Colombia, 60–74.

Pinto, R., Mateus, S. & Montenegro, O. *et al.* (1986) Evaluación de prácticas y pérdidas postcosecha de frijol en una zona productora. *Revista Acogranos (Colombia)*, **4**, 198–202.

Rolston, M.P. (1978) Water impermeable seed dormancy. *Botanical Review*, **44**, 365–396.

Schwartz, H.F. & Morales, F.J. (1989) Seed pathology. In: H.F. Schwartz & M.A. Pastor-Corrales (eds) *Bean Production Problems in the Tropics.* 2nd edition. 413–431. Centro Internacional de Agricultura Tropical, Cali, Colombia.

Shellie-Dessert, K.C. & Bliss, F.A. (1991) Genetic improvement of food quality factors. In: A. van Schoonhoven & O. Voysest (eds) *Common Beans: Research for Crop Improvement.* 649–677. CAB International, Wallingford, UK.

Soares, L.M.V. & Rodríguez-Amaya, D.B. (1989) Survey of aflatoxins, ochratoxin A, zearalenone, and sterigmatocystin in some Brazilian foods by using multi-toxin thin-layer chromatographic method. *Journal of the Association of Official Analytical Chemists*, **72**, 22–26.

Stathers, T.E. (2002) Pest management – Inert dusts. In: P. Golob, G. Farrell & J.E. Orchard (eds) *Crop Post-Harvest: Science and Technology. Volume 1 Principles and Practice.* Blackwell Science, Oxford, UK.

van Huis, A. (1991) Biological methods of bruchid control in the Tropics: a review. *Insect Science and its Application*, **12**, 87–102.

van Schoonhoven, A. (1976) Pests of stored beans and their economic importance in Latin America. In: *Proceedings of the Symposium on Tropical Stored Product Entomology*, 15th International Congress of Entomology, 19–27 August 1976, Washington DC, 691–698. Entomological Society of America, College Park MD, USA.

van Schoonhoven, A. (1978) Use of vegetable oils to protect stored beans from bruchid attack. *Journal of Economic Entomology*, **71**, 254–256.

Chapter 7

Cowpea: United States of America

G. N. Mbata

Cowpeas (*Vigna unguiculata*) are believed to have originated in India and have been spread by man to other parts of Asia and Africa and then in the seventeenth century, by way of the West Indies, to the Americas. They are now grown in warm and tropical climates around the world and about 80 000 ha of the crop is cultivated in the southern parts of the USA, particularly in California, Georgia and Texas (Fery 1985). There is considerable variation in the appearance of cowpea according to variety; one of the most popular varieties is the blackeyed bean (Plate 7). About 60 000 t of dry cowpeas are produced annually in the USA, of which 80% originate from the state of California. They are consumed locally and also exported to over 40 countries.

Cowpeas are moved from farm to warehouse. Scheduled fumigations with methyl bromide or phosphine protect the commodity from insect infestation. There is a zero tolerance of insects on cowpea; this results in stocks being fumigated at least four times a year to ensure that they are pest-free. Currently, both phosphine and methyl bromide are used for pest management, though methyl bromide is to be phased out and alternative procedures established. It is extremely rare for the crop to be marketed within the year of production, resulting in an overlap of harvest in store and subsequently an increase in the cost of pest management.

Historical perspectives

The USA is a leading producer and exporter of several types of dry edible beans which includes Navy, Great Northern, Pinto, light and dark red kidney, large and baby lima, small white, cowpea (mainly blackeyed beans), garbanzo and others. The states of North Dakota, Michigan, Nebraska, Minnesota and California are the largest producers of edible beans. While California ranks fifth in production of total edible beans in the USA, it ranks first in the production of cowpea (Table 7.1) where production can be over 50 000 t in a good year. Even though the blackeye varieties are the main type of cowpea produced in the USA, small quantities of other varieties are cultivated.

In California, the cowpea crop thrives in the summer fog along the coast. Dry cowpeas are processed in California and Texas, while Georgia is the leading processor of fresh cowpea, which are marketed under various names such as southern peas, southern field peas, acre peas and crowder peas. The dry blackeye varieties are marketed as blackeye beans or blackeye peas (Fery 1985).

Annual per capita consumption of cowpea in the US is between 0.1 and 0.13 kg (Table 7.2). This gives an estimated total domestic consumption of between 28 400 and 37 000 t in 2001. On average, 10% of the total US cowpea production is exported to over 40 countries worldwide (Table 7.2).

Physical facilities

Cowpea is assembled in storage centres where there are typically two or more blocks of warehouses. Each block consists of between two and eight warehouses (typical dimensions are 16.5–20.5 m wide, 35–45 m long and 9.5–13.0 m high). In the past, these structures were constructed entirely from wood, but are now built of metal or with wooden frames and steel sides (Fig. 7.1a). Some centres receive cowpea directly into corrugated steel silos (Fig. 7.1b) from where they pass into the warehouse for cleaning. Some farmers deliver the harvested cowpea in trucks. They are unloaded by a dumper, raised by bucket elevator and then flow by gravity into a fumigation chamber or into a storage silo

Table 7.1 Cowpea production and movement in California (1981–2000).

Year	Initial stock in August (tonne)	Annual harvest (tonne)	Total stock (tonne)	Annual dealer price (US$/tonne)	% of crop sold
1981–82	2 237	44 197	46 434	613.11	95.3
1982–83	2 186	52 820	52 820	456.15	76.7
1983–84	12 811	31 358	44 170	596.59	89.4
1984–85	4 677	46 923	51 600	461.85	82.2
1985–86	9 202	39 516	48 717	458.31	97
1986–87	1 322	36 812	38 134	695.73	95
1987–88	2 034	56 336	58 369	512.21	79
1988–89	12 354	42 869	55 223	564.14	75
1989–90	13 828	40 502	54 331	623.74	86
1990–91	7 422	46 041	53 464	631.80	80
1991–92	10 778	46 848	57 626	468.34	75
1992–93	14 489	25 895	40 384	514.17	81
1993–94	7 829	27 835	35 664	855.84	99
1994–95	458	39 365	39 823	802.34	87
1995–96	5 389	50 189	55 578	580.27	63
1996–97	20 352	25 531	45 883	650.09	78
1997–98	10 025	35 270	45 294	615.08	89.1
1998–99	4 951	28 831	33 782	753.16	83
1999–00	5 639	40 105	45 744	471.88	52
Means					
All years	7 789	39 855	47 643	596.00	81.50
Last 10 years	8 733	36 591	45 324	634.36	77.50
Last 5 years	9 271	35 985	45 256	614.10	72.20

Table 7.2 US cowpea: supply and utilisation 1990–2001. Source: National Agricultural Statistics Service, USDA and Bureau of the Census, US Department of Commerce.

Year	Supply				Utilisation				Ending stock (tonne)	Average dealer price US$/tonne
	Production (tonne)	Imports (tonne)	Carry over stock (tonne)	Total (tonne)	Exports (tonne)	Seed use* (tonne)	Domestic use (tonne)	Per capita use (kg)		
1990	47 561	508	27 504	75 573	5 033	1 576	35 892	0.13	33.	624
1991	50 000	203	33 045	83 248	6 355	1 474	34 875	0.12	41	466
1992	29 319	102	40 569	69 990	5 491	915	37 875	0.13	26	507
1993	31 606	1 575	25 775	58 956	7 575	966	32 486	0.11	18	840
1994	42 785	152	17 946	60 883	4 830	1 322	37 672	0.13	17	790
1995	55 437	51	17 031	72 519	2 034	1 627	35 028	0.12	34	578
1996	28 811	508	33 859	63 178	3 355	915	35 486	0.12	23	641
1997	41 006	51	23 386	64 443	3 508	1 220	32 944	0.11	27	610
1998	33 181	0	26 792	59 973	8 846	1 118	31 774	0.10	18	738
1999	66 159	51	18 302	84 511	6 355	2 084	34 113	0.11	42	460
2000	19 106	51	41 942	61 099	5 135	610	31 571	0.10	24	509
2001†	35 569	51	23 793	59 412	5 084	1 169	30 808	0.10	22	492

*Seed use calculated as acres planted multiplied by the estimated seeding rate per acre.

†California inventory on hand as of 30 November is used as ending stock. To this stock was added 10 168 t to account for the huge crop in Texas in 1999. Courtesy of Gary Lucier, USDA Economic Research Service.

(a)

(b)

Fig. 7.1 Storage structures for cowpea. (a) Typical cowpea warehouses and (b) steel silo between two warehouses used to receive cowpea and hold before cleaning and packing (California, US).

Fig. 7.2 Fumigation chamber for cowpea (California, US).

Fig. 7.3 Machines for cleaning, sorting and screening cowpea (California, US).

if the cowpeas are to be cleaned directly. Three types of fumigation chambers are used by warehouse managers; a concrete chamber (18.5 m³), a chamber made from PVC and wood (Fig. 7.2) or converted rail cars. The fumigation containers are fitted with recirculation fans and exhaust systems. Finally, in the warehouses cowpea are cleaned (Fig. 7.3) and loaded into 1.23 m³ steel or wooden tanks, paper cartons, polypropylene jumbo sacks of the same size, or into 25 or 50 kg paper bags (Fig. 7.4). The cowpeas held in steel or wooden tanks are transferred into maxi or paper bags before marketing although cowpea may be transport between sites in maxi-bags (Fig. 7.5).

Objectives of storage

The USA processes an average of 60 000 t of cowpea annually, mostly in California, which produces more than 80% of the entire US stock. On average California stores about 45 000 t, while Texas and other states store another 14 000 t. Harvests commonly overlap in store, as cowpea stocks are seldom completely sold in the year of harvest. Thus, there is always a carry-over stock every year, and this situation is likely to persist. In 2001, some stores still had remnants of the 1998 harvest.

(a)

(b)

(c)

Fig. 7.4 Containers for cowpea during storage. (a) 100 lb paper sacks, (b) 4 × 4 ft steel, wooden or paper containers, and (c) 4 × 4 ft polypropylene maxi-bags (California, US).

Fig. 7.5 Loading a maxi-bag of cowpea onto a truck (California, US).

Harvesting of cowpea takes place between August and November, peaking in September and October. Most harvested cowpea are moved into warehouses from surrounding farms and when taken into storage usually have a moisture content of 14–16%. More than 90% of cowpea are variety 46, which is preferred for its large, smooth seed; the remainder are varieties 5 or 27. All three varieties are very susceptible to insect infestation in storage.

Marketing of grain in the US is a free enterprise with little government control. The government serves as an unbiased third party by establishing standards for marketing. Three US agencies provide services needed for grain inspection (Cuperus & Krischik 1995):

- USDA–GIPSA (United States Department of Agriculture – Grain Inspection, Packers and Stockyard Administration), a non-regulatory agency that inspects and grades grain based on established standards and procedures;
- USDA–APHIS (United States Department of Agriculture – Animal Plant Health Inspection Service), a regulatory agency that verifies imported grain or animals to be free of quarantine pests and weeds; and
- FDA (Food and Drug Administration), a regulatory agency that prevents contaminated grain (such as grain contaminated with mycotoxin or pesticides or with high levels of insect damaged kernels) from being marketed for human consumption.

Table 7.3 Grades and grade requirements for cowpea. Source: USDA, GISPA (Grain Inspection, Packers and Stock Administration).

Percentage maximum limits of

| | | | Foreign materials | | | | | |
Grade	General appearance	Moisture*	Total defects (DK, FM, CCL and SP)†	Total damage	Total (includes stones)	Stones	Contrasting classes‡	Classes that blend§
US No. 1	The special off grade 'colour' may be applied after the removal of total defects	18.0	4.0	2.0	0.5	0.2	0.5	5.0
US No. 2		18.0	6.0	4.0	1.0	0.4	1.0	10.0
US No. 3		18.0	8.0	6.0	2.0	0.6	2.0	15.0
US substandard		18.0						
US Sample grade								

*Beans with more than 18.0% moisture are graded as high moisture.
†DK = damaged kernel; FM = foreign matter; CCL = contrasting classes; SP = splits.
‡Beans with more than 2.0% contrasting classes are graded 'mixed beans'.
§Beans with more than 15.0% classes that blend are graded 'mixed beans'.

The USDA–GIPSA and marketers designed a procedure for grading blackeye cowpea (USDA–GIPSA 1997). Assigning of grades to cowpea is based on percent total defects (including damaged grain, foreign materials in grain, contrasting classes and splits), moisture content, general appearance and the quantity of other beans. Based on these parameters, cowpea that are stored in the USA can be assigned three premium grades: 1, 2 and 3 (Table 7.3). Cowpea may also be assigned 'US sample grade' if it is musty, sour, materially weathered or insect-damaged or if heating has occurred. US sample grade may also apply if the stored cowpea have any commercially objectionable odour, or contain insect webbing, insect filth, animal filth, any unknown foreign substance, broken glass or metal fragments, which otherwise renders it of low quality. Cowpeas stored in bulk that do not meet the requirements for grades 1 to 3 or sample grade are regarded as US substandard. Other grades of cowpea are 'high moisture grade' for cowpea with moisture content higher than 18%, or 'mixed beans' for bulk cowpea with more than 2% of contrasting classes of other beans, or more than 15% of classes that blend.

Major sources of quality decline

The main cowpea pests in the USA are beetles from the family Bruchidae. Infestation of cowpea by bruchids such as *Callosobruchus maculatus* commences in the field and is carried into storage (Vol. 1 p. 99, Hodges 2002). Female *C. maculatus* seek out mature pods and stick eggs onto them. The early maturing pods form a source of infestation for the late maturing pods. It has been suggested that on emergence *C. maculatus* may fly from warehouses to farms where they infest mature cowpea. Neglected seeds in farms are the primary source of field infestation by another bruchid beetle, *Acanthoscelides obtectus* (Larson & Fisher 1925).

Infestation of cowpea in the field by *C. maculatus* can be low and thus difficult to detect. Field infestation of cowpea by *C. maculatus* in 42 United States farms resulted in a mean two-weekly emergence of 2.33 adults per bushel (35.2 L) within the first few weeks of storage (Hagstrum 1985). This is equivalent to one damaged seed per 100 000 undamaged seeds. Population levels remained very low for the first 18 weeks of storage due to low winter temperatures. However, following this period, if cowpea is not fumigated in spring the weevil multiplies very quickly.

Besides *C. maculatus*, another important storage pest of cowpea is Indian meal moth, *Plodia interpunctella* (Vol. 1 p. 104, Hodges 2002).

Commodity and pest management regimes

Successful storage is accomplished by starting with

whole, insect-free seeds in the storage facility. Warehouse managers know that infested cowpeas brought into the storage without disinfestation will put the remaining stock at risk. Thus steps are taken by managers to maintain insect-free commodities from the time cowpeas are received from the farmers until they are sold. Cowpeas received from farms are either fumigated on arrival, if they are not to be cleaned soon, or fumigated immediately after cleaning.

Cleaning and sorting of cowpea

Commodities from farms will contain grain dust and foreign materials such as stones, dirt, twigs, leaves, weed seeds, cereal grains, lentils, peas and materials other than cowpea. Cowpeas may be damaged by insects, disease, weather, or be mechanically broken or scarred. The screening process is mechanised (Fig. 7.3) and it involves three stages. The first stage is the cleaning stage, in which the fines, such as grain dust, are eliminated. Twigs, leaves and related materials are also screened off and the cowpeas segregated according to size. Next is the stage where gravity separates out the stones, and the final stage involves using air to blow out broken and damaged cowpea seeds. The final product is a near-uniform commodity.

Fumigation

The storage facilities themselves are never fumigated, only the cowpeas. The initial fumigation of stored cowpea takes place between September and November when the commodity comes in from the field. If the cowpeas are in storage over the summer, three fumigations take place: at the beginning, middle and end of summer.

It is difficult to detect pockets of *C. maculatus* infestation so that store managers generally lack objective information on which to make decisions about the need for fumigation. Thus fumigation is a routine rather than decision-based activity. Even though pre-emptive fumigation is done to prevent the development of *C. maculatus* in cowpea, samples are examined to make sure cycles of the bruchid do not develop in the commodity.

Fumigation is either done with methyl bromide or phosphine gas. About 25% of warehouse managers use only phosphine for fumigation, 25% use only methyl bromide and the remainder use methyl bromide for cowpea to be consumed and phosphine for cowpea for planting, as methyl bromide reduces the germination rate. Closed-loop fumigation, i.e. the continuous or intermittent recirculation of the fumigant, is the fumigation method used in sealed systems. Methyl bromide is a low boiling-point liquid pressurised in cylinders, expanding to a gaseous form when released. It is applied in fumigation chambers, at a dosage rate of 31 g m^{-3}, where fans operate to circulate and achieve an even distribution of the fumigant. Methyl bromide is chosen by some warehouse managers because it is fast acting and very effective. However, as methyl bromide reduces seed viability and is scheduled to be phased out under the terms of the Montreal Protocol, due to its negative environmental impacts (Vol. 1 p. 328, Taylor 2002), some cowpea warehouse operators use phosphine gas for the entire stock. The solid formulations of aluminium phosphide or magnesium phosphide, marketed as 'Phostoxin' or 'Magtoxin', respectively, generate the gas phosphine. Phosphide tablets, at an application rate of approximately three 3 g tablets per 2 m^3 are dispensed on top of the grain mass and the cowpea exposed to phosphine for about seven to ten days. The chamber for the fumigation with phosphine is equipped with a recirculation fan and an exhaust system. The withholding period after phosphine fumigation, before the stock can be moved or staff can enter the fumigation chamber, is about 48 hours if the ambient temperature is warm, or between 72 and 120 hours when temperatures are low.

Fogging

Storage moth pests, particularly *Plodia interpunctella*, are common in cowpea warehouses. These moths are kept in check by fogging the facility every two weeks. Fogging is usually done with dichlorvos (Vol. 1 pp 251–254, Birkinshaw 2002). Timed dispensers are used to deliver the insecticide into the environment of the warehouse. There is a withholding period of three days after fogging.

Protection of organic cowpea

Only a few warehouses process dry cowpea for the organic market. Organic cowpea are cleaned and sorted as described above but are treated in a fumigation chamber with an atmosphere containing 60–80% carbon dioxide. Current protocols recommend a minimum of 15 days at 60% carbon dioxide (Vol. 1 p. 332, Taylor 2002).

Economics of operation of the system

Dealers' prices for cowpea vary from year to year. In low

Table 7.4 Cost (US$) of handling, processing and protecting dry cowpea. Source: Cal-bean and Grain Co-op Inc, Pixley, CA and Rhodes Bean and Supply Co-op, Tracy CA.

Process	Cost (US$)	Frequency (per season)
Cleaning	66.9/tonne of field weight	Once
Fumigation	7.86/tonne	4–5 times
Insurance	4.92/tonne cleaned weight	Once
Storage	14.74/tonne of cleaned weight	Once
Miscellaneous costs		
Bagging		
100 lb paper sacks	0.5 each	
100 lb burlap sacks	0.6 each	
Tote sacks	10.5 each	
Bag markings	0.2 each line	
Loading fee	35 per container/railcar (102 tonne)	
Loading fee	25 per truckload	

production years, prices may rise to $787 t^{-1} with lower prices when production is high. Not all cowpeas can be sold immediately after harvest. Prices may be low during harvest, and it may be cost effective to hold cowpeas in storage waiting for prices to rise. Storage of cowpea involves risk because the commodity could be destroyed by bruchid beetles. In order to prevent this, cowpea are stored in warehouses managed by trained personnel and this is a major expense to the system. The storage facilities represent a fixed cost while variable costs include cleaning the commodity, fumigation, insurance and miscellaneous costs such as shipment (Table 7.4). The cost of cleaning, initial fumigation, insurance and storage are billed when the commodity is received from farmers, and any subsequent fumigations billed during the summer months. If the commodity stays in storage for more than one year, storage is billed again the following fall, as are subsequent fumigations. Packaging and loading are additional costs, as are the materials used in bagging (Table 7.4). Cowpeas can be marketed in bags, as truckloads, or container loads (railcars). Cowpeas for export are shipped in containers while those for local markets are shipped either in bags or in bulk as truckloads.

Future developments

A major concern of the cowpea industry in the USA is the problem of pest control. The most popular cowpeas processed as dry beans are varieties 5, 27 and 46. Even though cowpea varieties resistant to *C. maculatus* have been developed (Singh *et al.* 1985), these are yet to be adopted for large-scale cultivation in the US. The genotype and stage of pod maturity affects the suitability of cowpea pods for oviposition by *C. maculatus* (Fitzner *et al.* 1985) and ten varieties have been found that ex-

hibited total intact pod mortality, a mortality value from egg-hatch to adult emergence that is greater than 95% (Kitch *et al.* 1991). Cultivation of the varieties that are less suitable for oviposition and subsequent development of *C. maculatus* has the potential to reduce initial infestation of cowpea in the field, but none of the varieties showing pod resistance to bruchids is popularly cultivated in the USA. If varieties that show some level of resistance to *C. maculatus* were adopted then the number of fumigations required to protect cowpea during storage might be reduced.

Methyl bromide is one of the two widely used fumigants for the disinfestation of cowpea. The reasons given by warehouse managers for continued use of methyl bromide, rather than changing to phosphine, are its fast action and that it does not cause explosions or react with copper. However, the enforced phasing out of methyl bromide will encourage the search for alternative treatments, and may lead to increasing use of modified atmosphere storage. The potential for modified atmospheres low in oxygen, and rich in nitrogen concentrations up to 90%, or carbon dioxide up to 80%, has been demonstrated for the fumigation of cowpeas (Mbata *et al.* 1994). Modified atmospheres are used in the disinfestation of cowpeas marketed as organic.

Another disinfestation technique with potential for the organic market is the use of freezing temperatures. The method has been recommended since the early 1920s but detailed information on the response of cowpea weevil to commercial freezer temperatures, at $-18°C$, has only recently become available. Although complete kill of all life stages was obtained in laboratory trials, initial work suggested that rapid cooling would be necessary to prevent the development of acclimation by the pest to low temperatures (Johnson & Valero 2000).

One of the major recent developments in cowpea research was the isolation, characterisation and synthesis of the sex pheromone of *C. maculatus* (Qi & Burkholder 1982; Ramaswamy *et al*. 1995; Phillips *et al*. 1996; Shu *et al*. 1996; Mbata *et al*. 2000). The female releases a sex pheromone comprising at least two of the following components: 3-methylene heptanoic acid, (*Z*)-3-methyl-3-heptanoic acid, (*E*)-3-methyl-3-heptenoic acid, (*Z*)-3-methyl-2-heptenoic acid and (*E*)-3-methyl-2-heptenoic acid. Each of the components is active for *C. maculatus* males although combinations produced stronger excitation (Phillips *et al*. 1996) and a mixture of (*Z*)-3-methyl-3-heptenoic acid and (*Z*)-3-methyl-2-heptenoic acid elicits strong sexual excitation in males (Shu *et al*. 1996). Traps baited with this mixture lure males (Mbata *et al*. 2000). Efforts are continuing to develop a pheromone-based trap for monitoring *C. maculatus* infestation. The use of suction light traps equipped with a black light was found to attract both sexes of *C. maculatus* (Keever & Cline 1983). This trap can be used in monitoring *C. maculatus* populations in warehouses. Monitoring populations could eliminate time-based fumigation schedules that are currently in operation, and fumigation will then only need to be done if insects are detected in the commodity or caught in traps.

References

Birkinshaw, L.A. (2002) Insect control – insects and arachnids. In: P. Golob, G. Farrell & J.E. Orchard (eds) *Crop Post-Harvest: Science and Technology. Volume 1 Principles and Practice.* Blackwell Science, Oxford, UK.

Cuperus, G. & Krischik, V. (1995) Stored product integrated pest management. In: V. Krischik, G. Cuperus & D. Galliart (eds) *Stored Product Management.* US Department of Agriculture (Federal Grain Inspection Service)/Oklahoma Extension Service, Division of Agricultural Sciences and Natural Resources, Oklahoma State University, Stillwater OK, USA.

Fery, R.L. (1985) Improved cowpea cultivars for the horticultural industry in the USA. In: S.R. Singh & K.O. Rachie (eds) *Cowpea Research, Production and Utilization.* John Wiley & Sons, New York, USA.

Fitzner, M.S., Hagstrum, D.W., Knauft, D.A., Buhr, K.L. & McLaughlin, J.R. (1985) Genotypic diversity in the suitability of cowpea (Rosales: Leguminosae) pods and seeds for cowpea weevil (Coleoptera: Bruchidae) oviposition and development. *Journal of Economic Entomology*, **78**, 806–810.

Hagstrum, D.W. (1985) Preharvest infestation of cowpea by the cowpea weevil (Coleoptera: Bruchidae) and population trends during storage in Florida. *Journal of Economic Entomology*, **78**, 358–361.

Hodges, R.J. (2002) Pests of durable crops – insects and arachnids. In: P. Golob, G. Farrell & J.E. Orchard (eds) *Crop Post-Harvest: Science and Technology. Volume 1 Principles and Practice.* Blackwell Science, Oxford, UK.

Johnson, J.A. & Valero, K.A. (2000) Control of cowpea weevil *Callosobruchus maculatus* using freezing temperatures. USDA–ARS, Parlier CA. http://www.epa.gov/ozone/mbr/airc/2000/90johnson.pdf.

Keever, D.W. & Cline, L.D. (1983) Effect of light trap height and light source on the capture of *Carthartus quadricollis* (Guerin-Meneville) (Coleoptera: Cucujidae) and *Callosobruchus maculatus* (F.) (Coleoptera: Bruchidae) in a warehouse. *Journal of Economic Entomology*, **76**, 1080–1082.

Kitch, L.W., Shade, R.E. & Murdock, L.L. (1991) Resistance of cowpea weevil (*Callosobruchus maculatus*) larvae in pods of cowpea (*Vigna unguiculata*). *Entomologia Experimentalis et Applicata*, **60**, 183–192.

Larson, A.O. & Fisher, C.K. (1925) The possibilities of weevil development in neglected seeds in warehouses. *Journal of Economic Entomology*, **17**, 632–637.

Mbata, G.N., Reichmuth, C. & Ofuya, T. (1994) Comparative toxicity of carbon dioxide to two *Callosobruchus* species. In: E.J. Highley, H.J. Banks & B.R. Champ (eds) *Proceedings of the 6th International Working Conference on Stored-Product Protection*, Canberra, Australia, Vol. 1, CABI, Wallingford, UK.

Mbata, G.N., Shu, S. & Ramswamy, S.B. (2000) Sex pheromones of *Callosobruchus subinnotatus* and *C. maculatus* (Coleoptera: Bruchidae): congeneric responses and role of air movement. *Bulletin of Entomological Research*, **90**, 147–154.

Phillips, T.W., Phillips, J.K., Webster, F.X., Rong, T. & Burkholder, W.E. (1996) Identification of sex pheromone from cowpea weevil, *Callosobruchus maculatus*, and related with *C. analis* (Coleoptera: Bruchidae). *Journal of Chemical Ecology*, **22**, 2233–2249.

Qi, Y. & Burkholder, W.E. (1982) Sex pheromone biology and behavior of the cowpea weevil, *Callosobruchus maculatus* (Coleoptera: Bruchidae). *Journal of Chemical Ecology*, **8**, 527–534.

Ramsawamy, S.B., Shu, S., Monroe, W.A. & Mbata, G.N. (1995) Ultrastructure and potential role integumentary glandular cells in adult male and female *Callosobruchus subinnotatus* (Pic) and *C. maculatus* (Fabricius) (Coleoptera: Bruchidae). *International Journal of Insect Morphology and Embryology*, **24**, 51–61.

Shu, S., Koepnick, W.L., Mbata, G.N., Cork, A. & Ramaswamy, S.B. (1996) Sex pheromone production in *Callosobruchus maculatus* (Coleoptera: Bruchidae): electroantennographic and behavioral responses. *Journal of Stored Products Research*, **32**, 21–30.

Singh, B.B., Singh, S.R. & Adjadi, O. (1985) Bruchid resistance in cowpeas. *Crop Science*, **25**, 736–739.

Taylor, R.W.D. (2002) Fumigation. In: P. Golob, G. Farrell & J.E. Orchard (eds) *Crop Post-Harvest: Science and Technology. Volume 1 Principles and Practice.* Blackwell Science, Oxford, UK.

USDA–ERS (2001) Dry edible beans: US production by state and class for year 2000. *Vegetable and Specialties Situation and Outlook Report VGS-383*. April. 2001. USDA Economic Research Service, Washington DC, USA.

USDA–GIPSA (1997) *United States Standards for Beans*. USDA, Washington DC. http://www.usda.gov/gipsa/reference-library/standard/beans.htm.

Chapter 8
Miscellaneous Oilseeds

G. Farrell, N. D. G. White and D. S. Jayas

This chapter considers a range of seeds from which oil is commonly extracted, in particular soyabean, cottonseed, sunflower, palm kernel, sesame, linseed, castor and safflower. In addition, a detailed account is presented of canola (Plate 8), a variety of rapeseed, in Canada. These crops, together with peanut and copra, comprise the major worldwide oilseed production over the past 30 years. The major references to oilseeds have been given in the text, but details of minor oilseeds not covered in this discussion can be found in the FAO publication *Minor Oil Crops* (Axtell & Fairman 1992).

Oilseeds may be consumed either directly (soyabean, sesame) or indirectly when pressed to extract oil (cottonseed, palm kernel, linseed, castor, safflower) or both (sunflower, safflower, sesame, linseed). Often the press cake remaining after extraction is used as animal feed although castor cake is not used in this way due to the presence of the toxin, ricin. The handling and storage of oilseeds follows similar patterns to those used for cereal grains and pulses. Small-scale subsistence farmers in developing countries employ methods appropriate to the size of their harvest, their capacity to handle it and the requirements of the market. These methods may include stooking and stacking in the field at harvest, hand threshing, and storage in sacks, drums or pots at the homestead. Crop management in the field is onerous and labour intensive and can result in both qualitative and quantitative losses. In addition, pest management in the store is often lacking, or applied on an *ad hoc* basis in response to the presence of insects or rodents.

Transport of the commodity is usually on bicycles, handcarts, headloads or by using animal traction. In contrast, the management of oil crops in more developed parts of the world (where large-scale farming practices are common) is highly mechanised at all stages along the production chain. However, this access to machinery and information offers a choice to growers in developed countries – whether to deliver the crop to a grain handler, or whether to invest in the infrastructure and skills that will allow storage on-farm to modern standards and quality requirements. Given that oilseeds are much more difficult to store successfully than cereal grains, mainly because they must be kept at lower moisture contents, choosing to store oilseeds on-farm is a strategy that carries considerable technical and financial risk. Nevertheless, wherever oilseeds are stored the important qualities of the crop remain the same: low free fatty acids, high oil content and no residues of unregistered chemicals. In addition, the commodity must be free of insect pests, moulds and mycotoxins. All of these factors are affected by storage conditions or processes applied in store.

Worldwide Oilseeds

Historical perspectives

Oils and fats are essential components of human nutrition, and most are derived from plant sources such as oilseeds (in which energy is stored in the form of oil). Production of oilseeds has increased rapidly to meet the needs of the growing world population. In addition, technical advances have led to improvements in the quality and versatility of oilseed crops. The growing of oilseeds has been documented since ancient times. For example, sesame was an important crop in Persia 4000 years ago (Weiss 1983) and soyabean was domesticated in the eleventh century BC in China (Salunkhe *et al.* 1991).

Storage and handling methods are similar to those for grain. Historically, oilseeds were processed soon after harvest or stored in pots or jars. As production and market opportunities increased from the Middle Ages onwards, wooden warehouses were built that catered for grain and oilseeds stored in bags. Towards the end of the nineteenth century these warehouses were replaced by specialised containers, particularly in North America, where the new bin designs could hold several hundred tonnes. Handling equipment became increasingly mechanised so that oilseeds could be loaded quickly onto boxcars. Large silos were constructed at ports to store the commodity for export by ship. Thus bag storage gave way to vertical bulk storage (Patterson 1989). Techniques to maintain oilseed quality in store have also developed over time, from early recognition in the Roman era of the value of airtight storage to modern management methods using pesticides and microprocessor controlled aeration that are essential for maintaining the quality of oilseeds stored for long periods.

World trade in oilseeds has seen dramatic change in the last forty years, as markets have expanded and contracted (FAO 2002) (Fig. 8.1, Table 8.1). The major exporters are the USA, Canada, China, Brazil and Argentina, with Russia, Japan and the European Union as the major importers. Russia was previously an exporter but now imports oilseed products (apart from sunflower). Canada has become a major exporter. However, in developing countries, the shortage of local vegetable oils has increased their import, and placed a large burden on the hard currency reserves of, for instance, India and Pakistan. Thus, oilseeds and their products have an important role in the economies of many countries (Salunkhe *et al.* 1991).

Physical facilities

Since oilseeds are grown for their oil, a key function of storage must be to maintain the oil content of the harvested product. In general, oilseed storage facilities differ little from those used for cereals. However, many oilseeds are very sensitive to heat and therefore need better-quality store construction to exclude moisture during storage that could lead to the growth of mould and insect populations and consequently heating of the grain. The small size of some oilseeds requires well-designed and constructed stores and handling facilities to prevent leakage. Wall to roof and wall to floor joints, openings around hatches and augur entry points should be sealed properly (Moysey 1973). Escaped seeds will block ducts and prevent air flow and the commodity may become impossible to cool or dry (Armitage 1997).

Oilseeds should be dried quickly after harvest to temperatures below which fungi, mites and insects cannot develop. For canola in the UK, the extra cost of

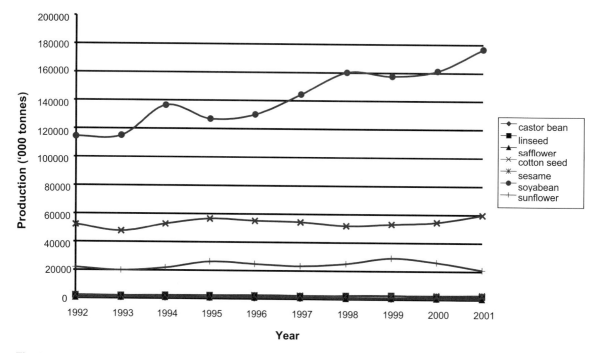

Fig. 8.1 Production trends for some oilseeds (from FAOStat Agriculture Database, FAO 2002).

Table 8.1 Production and export of some oilseeds. Source: FAO (2002).

Country	Production ('000 tonnes) (2001)							Export ('000 tonnes) (2000)						
	castor bean	cotton seed	linseed	safflower	sesame	soyabean	sunflower	castor bean	cotton seed	linseed	safflower	sesame	soyabean	sunflower
Argentina	—	—	—	36	—	26 737	3 188	—	30	—	—	—	4 123	283
Australia	—	1 765	—	40	49	—	—	—	621	—	12	—	—	—
Bangladesh	—	—	48	—	—	—	—	—	—	—	—	—	—	—
Belgium	—	—	—	—	—	—	—	—	30	65	0.8	—	97	—
Benin	—	—	—	—	—	834	—	—	30	—	—	—	—	—
Bolivia	87	—	—	—	—	—	—	—	—	—	—	—	216	—
Brazil	—	2 630	—	—	—	37 675	—	—	—	—	—	—	11 517	—
Canada	—	—	702	—	—	2 040	—	—	—	608	0.8	—	771	—
China	—	15 960	560	30	791	15 450	2 000	—	44	—	7	103	211	—
Côte d'Ivoire	—	—	—	—	—	—	—	—	—	—	—	—	—	—
Czech Republic	—	—	—	—	—	—	—	—	—	3	2	—	—	—
Ecuador	4	—	—	—	—	—	—	—	—	—	—	—	—	—
Egypt	—	—	30	—	—	—	—	—	—	—	—	—	—	—
Ethiopia	14	—	64	37	—	—	—	—	—	—	—	31	—	—
France	—	—	36	—	—	—	1 621	—	—	20	—	—	—	527
Germany	—	—	175	—	—	—	—	—	—	27	—	—	—	—
Greece	—	1 356	—	—	—	—	—	—	195	—	—	—	—	—
Guatemala	—	—	—	—	—	—	—	—	—	—	—	18	—	—
Hungary	—	—	—	—	—	—	668	—	—	—	—	—	—	280
India	850	5 250	244	168	730	5 600	770	20	—	—	1.5	183	—	—
Indonesia	4	—	—	—	—	863	—	—	—	—	—	—	—	—
Italy	—	—	—	—	—	885	—	—	—	—	—	—	—	—
Mexico	—	—	—	148	41	—	—	—	—	—	6	14	—	116
Moldova	—	—	—	—	—	—	—	—	—	—	—	—	—	—
Myanmar	—	—	—	—	426	—	—	—	—	—	—	20	—	—
Nigeria	—	—	—	—	69	—	—	—	—	—	—	30	—	—
Pakistan	—	5 488	1.2	—	51	—	—	—	—	—	—	—	—	—
Paraguay	18	—	—	—	—	3 585	—	2.3	—	—	—	—	1 796	—

(Continued.)

Table 8.1 (*Continued.*)

Country	Production ('000 tonnes) (2001)							Export ('000 tonnes) (2000)						
	castor bean	cotton seed	linseed	safflower	sesame	soyabean	sunflower	castor bean	cotton seed	linseed	safflower	sesame	soyabean	sunflower
Philippines	4	—	—	—	—	—	—	—	—	—	—	—	—	—
Romania	—	—	—	5.5	—	—	700	—	—	5	2	—	46	105
Russian Federation	—	—	—	—	—	—	2 700	—	—	5	2	—	46	1 115
Slovakia	—	—	—	—	—	—	—	—	—	—	—	—	—	82
Slovenia	—	—	—	—	—	—	—	—	—	—	—	—	—	62
South Africa	5	—	—	—	—	—	871	—	—	—	—	—	—	—
Spain	—	—	—	—	—	—	—	—	—	—	—	—	—	—
Sudan	—	—	—	—	300	—	—	—	—	—	—	139	—	—
Sweden	—	—	—	—	—	—	—	—	—	6	—	—	—	—
Syria	—	—	—	—	39	—	—	—	32	—	—	—	—	—
Tanzania	9	—	—	—	39	—	—	0.5	—	—	—	—	—	—
Thailand	—	—	—	—	—	—	—	—	—	9	2	17	—	—
The Netherlands	—	—	—	—	—	—	—	—	21	—	—	—	969	—
Togo	—	—	—	—	102	—	—	—	—	—	—	—	—	—
Turkey	—	1 909	—	—	—	—	—	—	35	—	—	—	—	—
Turkmenistan	—	1 800	—	—	—	—	—	—	—	—	—	—	—	—
Uganda	—	—	—	—	—	—	—	—	—	—	—	—	—	—
Ukraine	—	—	—	—	—	—	2 245	—	—	—	—	—	—	837
United Kingdom	—	—	39	—	—	—	—	—	—	67	—	—	—	—
United States of America	—	11 207	291	110	—	78 668	1 579	—	224	27	37	—	27 192	125
Uzbekistan	—	3 300	—	—	—	—	—	—	—	—	—	—	—	—
Venezuela	—	—	—	—	—	—	—	—	—	—	—	21	—	—
Viet Nam	5	—	—	—	—	—	—	—	—	—	—	—	—	—
Zimbabwe	—	—	—	10	—	—	—	—	42	—	—	—	—	—
World total	1 321	59 726	2 361	610	3 150	176 639	20 903	24	1 391	850	72	657	47 362	3 983

—Indicates no record or that the value is less than 0.2.

drying may be compensated for because reducing the moisture content increases the relative amount of oil in the seed; it is the oil content that is measured on a wet weight basis when fixing the price of the batch (Armitage 1997). Drying equipment designed for cereals may not be suitable for some oilseeds, because oilseeds differ from cereals in being more sensitive to temperatures and resistant to air flow. Air flows suitable for drying wheat may be impeded in linseed and rapeseed, thus leading to overheating and deterioration of the oilseed (Fig. 8.2). Under ideal conditions in the UK, if the cooling or drying system was designed for cereals the depth of canola in the store should be about one-third (Armitage 1997). Temperature sensors should be incorporated into stores so that untoward fluctuations are detected promptly (Moysey 1973).

Concrete floors should be laid on vapour barriers to prevent water vapour moving from the underlying soil through a concrete floor. This water movement can raise the moisture content of the lower 15 cm of grain by as much as 2% per month (Anon. 1954). It is particularly important as most oilseeds, except for soyabean, have to be stored at lower moisture contents than cereals. High oil contents enable some fungi, such as *Eurotium amstelodami,* to begin development on oilseeds at moisture contents in equilibrium with 65% r.h. (Wallace 1973; Sauer *et al.* 1992), whereas on cereals this does not occur until moisture contents are in equilibrium with > 70% r.h.

Painting the outside of steel stores white can lead to a reduction in the temperature of the commodity within. The temperature drop may be small (in the order of 3–4°C cooler than unpainted bins) but may be significant in the reproduction of insects or commodity respiration and internal heating. Again, oilseed stores would benefit (Moysey 1973).

Objectives of storage

Oilseeds, like cereal grains, are subject to grading schemes and quality criteria that vary from country to country, according to local requirements and the use to which the seed will be put. Most care is needed in the handling of oilseeds intended for breeding stock or seed, because physical damage reduces viability. The marketing system can affect the storage period. A prime example is the reluctance of confectioners in North American to accept sunflowers that have been stored for more than one year, because confectionery quality standards are higher than those for oil (Canadian Grain Commission 2001). In the USA sunflowers stored for production of oil can be kept for more than one season (Warrick 2001). Aspects of the harvest, storage and marketing of sunflowers in three countries are illustrated in Table 8.2.

Generally, castor seed for export must be clean, with no more than 2–5% damaged or broken seeds and moisture content less than 9%. Regulations restrict carriage of castor in the same transport as food or feedstuffs because castor seed contains the poison, ricin (Weiss 1983). Storage of substantial quantities of castor seed on-farm should be avoided because the seeds are large and occupy considerable space in relation to their weight. However, storage for short periods may be inevitable while full loads are being made up for rail or road transport. Similarly, the storage of safflower intended for oil extraction is uneconomic and should be avoided. If possible, harvesting should be timed such that the delivery of threshed seed to the mill is rapid (Weiss 1983).

It is advantageous for economic and marketing reasons to position processing equipment near to producers and end-users. This limits transport costs and avoids extended storage periods. For these reasons, much of the soyabean crop in developed countries is processed

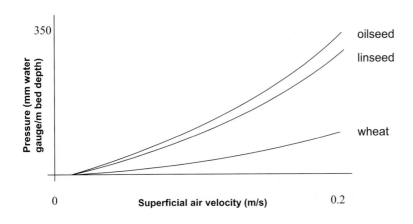

Fig. 8.2 Pressure resistance curves for crops (modified from HGCA 1998).

Table 8.2   Comparison of sunflower production practices in US, Germany and Ukraine.

Practice	US†	Germany†	Ukraine†	Constraints to Ukraine oilseed production‡
Harvest	Desiccants may be used to accelerate drying before harvest.* Seed is then passed through a grain dryer to 9.5% m.c. then stored on farm or at a local elevator	Grain is tested on-farm for moisture and quality. If harvested early, the farmer will take the crop to a local elevator for drying and storage when on-farm facilities are not available.	Desiccants delivered from the air are used in the main production areas. The crop is combined when the m.c. is 8–10%	High harvest losses (7% Ukraine, 2% Western Europe); high storage losses (7% Ukraine, 1% Western Europe)
Storage and sale	Seed is cleaned, stored below 10% m.c. and the bins aerated regularly	Few farmers have aeration facilities, consequently much of the crop is sent to local elevators	Most farms lack drying facilities and farmers may be unable to send the crop to local elevators. Therefore the seed is generally sold direct to the oilseed crushers, traders or input suppliers, who take responsibility for storage and shipping	High losses in state-owned elevators (14% Ukraine, 7% Western Europe); high costs at seaports (US$12/tonne in Ukraine, US$6/tonne in Western Europe); high trade margins (25% Ukraine, 5% Western Europe)

Source: †Ronco Consulting (1998); ‡Striewe (2000).
*Gramoxone Super (paraquat) sprayed onto the crop just before harvest to remove moisture and make maturity more uniform (Warrick 2001).

relatively close to the production areas or near to areas where livestock or poultry are raised since the prime use of soyabean is as an ingredient in animal feedstuffs.

Major sources of quality decline

The major influence on the quality of stored oilseed is moisture. Moisture contents for safe storage depend inversely on temperature and oil content. For example, canola should be stored cool and dry, preferably below 25°C (below 20°C is even better), and at less than 7% moisture content. Under these conditions fungal contamination will not occur, the formation of free fatty acids will be minimised and colour will be retained. This environment also restricts insect development, and insect control will not be required if the oilseed is stored for a few weeks (Banks 1998). However, canola below 6% moisture content breaks easily, but the maximum trading moisture content in the UK is usually 9%, and so there may be a narrow range of moisture content for 'safe' storage (Armitage 1997).

Poor storage may lead to a number of problems, of which heating is particularly important. Heating leads to increased free fatty acid content and loss of market value. Water ingress gives rise to crusting of the seed surface and mould growth. In extreme cases, the tem-

perature of the bulk can rise to a level at which chemical oxidation occurs and the seed mass may smoulder or ignite (Banks 1998). Moisture is particularly damaging in cottonseed storage, since temperature of the seed mass may rise as a result of respiratory activity (Thomson 1979). Once generated, heat is not rapidly lost because of the insulating effect of linters on the cottonseeds. As with other oilseeds, heating leads to the decomposition of glycerides to free fatty acids that increases oil-refining costs, and so cottonseed storage containers must be well ventilated or aerated mechanically (Salunkhe *et al.* 1991).

Sesame seed is relatively delicate, and damage during threshing can lead to immediate loss of viability – even microscopic cracks in the testa will make the seed nonviable (Weiss 1983). Cleanliness of oilseeds is important, since contaminants not removed during processing may cause heating that leads to discoloured or rancid oil, particularly with sesame and sunflower.

Although oilseeds may be attacked by insects and vertebrates in store, loss data are not often available and so losses worldwide are hard to assess. However, some work has been done on individual crops. Stored castor seed is relatively free of pest attack, if the testa is unbroken (Weiss 1983), and so control measures are seldom necessary. Cosmopolitan pests of castor include

Ephestia cautella, Lasioderma serricorne and *Tribolium castaneum.*

Safflower is resistant to most storage pests when the seed is intact and dry. Nevertheless, since few grain bulks are composed wholly of such seed, a range of insects may be found in store including *L. serricorne, Stegobium paniceum* and several *Trogoderma* spp (Bass 1968; Weiss 1983). New thin-hulled varieties are more susceptible to damage than those with thicker hulls, and are therefore more likely to support increased numbers of insect pests. In addition, *L. serricorne* has been found feeding on maturing safflower heads in the field. It may be necessary to treat safflower grown for planting to prevent the introduction of the pest into the storage system (Weiss 1971).

Sunflower is susceptible to attack by a range of cosmopolitan insect and vertebrate pests, though varieties with large, hard seeds are less liable to damage than types with thin hulls and high oil content. The most common storage pests include *Araecerus fasciculatus, Ephestia* spp, *Oryzaephilus mercator, Plodia interpunctella, Sitophilus* spp, *Tribolium* spp and mites of the genus *Tyroglyphus* (Salunkhe & Desai 1986).

Insecticides and sealed storage are recommended for pest management in sunflower breeding stock, but commercial crops are seldom protected in this way, since normal hygiene practices are usually sufficient. Smallholder crops usually suffer most from storage pests, and, as with sesame, farmers may have to save 50% more than is necessary to ensure an adequate stock for the following year's crop (Weiss 1983).

Losses in stored sesame seed and cake are said to be substantial, although there is little published information to back up this claim (Weiss 1983). Losses in sesame stored in Uttar Pradesh, India, caused mainly by *T. castaneum* and *Corcyra cephalonica* have reportedly reached 4.4% (Srivastava 1970). *Tribolium castaneum* and *E. cautella* are cosmopolitan pests of sesame, often found in ship's cargoes. Other oilseed insect pests are *C. cephalonica* in India, Africa and Brazil, *Cryptolestes pusillus* and *O. mercator* in Africa, *Carpophilus* spp and *E. cautella* in India and Pakistan, and *Tribolium confusum* in Uganda and the Central African Republic (Weiss 1983).

Canola can become infected in store, particularly at the grain surface (Banks 1998). In Australia, *T. castaneum* and *Trogoderma variabile* are potentially serious pests of canola (Rees *et al.* 1999). In the UK, mites are increasingly important on canola, favoured by the mild,

moist climate and storage moisture contents of about 9% (Armitage 1997).

Stored linseed is not as vulnerable to insect infestation as cereals although, in Canada, *Oryzaephilus* and *Tribolium* spp have been found in stores. Fungus beetles and mites have been noted on relatively dry (10.1–13.5% m.c.) to damp (> 13.5% m.c.) linseed, and work in Manitoba suggested that linseed varieties varied in their susceptibility to *O. surinamensis* (Anon. 2001).

Commodity and pest management regimes

Appropriate post-harvest handling is key to effective and safe storage of oilseeds. Requirements depend both on the seed type and on the demands of the market. Oilseeds in store require better quality management than cereal crops, particularly sunflower which can heat up very quickly at above 10% m.c. if the bins are not aerated. Due to their vulnerability to changes in storage conditions, more frequent inspection of oilseeds than for cereals is advisable, and so oilseed stores should be designed to facilitate ease of inspection. The small size of some oilseeds, such that their surface area is proportionately greater than that of cereals, means that insecticide doses may have to be double those recommended for cereals (Armitage 1997).

*Castor (***Ricinus communis*)*

Clusters of castor seed harvested slightly green or with wet capsules should be dried before hulling. They can be spread in the sun in the tropics, but may require artificial drying in temperate regions. Care must be taken to avoid overlong exposure to sunlight. Most modern dryers are suitable, but castor capsules are easily damaged and require more care when loading than soyabeans or cereals. Uniform density in the dryer is necessary, since uneven loading leads to inefficient drying. Air movement through the load should be continuous to prevent packing of the capsules which causes channelling, in which permanent passages are formed in the seed bulk. In addition, excessive static pressures arise if high-moisture capsules are too tightly packed (Weiss 1983). In the USA, castor stores well for at least two years if the moisture content is 6% or lower (Brigham 1993).

Castor should not be stored in the open except for short periods, since heat and sunlight degrade quality, reduce the oil content and increase the risk of insect infestation. However, during three years of on-farm storage in India, in bamboo baskets or gunny bags, there was no significant change in oil content. After decortication, castor kernels stored well for up to 1.5 months under ordinary storage conditions, but quality declined markedly thereafter (Salunkhe & Desai 1986).

The viability of castor seed declines during storage even when temperatures are low. In Russia, germination of castor seed fell from 93% to 3% after six months storage at −5°C to −7°C (Blagdyr & Sevastyanova 1975). Seed is easily damaged through rough handling and so mechanical operations should be limited. Bagging of seed must be done with care; wooden scoops and shovels together with rubber conveyor belts are recommended. Castor transported by sea must be stowed away from boilers so that it is not degraded by heat.

Linseed *(*Linum usitatissimum*)*

In Canada, linseed should be cooled before storage, since freshly harvested seed can continue to respire at a high rate for up to six weeks after harvest, before becoming dormant. This, coupled with growth of fungi in air spaces above 70% relative humidity, may lead to heating. Heat can spread rapidly throughout the whole store from isolated areas. Linseed can be stored safely at seed moisture contents of 10% or lower in Canada and 9–11% in the USA (Berglund *et al.* 1999). In North America, a wet harvest or early autumn frost will increase the amount of heated linseed. In the absence of aeration, harvested crops should be inspected frequently for hot spots. Moisture migration is more likely to happen in the autumn and early winter, though it has been known in the spring (Berglund *et al.* 1999).

Linseed harvested using a combine harvester is often contaminated with large amounts of green weed-seed dockage (up to 10% in Canada), which should be removed before storage to limit heating and mould growth arising from increases in moisture content (Berglund *et al.* 1999). Dockage is best reduced by controlling broadleaf and grassy weeds pre-harvest (Anon. 2001).

In India, farmers store linseed in split bamboo or rice straw baskets, plastered with mud, or in earthenware pots. Traders and millers may store the crop for up to a year in gunny bags (Gill 1987).

Safflower *(*Carthamus tinctorius*)*

Safflower can be harvested at 12–14% m.c. with a combine harvester, but must then be dried, either artificially or in sacks, though artificial drying may be detrimental to

crops intended for planting (Weiss 1983). However, the crop is ideally harvested when the seed is below 8% m.c., and preferably about 5% m.c. Seed at such low moisture contents requires little attention in store and undamaged seed will not be at risk from stored product insects. At 4% to 7% m.c. and −10°C to −12°C, germination after eight years was over 80% (Bass 1968). Drying temperatures should not exceed 43°C to ensure high oil yield and no seed damage. In the USA, harvesting at 12% m.c. and drying with aeration, when the air temperature is warm and humidity low, results in the least harvest losses (Gregoire 1999). Grain bins are suitable for the bulk storage of safflower, provided that the moisture content is about 5%. When seed is required for oil extraction, aeration is not normally practised since it is not cost effective in the USA. It may, however, be justified for the production of certified seed.

Sunflower *(*Helianthus annuus*)*

Sunflowers are ready for harvest when the back of the flower head turns yellow or brown, and seed are at 10–12% m.c. Heads should be harvested promptly. The seed should be cleaned before storage to prevent heating of the grain bulk. Sunflower seed dries easily in sacks and, in the tropics, bags of the crop can be left in the shade with their necks open to dry. In temperate regions, cold-air drying is sufficient, and can inhibit insects if insecticide treatment is not possible or is undesirable. Sunflower stores well at below 9% m.c., with no loss in oil quality (Weiss 1983).

Sesame *(*Sesamum orientale = S. indicum*)*

Provided sesame is clean and dry, large-scale storage presents few problems. It is more economical to store on-farm than most oilseeds because the seed is small and has a high bulk density. Sesame can be moved by modern conveyors and augurs with little damage, provided the equipment is in good condition. In developing countries, sesame is stored in drums, jars or pots, with tight-fitting lids to exclude insects and rodents. In parts of East and West Africa, granaries of mud and wattle holding about 100 kg have been used. These may be grouped on raised wooden platforms, protected under a thatch roof (Weiss 1983). In terms of viability, sesame can be kept up to five years in humid and warm conditions in paper, cloth or polythene bags, and over ten years in the same containers if moisture contents are kept low with the help of silica gel (van Rheenen 1981). Germination rates may be undiminished after one year's storage at 18°C and 5–6% m.c. (Prieto & Leon 1976). Sesame will keep for five years at room temperature in the USA without loss of viability (Oplinger *et al*. 1997).

Soyabean *(Glycine max)*

At harvest, soyabean should not exceed 14–15% m.c.; at higher levels the beans will not store without drying, and below 12% mechanical damage occurs (Weiss 1983). Correct harvesting, whether by hand or mechanically, should produce sound, whole, clean beans. This reduces any subsequent storage problems. Drying in bags in the sun, or artificially, is acceptable, but cold air should be used in bin dryers. Hot-air dryers require close supervision to prevent heat damage, with 40°C as the maximum temperature. During on-farm storage over a season, the beans should be maintained at 10–12% m.c. At these levels soyabeans will usually keep without change in grade for two years (Salunkhe *et al*. 1991), though regular store inspection at the top of the bin is required to detect moisture migration. If this occurs, the bin contents must be turned. However, in normal commercial practice storage periods range from a few weeks to about six months.

For long-term storage, soyabean can also be kept in laminated bags made from cellophane, aluminium and polyethylene. In these bags, viability after 30 months at room temperature is nearly 100%, whereas the seeds lose all viability after 15 months in paper bags and after 25 months in polyethylene bags (AVRDC 1978). Laminated bags provide a better barrier to water vapour transmission than paper or polyethylene alone. When maintained in store at ambient temperatures, soyabean rapidly loses viability after about six months. Storage of soyabean in the humid tropics with limited facilities is problematic, since high humidity usually reduces quality and viability more quickly than high temperature.

Soyabean produced for seed purposes should be harvested as soon as the moisture content reaches 14%. Combine harvesters should be operated at slower speeds than normal, since there is a correlation between high drum speeds and poor germination. Handling and conveying operations should be done as gently as possible to minimise damage to the seed (Salunkhe & Desai 1986). Storage of high-value seed, such as genetic lines, germplasm or cultivars, requires sophisticated stores with refrigeration-humidification systems to maintain stocks at 10°C and 50% relative humidity. Under these conditions the seed will retain its condition for five to eight years (Delouche 1975).

Cottonseed *(*Gossypium *spp)*

Cottonseed may be stored for short periods after harvest, before ginning, in which case the stored product consists of the seed and associated lint. High temperature and moisture contents must be minimised during the storage of cottonseeds. As long as moisture content remains below 12%, viability, microflora and free fatty acid content are not significantly affected (Salunkhe *et al*. 1991). Cottonseed leaving the ginnery cannot be stored for long periods at high moisture contents, and so it is passed through hot-air dryers. The seed is then cooled before being sent for storage. Cottonseed at 16% m.c. can be stored successfully at 21°C. Storage at higher temperatures is unsafe at more than 11–12% m.c. (Salunkhe *et al*. 1991).

Palm oils *(*Elaeis guineensis*)*

Palms produce fruit containing two types of oil of commercial value: palm oil extracted from the flesh and palm kernel oil extracted from the kernels. The trees yield fruit continuously throughout the year. This is a major advantage over annual oil crops, and permits very efficient utilisation of labour and equipment for harvesting and oil extraction. However, during harvesting branches of fruit are cut off the palm and usually allowed to fall several metres to the ground. The resulting mechanical shock initiates enzyme activity that gives rise to a high free fatty acid content in the palm oil, and so processing should not be delayed (Godlin & Spensley 1971). Since the flesh is commonly processed within hours of harvesting storage facilities are not required.

Palm kernels taken from the fruits are difficult to store without loss of quality and have to be well dried if they are to be kept for any length of time (Hayma 1995).

Pest control operations

Oilseeds are often fumigated but when using methyl bromide (Vol. 1 p. 327, Taylor 2002), application rates are higher and treatment times longer than for cereals (Table 8.3). The reason for this is that methyl bromide is strongly sorbed by commodities with a high oil content (Table 8.4), although most of this gas desorbs slowly after the fumigation.

The dosage rates in Table 8.3 apply to fumigations under gas-proof sheets and in freight containers that are usually fully loaded. If this method is to be used for mite control, dosage rates should be doubled. Penetration of

Table 8.3 Methyl bromide dosage rates and exposure periods for various commodities at different temperatures. Source: EPPO (1993).

Commodity	Dosage (g m⁻³)			Exposure period (h)
	< 10°C	10–20°C	> 20°C	
Wheat, barley	50	35	25	24
Oilseed	75	50	35	48
Oilseed cake	120	85	60	48

Table 8.4 Fumigants remaining in the gas phase in a closed system at 25°C, 60% r.h. and with a 95% filling ratio. Source: Banks (1992).

Commodity	% remaining in gas phase			
	Phosphine		Methyl bromide	
	1 day	5 days	8 h	24 h
Wheat	78	56	32	18
Barley	74	55	48	35
Canola	64	28	31	18
Safflower	20	02	07	05

Note: These values include the initial loss and that subsequently lost to reaction.

methyl bromide into oilseed cake is poor and may be uneconomic using the recommended dosage rates. In such cases, the use of phosphine should be considered. Diapausing larvae of the beetle *Trogoderma granarium* and moth *Ephestia elutella* are highly tolerant of methyl bromide. Where these insects are present, dosages should be increased by one half and exposure periods increased to 48 h in order to achieve a *ct*-product of at least 300 mg L⁻¹ h⁻¹ (Vol. 1 p. 322, Taylor 2002).

In many situations when time is not a constraint, phosphine can be used instead of methyl bromide because there is, generally, little or no reaction of this chemical with commodities. However, work in Australia has shown that phosphine is sorbed by cottonseed and linseed, and should not be used to fumigate these oilseeds (van S. Graver & Annis 1994).

When using phosphine it is best to fumigate commodities at their recommended safe moisture contents. Thus oilseeds should be at lower moisture contents than cereals (Table 8.5).

Table 8.5 Some recommended moisture contents for successful fumigation of various commodities with phosphine. Source: van S. Graver & Annis (1994).

Commodity	Moisture content (%)
Copra	7.0
Peanuts (shelled)	7.0
Palm nuts	5.0
Wheat	13.5

Economics of operation of the system

There are two major costs associated with the storage of cereals and oilseeds: the cost of physically storing the commodity (warehousing, insurance and protection against special risks) and the interest payable or foregone on the capital invested in the stock (UNCTAD 1978). In addition, there may be management costs, though these are often insignificant in comparison. The temperate or tropical location of the store introduces climatic conditions as another factor, as does the maximum period for which the bulk may be held in store without the need for stock rotation (Patterson 1989). There is also a charge for turning the grain in the bin to prevent caking and deal with moisture migration. Recent costings for storage operations with oilseed appear not to be available but estimates by UNCTAD (1978) suggest $US1 t⁻¹ y⁻¹ for turning oilseed in temperate regions and $1.80 for oil cake, in 2003 prices probably about $5 and $9 respectively. The annual warehouse rental charge varied more widely, at $5–30 t⁻¹, in 2003 prices probably $15–60 t⁻¹. In the tropics, where manual labour is used to move oilseeds, individual assessments based on local labour rates must be used (Patterson 1989).

Future developments

Per capita consumption of oils and fats in industrialised countries has reached a point where substantial increases are not expected. However, in many parts of the developing world fat consumption remains low, often below recommended daily intakes (Salunkhe *et al.* 1991). To meet nutritional requirements, efforts are being made to increase production in deficit countries. This will necessitate improvements in quality and has implications for storage both on-farm, for home consumption, and at processing centres to meet the wider needs of consumers.

The withdrawal of approved admixture insecticides in developed countries will increase the demand for fumigation, a neglected area where oilseeds are concerned. Improved stores are being tested in Australia that will accept oilseeds at higher moisture contents and dry the commodity to normal safe limits in store. The phasing out of methyl bromide and the development of resistance to phosphine has led to increased research interest in controlled atmosphere storage using carbon dioxide. It may be possible to use this gas for treating oilseeds in transit, on railcars and on board ship, as well as in large-scale silos at ports and major cities.

Canola (rapeseed): Canada

Historical perspectives

Several cultivars of rapeseed are called canola, a mustard crop grown primarily for its seed (Plate 8) that yields about 40% oil and high-protein animal feed. Rapeseed has resulted from the hybridisation of similar brassicas (Cruciferae) and may be assigned to *Brassica napus* or *Brassica rapa*. Natural populations of *B. napus* probably originated in Europe and northwestern Africa while *B. rapa* originated in different areas of Asia and Europe (Daun 1993a). Oilseed rape was grown in India, China, and Japan 2000 years ago. In the thirteenth century, rape was grown widely in northern Europe and was the major source of lamp oil, but in the latter half of the nineteenth century it was largely replaced by other vegetable oils. In the twentieth century, the oil was recognised as an edible product and production increased.

Canola cultivars have slightly different agronomic characteristics depending on the species from which they are derived. Traditionally, canola has been unsuitable as a source of human or animal food due to the presence of two naturally occurring toxins, erucic acid and glucosinolates. However, selective breeding has resulted in a crop with < 2% erucic acid in its oil, which is used for human consumption, and < 30 micromoles of glucosinolates per gram in its meal, which is used as an animal protein supplement. Both winter and spring canola cultivars are available for different geographical locations (Daun 1993a, b) but only spring canola is grown in Canada. World production of canola and rapeseed peaked at 42.6m t in 1999–2000 (Anon. 1999). At that time, canola production was 2.4m t in Australia and 8.8m t in Canada while the European Union produced 11.4m t of rapeseed, primarily for industrial oil. The other major producers of rapeseed are China and India (FAS 1998). Canada is the major exporter of canola.

Commercial rapeseed production in western Canada began in 1942 using seed from Argentina as a source of oil for marine lubrication. After 1956, production expanded rapidly in Canada with the development of edible oil variants and as a crop to alternate and compete with wheat in western Canada (Daun 1993a). In 1985, the United States Food and Drug Administration granted canola oil the status of 'generally recognized as safe' for human consumption.

Physical facilities

Harvested canola is handled in bulk in Canada. It is stored in structures used for cereals (Fig. 8.3), usually for 3–12 months (harvested in August or September) on the farm, then transported by truck to primary grain elevators. From there it is moved by rail hopper car to domestic crushing plants or as whole seed to terminal elevators on the coasts for export. The seed is marketed through private grain handling companies. The bulk seed

Fig. 8.3 A bolted galvanised steel granary in Canada, typically used for cereal storage, holding canola seed that is being sampled using a probe.

stored on the farm is usually moved by screw auger and is placed in 3–6-m diameter bolted galvanised steel granaries often having fully perforated floors for aeration.

Objectives of storage

Canola is rarely stored on the farm for more than one year (Muir 1997) and the crop is generally held only until it can be sent for domestic oil extraction or for export. Frost, heat or drought stress near harvest can result in large amounts of green seed since these conditions can arrest maturation. Cutting the canola and leaving it in the field (swathing) for four days can reduce the proportion of green seeds (Cenkowski *et al.* 1989) and some maturation is possible if the canola is held in storage for at least five months (Thomas 1984).

During storage the seed must be kept cool and dry or rapid moulding will occur and hot spots will develop. This would decrease germination, discolour the oil due to heating, increase free fatty acids and damage oil flavour (Mills 1989). From storage the canola is sent to the crushing plant at the processor's convenience or sold by private or co-operative grain companies for export. In Canada, the oil is extracted chemically for use in margarine, cooking oil and a multitude of other food uses. During oil removal from the seed the enzyme myrosinase must be inactivated by exposure to high temperature, but this must not exceed 110°C or the oil cannot be hydrogenated (Daun 1993b). Canola oil is now the most widely consumed vegetable oil in Canada (Daun 1993b).

The canola meal that remains after seed crushing is sold immediately to animal feed mills.

Major sources of quality decline

Canola is more difficult to store without spoilage than cereals because it contains 43–45% of its dry mass as oil compared to only 1.9% for wheat (Hoseney 1986). The high oil content of all oilseed crops means they must be stored at lower moisture contents than cereals, at a moisture content in equilibrium with 65% r.h. At 25°C this corresponds to 8.3% m.c. (Fig. 8.4) and at 40°C to 7% m.c. (Pichler 1956; Muir 1973; Pixton & Warburton 1977; Sokhansanj *et al.* 1986; Jayas *et al.* 1988, 1995). In Canada, normal grades of canola must have an m.c. of 10% or less (Canadian Grain Commission 1999), although for safe long-term storage canola should probably be at 8% m.c. or less (Moysey 1973; Mills & Sinha 1980). Throughout the remainder of this account the information given applies as much to rapeseed as it does to canola.

Seed maturity and damage

It is important to grow canola with as few weed species as possible because weed seeds are often immature when canola is harvested and can have a high moisture content, leading to fungal spoilage. Occasionally, canola can be harvested with a high percentage of immature seeds that appear green rather than yellow when crushed. The

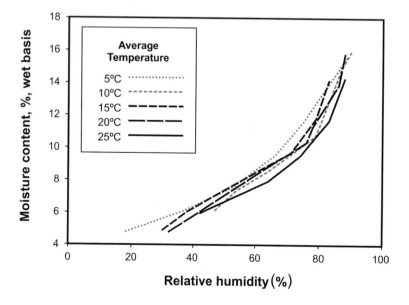

Fig. 8.4 Experimental equilibrium moisture data at different temperatures averages for three samples of canola (data from Sokhansanj *et al.* 1986).

Table 8.6 USDA and Canadian Grain Commission (CGC 2000, unpublished) grade and grade requirements for canola.

Grading factor (maximum % limit)	US and Canadian grades				Additional Canadian criteria		
	1	2	3		1	2	3
Damaged kernels (%)							
Heat damaged	0.1	0.5	2.0	Excreta (%)	0.02	0.02	0.02
Distinctly green	2.0	6.0	20.0	Insect excreta (%)	0.1	0.2	0.3
Total	3.0	10.0	20.0				
Conspicuous admixture (%)							
Ergot	0.05	0.05	0.05				
Sclerotinia	0.05	0.10	0.15				
Stones	0.05	0.05	0.05				
Total	1.0	1.5	2.0				
Inconspicuous admixture† (%)	5.0	5.0	5.0				
Maximum count limits of other material							
Animal filth	3	3	3				
Glass	0	0	0				
Unknown foreign substance	1	1	1				

Sample grade: canola that: (1) does not meet the requirements for Nos 1, 2, or 3; or (2) has a musty, sour or commercially objectionable foreign odour; or (3) is heating or otherwise of distinctly low quality.

*Conspicuous admixture is all matter other than canola that is readily distinguishable from canola and which remains in the sample after the removal of machine-separated dockage.

†Inconspicuous admixture: any seed which is difficult to distinguish from canola. This includes, but is not limited to, common wild mustard (*Brassica kaber* and *B. juncea*), domestic brown mustard (*B. juncea*), yellow mustard (*B. hirta*) and seed other than the mustard group.

maximum acceptable proportion of green seeds for no. 1 and no. 2 grades in Canada and the USA is 2% and 6%, respectively (Table 8.6). The presence of green seeds increases the refining costs of canola oil and can shorten the shelf life of products containing such oil (Mills 1996).

If canola is frost-damaged then heating and spoilage usually occurs in the first few months after harvest. It is at this time that the canola must be carefully monitored to detect the onset of spoilage (Mills *et al.* 1984). A bin containing more than 31 t of severely frost-damaged canola in Canada was reported to have suffered rapid spoilage. The grain temperature was 102°C after three months of storage when the outdoor temperature was only 2°C (Mills *et al.* 1984). Canola plants left standing under snow for the winter and then harvested yield seed that stores less well than that harvested in autumn (Daun *et al.* 1986). However, because canola seeds grow in pods they are less exposed to weather conditions than cereal seeds. When canola seed has been heat-damaged by hot spots during storage the seed has low to zero viability, a tobacco-like odour, low pH in deionised water, high electrolyte leakage from the seed into soak water, a high proportion of jet-black seeds and extremely high free fatty acid levels in the oil (Mills & Kim 1977).

Typical dockage associated with harvested canola includes buckwheat seeds (*Fagopyrum esculentum*), wild oats (*Avena fatua*), and other volunteer cereal grain, other weeds, which have a higher moisture content than the canola and broken canola pods (Prasad *et al.* 1978). The dockage accumulates near the inner walls of the structure and even the use of grain spreaders does not always distribute the dockage and chaff more uniformly in a bulk (Jayas *et al.* 1987a). Canola seeds lying in pods in a field after cutting can be 5°C warmer than ambient air on a sunny day (Prasad *et al.* 1978) and this must be considered when placing seed into storage. Mechanical damage to seeds during harvest is uncommon above 7% m.c. (Harner 1989). There have been reports in the literature of 'sweating' by freshly harvested canola for several weeks (Thomas 1984; Mills 1989, 1996). Such seed is metabolically very active, producing moisture and carbon dioxide by respiration. Recent, unpublished data indicate that mature, cleaned seed does not 'sweat', so the activity is probably due to hot harvests, green seed and moist weed seeds.

Beginning soon after harvest, canola should be monitored and aeration or grain turning employed as soon as spoilage is detected (Mills 1989). Monitoring carbon dioxide concentrations can be a reliable method of detecting spoilage (White *et al.* 1982) with levels in farm storage bins being up to 5.7% in air compared to ambient air levels of 0.03% (Muir *et al.* 1985). Carbon dioxide levels in excess of 1% indicate at least localised spoilage in bulk seed (Sinha *et al.* 1981).

Insects and mites

Insects are not a major problem in stored canola because of its high oil and low carbohydrate content. Nevertheless, infestations of *Tribolium* spp, *Cryptolestes* spp, or Pscoptera (Sinha & Wallace 1977) are common and these insects appear to be feeding on cereal dockage and/or fungi.

Insects that can feed and multiply slowly on whole and crushed canola include the psocid *Liposcelis bostrychophilus* (Sinha 1988), the beetles *Oryzaephilus mercator*, *Tribolium castaneum*, *Tribolium confusum* and *Oryzaephilus surinamensis* (Sinha 1972, 1976), whereas *Cryptolestes ferrugineus* does not survive on whole canola or the crushed seed (Sinha 1972). Canola meal does not appear to be a very suitable medium for the development of storage pests, since it typically contains 5.3% oil by mass and 38% protein (Daun 1993b). At 30°C and 65% r.h., of 16 insect and two mite species tested only four *Tribolium* species produced any progeny and less than would have been found on wheat meal under the same conditions (White & Jayas 1989).

Mites can be a severe problem on stored damp canola, especially above 10% m.c., feeding directly on the seed (Hudson *et al.* 1991) or on associated fungi (Sinha 1968; Sinha & Wallace 1973; Mills 1976). In canola stored in western Canada for six years, *Acarus* spp, *Lepidoglyphus destructor*, *Aeroglyphus robustus*, the predators *Cheyletus eruditus* and *Blattisocius keegani*, and the fungus-feeder *Tarsonemus granarius* were common (White & Sinha 1979; Sinha 1984). In various studies, the dominant species in Canada and the United Kingdom were *L. destructor*, *Acarus siro* and *C. eruditus* (Armitage 1980; Sinha 1984); in southern France they were *Tyrophagus putrescentiae*, *C. eruditus* and *A. siro* (Fleurat-Lessard 1973). All these mite species could carry fungal spores on or in their bodies, so spreading infection (Sinha 1966; Sinha & Wallace 1966; Hughes 1976).

Microflora

Fungi are a serious problem on canola stored at > 8% m.c.

when stored for long periods. The fungi kill the seeds, cause seed clumping, and produce hot spots due to their respiratory by-products of heat and moisture (Mills 1989; Sauer *et al.* 1992). At low r.h. (65–70%), spoilage is often initiated by *Eurotium amstelodami*. Increased localised moisture can then lead to the growth of storage fungi such as *Aspergillus candidus*, *Aspergillus versicolor* and numerous *Penicillium* spp (Mills & Sinha 1980).

Measurement of canola deterioration times at 5° to 25°C and 6% to 17% m.c has shown deterioration to be fastest at the higher temperatures and moisture contents (Burrell *et al.* 1980). Prior to visible moulding, seeds become clumped together with fungal mycelia and this characteristic is considered the earliest indicator of the need to dry canola. For example, at 25°C and 10.6% m.c., clumping occurred on the 11th day and visible mould on the 21st day, although even by the 40th day there was no germination loss. Canola harvested at more than 10% m.c. should be dried within 1–2 weeks to avoid spoilage (Burrell *et al.* 1980). Recent research in Australia has indicated that even dry canola stored in welded or rolled steel bins can produce hazardous levels of both carbon dioxide and carbon monoxide, so people entering such structures should ensure that the headspace is adequately ventilated (Reuss & Pratt 2001).

The periods for which canola seed, harvested at various temperatures and moisture contents, can be stored without aeration are presented in Fig. 8.5. Safe storage guidelines for canola meal are given in Fig. 8.6. Quality losses are shown as elevated concentrations of free fatty acids, loss of seed viability, discoloration and fungal incidence (Mills & Sinha 1980; White & Jayas 1989). When stored moist, canola seeds mould very rapidly and a great deal of metabolic heat is produced by the fungi. This can lead to bin-burnt or fire-burnt (charcoal black) seed producing dark oil, or in some cases to actual ash (Mills 1989).

Commodity and pest management regimes

Physical properties of canola

The physical properties of canola and wheat as individual seeds and as bulks are given in Table 8.7. Canola seeds are spherical and so pack more tightly than cereals, reducing porosity to 33–35%. As the porosity of canola bulks is somewhat lower than the equivalent cereal bulks, two to three times more static pressure is needed to force air through them (Fig. 8.7) (Jayas *et al.* 1987b). This problem can be exacerbated if canola is wetted be-

Fig. 8.5 Laboratory and commercial data on the maximum periods (up to 147 days) for the safe storage of canola binned at various temperatures and moisture contents at a depth of 2 m in the centre of a bin 5 m in diameter. Laboratory tests: solid triangle with 77 denotes signs of spoilage within 77 days; open triangle denotes no indication of spoilage within 147 days. Commercial bins: solid circle with 144 denotes spoilage after 144 days; open circle denotes no spoilage detected (Mills & Sinha 1980).

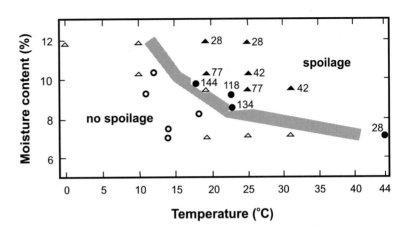

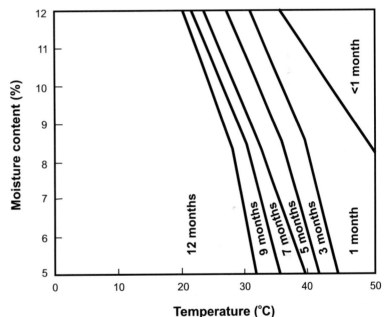

Fig. 8.6 Estimated safe storage times for canola meal based on initial colour change from yellow-green to brown and on fat acidity values below 88 mg KOH/100 g canola meal (White & Jayas 1989).

cause gums in the seed coat make it stick together. The presence of chaff and fines have a marked effect on the porosity of canola bulks. Since the bulk densities of chaff (383 kg m^{-3}) and fines (570 kg m^{-3}) are somewhat lower than canola (700 kg m^{-3}) an increase in the proportion of these contaminants decreases bulk density, increasing porosity linearly (Jayas *et al.* 1989).

Various structural surfaces (wood, steel, concrete) offer less resistance to canola flow than to wheat movement (Table 8.7) at low moisture contents. Since canola seeds are small, have low friction coefficients, and flow very easily, care must be taken to ensure storage bins do not leak seed.

Aeration and near-ambient drying

The purpose of aeration is to cool canola to temperatures that will prevent spoilage. Aeration air flow rates are too low, typically 1–2 L s^{-1} m^{-3}, to result in any significant drying (Jayas & Muir 2001). Insects and mites in stored canola can be suppressed by this type of aeration. In the United Kingdom, aeration of seeds at 9% m.c. at air temperatures below 5°C for four months decreased the numbers of mites (*Acarus* spp) from 36 000 kg^{-1} to 4000 kg^{-1} (Armitage 1984).

For maximum efficiency, aeration should start as soon as bins begin to be filled (Friesen & Huminicki 1986).

Table 8.7 Physical characteristics of canola and wheat.

	Canola	Wheat
Dimension of seeds (mm)	Width 1.0–2.1 × length 1.1–2.5[a]	Width 1.7–4.0 × length 4.2–8.6[b]
Shape	spherical	ellipsoidal
Kernel density (kg dm^{-3})	1.09–1.13[c]	1.37–1.38[c]
Porosity of bulk seeds (% air)	33–35[c]	38–40[c]
Specific heat of the dry mass (J kg^{-1} K^{-1})	1553–1569[d]	1185–1260[e]
Equations for calculating thermal conductivity ($K = W\ M^{-1}\ K^{-1}$)	$K = 0.1600 + 0.043M$	$K = 0.1337 + 0.25M$
where M = moisture content (kg H$_2$O/kg dry matter)	$0 \times M \times 0.30$[a]	$0.05 \times M \times 0.22$[f]
Bulk density (standard) (kg m^{-3})	670[c]	768[c]
Angle of repose (°)		
Filling	24[c]	26[c]
Emptying	26[c]	23[c]
Mean coefficient of sliding friction (tangent of slope angle when sliding began)[c]	at 8.1% m.c.	at 12.7% m.c.
Galvanised steel	0.28	0.32
Steel-trowelled concrete	0.29	0.42
Wood-floated concrete	0.38	0.47

[a]Pabis *et al.* (1998); [b]Grochowicz (1971); [c]Muir & Sinha (1988); [d]Muhlbauer & Scherer (1977); [e]Mohsenin (1980); [f]Gadaj & Cybulska (1973).

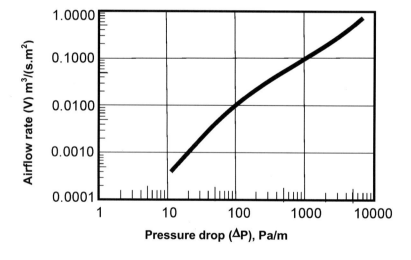

Fig. 8.7 Pressure drop across a spout-filled canola column 6.5% m.c. (Jayas *et al.* 1987b).

At flow rates of 1 L s^{-1} m^{-3} about 200 h are needed to move a cooling front through a farm grain bin; at twice that air flow, the time is halved. Air is usually blown into the bulk from the bottom of the bin so completion of the cooling can be monitored from the top of the bulk. In *B. napus* bulks, aeration fans should operate at 200 Pa at a seed depth of 3.4 m and 500 Pa at 8.3 m deep. For the slightly smaller *B. rapa* seeds, the static pressures should be 300 Pa and 750 Pa at the same depths (Jayas *et al.* 1987b).

Near-ambient drying uses considerably higher air flow rates than aeration and this gives a moderate drying effect even though heat is not added to the system, apart from any that may be generated by the fan blowing the air (Vol. 1 p. 158, Boxall 2002; Jayas & Muir 2001). Static pressures of 1000 and 2000 Pa at 3 m and 4.3 m depths, respectively, are needed for *B. napus*; while using the same fans these pressures should occur at depths of 2.6 m and 3.6 m for *B. rapa* (Thomas 1984). The cooling and drying effect of near-ambient drying, in bins with perforated floors, have been modelled by Muir and Sinha (1986) and validated by Muir *et al.* (1991). To achieve the required air flow rates, a sufficient area of floor perforations is required and a fully perforated floor is recommended (Friesen & Huminicki 1986). Fully perforated floors are superior to ducts or partially perforated floors because they give more even air distribution; for best performance the top of the bulk should also be levelled (Alagusundaram *et al.* 1994; Pabis *et al.* 1998). Forcing air horizontally can reduce energy use or more canola can be dried using the same fan (Jayas *et al.* 1987c). In western Canada, near-ambient drying of canola with initial moistures of 8.8–12.8%, for two years in the period December–February, reduced the moisture content by 2.3% (Sinha *et al.* 1981).

Typically, near-ambient drying is done 24 h a day regardless of sporadic high ambient relative humidity (Friesen & Huminicki 1986). Under these conditions, thin layers of the seed will become re-wetted as a moisture front moves through the bulk. The re-wetting rate does not change when air flows are between 0.25 and 0.43 m s^{-1} but it is slower at 0.10 m s^{-1} (Shatadal *et al.* 1990). For every 6°C or 1% m.c. reduction, below 25°C and 9% m.c., the storage life of canola will double (Harner 1989).

Heated-air drying

Canola to be used as seed should not be heated above 45°C (especially when damp) or above 65°C for other use (Canadian recommendations; Friesen 1980). Generally, the maximum air plenum temperature for drying canola depends on seed moisture content, seed use, type of dryer and expected storage period. As canola presents more resistance to air flow than cereal bulks, the fan operating on a dryer at the same speed used for cereal grain will produce a higher static pressure but less air flow. This can cause temperatures in the hot air plenum to rise and thus dryer controls are important (Nellist & Bruce 1995). Manufacturers' guidelines should be followed.

Drying times for bin-drying canola to 10% m.c. at 40° and 45°C are given in Table 8.8 which demonstrates the relationships between moisture content, grain depth and drying time. If the moisture content of canola is above 17% the seeds should be cleaned (Daun 1993a) and dried in two passes, first, using high-temperature batch or continuous dryers to lower the moisture to 12% and then using near-ambient drying (Harner 1989) to the final moisture. This reduces damage to the seed. In Europe, canola may be harvested at moisture contents as high as 30% and is often dried at 50°C to safeguard germination or 65°C for quality oil (Bailey 1980).

Pesticides and fumigants

If canola bulks become infested with insects or mites and cannot be cooled or dried then they can be disinfested using the fumigants phosphine or carbon dioxide (Jayas *et al.* 1993). To do this the storage structure must be sufficiently airtight. In the case of CO_2 fumigation, the fact that canola seed adsorbs CO_2 more rapidly and to a greater extent than cereals must be taken into account when calculating gas dosages and pressure stresses on bins (Cofie-Agblor *et al.* 1998).

The insecticide pirimiphos-methyl may be applied to bulk seed in Europe to control mites but residues occur in the oil and cannot be completely removed by oil extraction and processing (Good *et al.* 1977), with over 1.5 ppm remaining. Canola is often stored in bins that have held cereals previously and in which contact insecticides have been used to treat insect infestation. It is not recommended that canola be put in such bins since the highly lipophilic organophosphate insecticides such as malathion or bromophos enter the seeds at the outer layer of the bulk and can raise the overall residue level to exceed recommended tolerances in Canada (Watters & Nowicki 1982) of 0.2 ppm for bromophos (White & Leesch 1996), or 0.1 ppm for malathion (White & Nowicki 1985).

Main recommendation for conserving canola

To ensure safe storage of canola the following actions should be taken (Mills 1989):

- Clean the seeds as soon as possible.
- Bin the seed at least 1.5 percentage points below 10% m.c. for long-term storage.
- Use an efficient deflector under the auger to spread the heavier, moister, green material and fines away from the core of the bulk.
- Use an aeration unit to cool the seed to 0–5°C quickly for winter storage in western Canada.
- Monitor the seed temperature every few days in autumn and every two weeks during the winter.

Economics of operation of the system

Western Canada has dryland farming and canola yields are typically about 1.3 t ha^{-1} (Canada Grains Council

Dryer size	Initial moisture (% w.b.)	Depth (m)	Volume (m³)	Time (h)
Air temperature 40°C				
3.75 kW motor	22	0.6	16	17
	20	0.9	24	21
	18	1.2	32	26
	16	1.35	36	25
	14	1.5	40	22
5.4 kW motor	22	0.9	24	19
	20	1.05	28	20.5
	18	1.2	32	21
	16	1.5	40	25
	14	1.8	48	23
9.3 kW motor	22	1.05	28	21
	20	1.2	32	20
	18	1.5	40	23
	16	1.8	48	21
	14	2.1	56	19
Air temperature 45°C				
3.75 kW motor	22	1.5	40	36
	20	1.8	48	32
	18	2.25*	60	30
	16	2.25*	60	45
	14	2.25*	60	20
5.4 kW motor	22	1.5	40	32
	20	1.8	48	28
	18	2.4	64	30
	16	2.7*	72	34
	14	2.7*	72	32
9.3 kW motor	22	1.65	44	31
	20	1.95	52	30
	18	2.4	64	26
	16	3.0*	80	29
	14	3.0*	80	22

Table 8.8 Drying chart for bin drying canola to a final moisture of 10% w.b. (Friesen 1980) at two air temperatures in a 5.8-m diameter bin with twin-screw stirring devices.

*Depths shown are maximum recommended because of static pressure limitations of the fan.

1998). In 1999, the price of canola oil was $Can 646 t⁻¹ and canola meal $Can 146 t⁻¹ (Canola Council of Canada 2000). At the same time, seed production costs were $Can 176 t⁻¹ and shipment and handling costs from mid-prairie to a west coast port $Can 70 t⁻¹ (which includes elevator charges, cleaning, transport, terminal handling and storage). This gives a total cost to the farmer of $Can 246 t⁻¹ (Canola Council of Canada 1999). Canola seed price was $Can 336 t⁻¹ in 1999, leaving a potential profit to the farmer of $Can 90 t⁻¹. This of course varies yearly with fluctuating commodity prices and is frequently much less. If east coast ports are to be used, costs are higher as transport by ship across the Great Lakes is required (B. Timlick, pers. comm.).

Future developments

Genetically modified (GM) varieties of canola, mainly glyphosate-resistant types (herbicide-resistant) are currently grown and comprise about 55% of Canadian canola production. Difficulties in marketing genetically modified crops around the world put exports at risk. It is likely that GM canola will not be grown widely if it affects sales, not only of seed, oil and meal, but also of any honey produced by bees collecting pollen from canola in western Canada. Numerous varieties of herbicide-tolerant canola have also been developed through reduced mutation; these are not considered to be 'genetically modified'.

References

Alagusundaram, K., Jayas, D.S., Friesen, O.H. & White, N.D.G. (1994) Airflow patterns in bulks of wheat, barley, and canola stored in bins with partially perforated floors: an experimental investigation. *Applied Engineering in Agriculture*, **10**, 791–796.

Anon. (1954) Moisture migration from the ground. *Housing Research Paper 28*. Housing Home Finance Agency, Washington DC, USA.

Anon. (1999) Grow canola.com the source for canola information. Canadian Canola Growers. http://www.statcomonline.com/canolafree.

Anon. (2001) Crop Production. Flax Council of Canada, Winnipeg, Manitoba, Canada. http://www.flaxcouncil.ca.

Armitage, D.M. (1980) The effect of aeration on the development of mite populations in rapeseed. *Journal of Stored Products Research*, **16**, 93–102.

Armitage, D.M. (1984) The vertical distribution of mites in stored produce. In: D.A. Griffiths, & C.E. Bowman (eds) *Acarology VI*, Volume 2.

Armitage, D. (1997) Avoid store flaws. *Crops*, 5 July 1997.

AVRDC (1978) Progress Report 1977. Asian Vegetable Research and Development Centre, Shanhua, Taiwan.

Axtell, B.L. & Fairman, R.M. (1992) Minor oil crops. FAO Agricultural Services Bulletin No. 94. Food and Agriculture Organization of the United Nations, Rome, Italy. http://www.fao.org/inpho/vlibrary/x0043e/x0043e00.htm

Bailey, J. (1980) Oilseed rape harvesting losses can be high. *Arable Farming*, May 1980, 59–61.

Banks, H.J. (1992) Uptake and release of fumigants by grain: sorption/desorption phenomena. In: S. Navarro & E. Donahaye (eds) *Proceedings of the International Conference on Controlled Atmosphere and Fumigation in Grain Storage*, June 1992, Winnipeg, Manitoba, Canada. Caspit Press, Jerusalem, Israel.

Banks, H.J. (1998) It is a brave storer who stores oilseed without aeration. Australian Oilseeds Federation, Wilberforce, New South Wales, Australia. http://www.ento.csiro.au/research/storprod/sga/sga_aug98_4.htm.

Bass, L.D. (1968) Report of the Plant Physiologist. National Seed Storage Laboratory. Fort Collins CO.

Berglund, D.B., Peel, M.D. & Zollinger, R.K. (1999) Flax production in North Dakota. North Dakota State University. Fargo ND. http://www.ext.nodak.edu/extpubs/plantsci/crops/a1038.htm.

Blagdyr, A.P. & Sevastyanova, L.B. (1975) Changes in sowing quality of sunflower and castor seed stored under low temperatures. *Selek. i. Semen.*, **1**, 59.

Boxall, R.A. (2002) Storage losses. In: P. Golob, G. Farrell & J.E. Orchard (eds) *Crop Post-Harvest: Science and Technology. Volume 1 Principles and Practice.* Blackwell Publishing, Oxford, UK.

Brigham, R.D. (1993) Castor: return of an old crop. In: J. Janick & J.E. Simon (eds) *New Crops*. Wiley, New York.

Burrell, N.J., Knight, G.P., Armitage, D.M. & Hill, S.T. (1980) Determination of the time available for drying rapeseed before the appearance of surface moulds. *Journal of Stored Products Research*, **16**, 115–118.

Canada Grains Council (1998) *Canadian Grains Industry Statistical Handbook.* Canada Grains Council, Winnipeg, Manitoba, Canada.

Canadian Grain Commission (1999) *Official Grain Grading Guide.* Canadian Grain Commission, Winnipeg MB, Canada. http://www.cgc.ca; Canadian Grain Storage web page http://res2.agr.ca/winnipeg/stored.htm.

Canadian Grain Commission (2001) *Grain Grading Guide.* Canadian Grain Commission, Winnipeg, Manitoba, Canada.

Canola Council of Canada (1999) Canola Production Centre. http://www.canola-council.org/cpc/99report/99cpc.htm.

Canola Council of Canada (2000) Statistics and markets. http://www.canola-council.org.

Cenkowski, S., Sokhansanj, S. & Sosulski, F.W. (1989) Effect of harvest date and swathing on moisture content and chlorophyll content of canola seed. *Canadian Journal of Plant Science*, **69**, 925–928.

Cofie-Agblor, R., Muir, W.E., Jayas, D.S. & White, N.D.G. (1998) Carbon dioxide sorption by grains and canola at two CO_2 concentrations. *Journal of Stored Products Research*, **34**, 159–170.

Daun, J.K. (1993a) Oilseeds – production. In: *Grains and Oilseeds Handling, Marketing, Processing.* Vol. 2. 4th edition. Canadian International Grains Institute, Winnipeg, MB, Canada.

Daun, J.K. (1993b) Oilseeds – processing. In: *Grains and Oilseeds Handling, Marketing, Processing.* Vol. 2. 4th edition. Canadian International Grains Institute, Winnipeg, MB, Canada.

Daun, J.K., Cooke, L.A. & Clear, R.M. (1986) Quality, morphology and storability of canola and rapeseed harvested after overwintering in northern Alberta. *Journal of the American Oil Chemists' Society*, **63**, 1333–1340.

Delouche, J.C. (1975) Seed quality and storage of soyabeans. In: D.K. Whigham (ed.) *Soyabean: Production, Protection and Utilization.* Proceedings of a Conference of Scientists of Africa, the Middle East and South Asia, 1974, Addis Ababa, Ethiopia. University of Illinois, Urbana IL, USA.

EPPO (1993) Methyl bromide fumigation of stored products to control stored-product insect pests in general. Standard 12. *OEPP/EPPO Bulletin*, **23**, 207–208.

FAO (2002) FAOStat Agriculture Database. Food and Agriculture Organization of the United Nations, Rome, Italy. http://apps.fao.org/page/collections?subset=agriculture.

FAS (1998) World agricultural production. Foreign Agricultural Service, United States Department of Agriculture, Washington DC, USA. http://www.fas.usda.gov/wap/circular/1998/98–04/toc.htm.

Fleurat-Lessard, F. (1973) Les acariens des stocks de graines de colzas. Centre Technique Interprofessionne des Oleagineux Metropolitans, Paris, France.

Friesen, O.H. (1980) *Heated-Air Grain Dryers.* Publication 1700, Agriculture Canada, Ottawa, ON, Canada. http://res2.agr.ca/winnipeg/stored.htm.

Friesen, O.H. & Huminicki, D.N. (1986) *Grain Aeration and Unheated Air Drying.* Manitoba Agriculture, Agdex 732–1. http://res2.agr.ca/winnipeg/stored.htm.

Gadaj, S.P. & Cybulska, W. (1973) Thermal conductivity of a layer of kernels (in Polish). *Roczniki Nauk Rolniczych*, **70**, 7–15.

Gill, K.S. (1987) *Linseed.* Indian Council of Agricultural Research, New Delhi, India.

Godlin, V.J. & Spensley, P.C. (1971) *Oils and Oilseeds.* Crop and Product Digests 1. Tropical Products Institute, Overseas Development Administration, London, UK.

Good, E.A.M., Stables, L.M. & Wilkin, D.R. (1977) The control of mites in stored oilseed rape. In: *Proceedings 1977 British Crop Protection Conference – Pests and Diseases.*

Gregoire, T. (1999) ProCrop (1999) Crop Production Database. Extension Service, North Dakota State University. Fargo ND, USA. http://www.ag.ndsu.nodak.edu/aginfo/procrop/saf/saffst09.htm.

Grochowicz, J. (1971) Machinery for cleaning and sorting of seeds (in Polish). PWRIL, Warsaw, Poland.

Harner, J.P. (1989) Handling and storage. Kansas State University Co-operative Extension Service, Manhattan KS.

Hayma, J. (1995) *The Storage of Tropical Agricultural Products.* 5th edition. Agrodok 31. Agromisa Foundation, University of Wageningen, The Netherlands.

HGCA (1998) Bulk storage drying of grain and oilseeds. *Topic Sheet 16.* Home-Grown Cereals Authority, London, UK.

Hoseney, R.C. (1986) *Principles of Cereal Science and Technology.* American Association of Cereal Chemists, St. Paul MN.

Hudson, R.D., Woodruff, J.M., Shumaker, G.A., *et al.* (1991) Canola a new crop for Georgia. Canola production guide. Georgia Extension Canola Committee, Tifton GA.

Hughes, A.M. (1976) The Mites of Stored Food and Houses. *Technical Bulletin 9*, Ministry of Agriculture, Fisheries and Food, Her Majesty's Stationery Office, London, UK.

Jayas, D.S. & Muir, W.E. (2001) Aeration systems. In: S. Navarro & R.T. Noyes (eds) *The Mechanics and Physics of Modern Grain Aeration Management.* CRC Press LLC, Boca Raton FL.

Jayas, D.S., Sokhansanj, S., Moysey, E.B. & Barber, E.M. (1987a) Distribution of foreign material in canola bins filled using a spreader or spout. *Canadian Agricultural Engineering*, **29**, 183–188.

Jayas, D.S., Sokhansanj, S., Moysey, E.B. & Barber, E.M. (1987b) Airflow resistance of canola (rapeseed). *Transactions of the American Society of Agricultural Engineers*, **30**, 1484–1488.

Jayas, D.S., Sokhansanj, S., Moysey, E.B. & Barber, E.M. (1987c) The effect of airflow direction on the resistance of canola (rapeseed) to airflow. *Canadian Agricultural Engineering*, **29**, 189–192.

Jayas, D.S., Kukelko, D.A. & White, N.D.G. (1988) Equilibrium moisture–equilibrium relative humidity relationship for canola meal. *Transactions of the American Society of Agricultural Engineers*, **31**, 1585–1588, 1593.

Jayas, D.S., Sokhansani, S. & White, N.D.G. (1989) Bulk density and porosity of two canola species. *Transactions of the American Society of Agricultural Engineers*, **32**, 291–294.

Jayas, D.S., White, N.D.G., Muir, W.E. & Sinha, R.N. (1993) Controlled atmosphere storage of cereals and oilseeds. *Journal of Applied Zoological Research*, **4**, 1–12.

Jayas, D.S., White, N.D.G. & Muir, W.E. (1995) *Stored-grain Ecosystems.* Marcel Dekker Inc., New York, USA.

Mills, J.T. (1976) Spoilage of rapeseed in elevator and farm storage in western Canada. *Canadian Plant Disease Survey*, **56**, 95–103.

Mills, J.T. (1989) *Spoilage and Heating of Stored Agricultural Products. Prevention, Detection and Control.* Agriculture Canada Publication **1823E**. Ottawa ON, Canada.

Mills, J.T. (1996) Storage of Canola. Government of Alberta, Canada. http://www.agric.gov.ab/crops/canola/storage1.html.

Mills, J.T. & Kim, K.M. (1977) Chemical and physiological characteristics of heat-damaged stored rapeseed. *Canadian Journal of Plant Science*, **57**, 375–381.

Mills, J.T. & Sinha, R.N. (1980) Safe storage periods for farm-stored rapeseed based on mycological and biochemical assessment. *Phytopathology*, **70**, 541–547.

Mills, J.T., Clear, R.M. & Daun, J.K. (1984) Storability of frost-damaged canola. *Canadian Journal of Plant Science*, **64**, 529–536.

Mohsenin, N.N. (1980) *Thermal Properties of Foods and Agricultural Materials.* Gordon & Breach Scientific Publications, New York, USA.

Moysey, E.B. (1973) Storage and drying of oilseeds. In: R.N. Sinha & W.E.Muir (eds) *Grain Storage: Part of a System.* AVI Publishing, Westport CT.

Muhlbauer, W. & Scherer, R. (1977) The specific heat of cereals (in German). *Grundlagen der Landtechnik*, **27**, 33–40.

Muir, W.E. (1973) Temperature and moisture in grain storage. In: R.N. Sinha & W.E.Muir (eds) *Grain Storage: Part of a System.* AVI Publishing Company, Westport CN, USA.

Muir, W.E. (1997) Grain Preservation Biosystems. University of Manitoba, Winnipeg MB, Canada.

Muir, W.E. & Sinha, R.N. (1986) Theoretical rates of flow of air at near-ambient conditions required to dry rapeseed. *Canadian Agricultural Engineering*, **28**, 45–49.

Muir, W.E. & Sinha, R.N. (1988) Physical properties of cereal and oilseed cultivars grown in western Canada. *Canadian Agricultural Engineering*, **30**, 51–55.

Muir, W.E., Waterer, D. & Sinha, R.N. (1985) Carbon dioxide as an early indicator of stored cereal and oilseed spoilage. *Transactions of the American Society of Agricultural Engineers*, **28**, 1673–1675.

Muir, W.E., Sinha, R.N., Zhang, Q. & Tuma, D. (1991) Near-ambient drying of canola. *Transactions of the American Society of Agricultural Engineers*, **34**, 2079–2084.

Nellist, M.E. & Bruce, D.M. (1995) Heated-air grain drying. In: D.S. Jayas, N.D.G. White & W.E. (eds) *Stored-grain Ecosystems.* Marcel Dekker Inc., New York, USA.

Oplinger, E.S., Punam, D.H., Kaminski, A.R., *et al.* (1997) *Sesame.* Alternative Field Crops Manual. University of Wisconsin/University of Minnesota. Madison WI/St. Paul MN. http://hort.purdue.edu/newcrop/afcm/sesame.html.

Pabis, S., Jayas, D.S. & Cenkowski, S. (1998) *Grain Drying: Theory and Practice.* John Wiley & Sons, Inc., New York, USA.

Patterson, H.B.W. (1989) *Handling and Storage of Oilseeds, Oils, Fats and Meal.* Elsevier Applied Science, London and New York.

Pichler, H.J. (1956) Sorption isotherms for grain and rape. *Journal of Agricultural Engineering Research*, **2**, 159–165.

Pixton, S.W. & Warburton, S. (1977) The moisture content/equilibrium relative humidity relationship and oil composition of rapeseed. *Journal of Stored Products Research*, **13**, 77–81.

Prasad, D.C., Muir, W.E. & Wallace, H.A.H. (1978) Characteristics of freshly harvested wheat and rapeseed. *Transactions of the American Society of Agricultural Engineers*, **21**, 782–784.

Prieto, S. & Leon, R.S. (1976) Influences of storage conditions and periods on germination of sesame seeds. Centro de Investigaciones Agropecuarias Región Centro Occidental (Venezuela), **6**, 35–40.

Rees, D., Boyd, B., Brown, B., *et al.* (1999*) Stored products and structural pests: commodity and pest ecology.* CSIRO Entomology Report of Research 1997–1999. CSIRO, Canberra ACT, Australia.

Reuss, R. & Pratt, S. (2001) Accumulation of carbon monoxide and carbon dioxide in stored canola. *Journal of Stored Products Research*, **37**, 23–34.

Ronco Consulting, (1998) Making Ukrainian Grain Production Profitable. Ronco Consulting Corporation, Washington DC. http://www.roncoconsulting.com/bulletins/asp/3/table2.html.

Salunkhe, D.K. & Desai, B.B. (1986) *Postharvest Biotechnology of Oilseeds.* CRC Press, Boca Raton FL.

Salunkhe, D.K., Chavan, J.K., Adsule, R.N. & Kadam, S.S. (1991) *World Oilseeds.* Van Nostrand Reinhold, New York, USA.

Sauer, D.B., Meronuck, R.A. & Christensen, C.M. (1992) Microflora. In: D.B. Sauer (ed.) *Storage of Cereal Grains and Their Products.* 4th edition. American Association of Cereal Chemists, St. Paul MN.

Shatadal, P., Jayas, D.S. & White, N.D.G. (1990) Thin-layer rewetting characteristics of canola. *Transactions of the American Society of Agricultural Engineers*, **33**, 871–876.

Sinha, R.N. (1966) Development and mortality of *Tribolium castaneum* and *T. confusum* on seed-borne fungi. *Annals of the Entomological Society of America*, **59**, 192–201.

Sinha, R.N. (1968) Adaptive significance of mycophagy in stored product arthropods. *Evolution*, **22**, 785–798.

Sinha, R.N. (1972) Infestibility of oilseeds, clover, and millet by stored-product insects. *Canadian Journal of Plant Science*, **52**, 431–440.

Sinha, R.N. (1976) Susceptibility of small bulks of rapeseed and sunflower seed to some stored-product insects. *Journal of Economic Entomology*, **69**, 21–24.

Sinha, R.N. (1984) Acarine community in the stored rapeseed ecosystem. In: D.A. Griffiths & C.E. Bowman (eds) *Acarology* VI. Vol. 2. Ellis Horwood Limited, Chichester, UK.

Sinha, R.N. (1988) Population dynamics of Psocoptera in farm-stored grain and oilseed. *Canadian Journal of Zoology*, **66**, 2618–2627.

Sinha, R.N. & Wallace, H.A.H. (1966) Association of granary mites and seed-borne fungi in stored grain and in outdoor and indoor habitats. *Annals of the Entomological Society of America*, **59**, 1170–1181.

Sinha, R.N. & Wallace, H.A.H. (1973) Population dynamics of stored-product mites. *Oecologia*, **12**, 315–327.

Sinha, R.N. & Wallace, H.A.H. (1977) Storage stability of farm-stored rapeseed and barley. *Canadian Journal of Plant Science*, **57**, 351–365.

Sinha, R.N., Mills, J.T., Wallace, H.A.H. & Muir, W.E. (1981) Quality assessment of rapeseed stored in ventilated and non-ventilated farm bins. *Sciences des Aliments*, **1**, 247–263.

Sokhansanj, S., Zhijie, W., Jayas, D.S. & Kameoka, T. (1986) Equilibrium relative humidity – moisture content of rapeseed (canola) from 5°C to 25°C. *Transactions of the American Society of Agricultural Engineers*, **29**, 837–839.

Srivastava, A.S. (1970) Important insect pests of stored oilseeds in India. *International Pest Control*, **12**, 18–20, 26.

Striewe, L. (2000) Grain and oilseed marketing in Ukraine. Iowa State University Ukraine Agriculture Policy Project. Iowa State University, Ames IA. USA.http://www.cper.kiev.ua/publicat/opapers/l1/ludwig1.htm.

Taylor, R.W.D. (2002) Remedial treatments in pest management. In: P. Golob, G. Farrell & J.E. Orchard (eds) *Crop Post-Harvest: Science and Technology. Volume 1 Principles and Practice.* Blackwell Publishing, Oxford, UK.

Thomas, P. (1984) Canola Growers Manual. http://www.canola-council.org.

Thomson, J.R. (1979) *An Introduction to Seed Technology.* Wiley, New York, USA.

UNCTAD (1978) *Storage Costs and Warehouse Facilities.* United Nations Conference on Trade and Development Secretariat, UNCTAD/CD/Misc. 75, United Nations, Geneva, Switzerland.

van Rheenen, H.A. (1981) Longevity of sesame seed under different storage conditions. In: A. Ashri (ed.) *Sesame: Status and Improvement.* FAO Plant Production and Protection Paper No. 29.

van S. Graver, J. & Annis, P.C. (1994) *Suggested Recommendations for the Fumigation of Grain in the ASEAN Region. Part 3. Phosphine Fumigation of Bag-stacks Sealed in Plastic Enclosures: An Operations Manual.* Australian Centre for International Agricultural Research, Canberra, ACT, Australia.

Wallace, H.A.H. (1973) Fungi and other organisms associated with stored grain. In: R.N. Sinha & W.E.Muir (eds) *Grain Storage: Part of a System.* AVI Publishing, Westport CT.

Warrick, W.E. (2001) Sunflower production guide. Texas Cooperative Extension. Texas A&M University, San Angelo TX. http://sanangelo.tamu.edu/agronomy.

Watters, F.L. & Nowicki, T.W. (1982) Uptake of bromophos by stored rapeseed. *Journal of Economic Entomology*, **75**, 261–264.

Weiss, E.A. (1971) *Castor, Sesame and Safflower.* Leonard Hill, London, UK.

Weiss, E.A. (1983) *Oilseed Crops.* Longman Tropical Agriculture Series, London, UK.

White, N.D.G. & Jayas, D.S. (1989) Safe storage conditions and infestation potential of canola meal by fungi and insects. *Journal of Stored Products Research*, **25**, 105–114.

White, N.D.G. & Leesch, J.G. (1996) Chemical control. In: Bh. Subramanyam & D.W. Hagstrum (eds) *Integrated Management of Insects in Stored Products.* Marcel Dekker, Inc., New York, USA.

White, N.D.G. & Nowicki, T.W. (1985) Effects of temperature and duration of storage on the degradation of malathion residues in dry rapeseed. *Journal of Stored Products Research*, **21**, 111–114.

White, N.D.G. & Sinha, R.N. (1979) Life history and population dynamics of the mycophagous mite *Tarsonemus granaries* Lindquist (Acarina: Tarsonenidae). *Acarologia*, **22**, 353–360.

White, N.D.G., Sinha, R.N. & Muir, W.E. (1982) Intergranular carbon dioxide as an indicator of deterioration in stored rapeseed. *Canadian Agricultural Engineering*, **24**, 43–49.

Chapter 9
Peanuts

M. Sembène, A. Guèye-NDiaye, C. L. Butts and F. H. Arthur

Peanut (*Arachis hypogaea*) is an herbaceous, annual legume, originating in South America (Plate 9). The pods of the peanut develop in the ground and hold one or more kernels that contain 38–50% oil and are also rich in protein. Peanut may also be referred to as groundnut, and after oil extraction the cake remaining is usually called groundnut cake. Currently, peanut is cultivated in subtropical savanna regions in sub-Saharan Africa, the USA, Middle East and Asia. It is consumed by humans as grain, flour, paste and oil. Groundnut cake and foliage are used as animal fodder and the oil is also used in soap production. This chapter considers the storage and handling of peanut in Senegal where, in a good year, about 1m t of pods are produced, and in the USA where the harvest can reach about 1.6m t. In Senegal, smallholder farmers deliver peanuts to a large state-dominated marketing system and in the USA large-scale farmers deliver to co-operative or commercial stores that use more modern facilities.

Peanut was introduced into Africa from Brazil by the Portuguese at the end of the fifteenth century (Adrian & Jacquot 1968; Pehaut 1976). Senegal has been the largest West African peanut producer since 1840 (Adrian & Jacquot 1968). The crop is nearly all rain fed and most of it marketed and stored as part of a campaign involving state, parastatal and private organisations. Only a small part of the total peanut production is stored by farmers and traders. Of the groundnut production in Senegal, 15% is used for domestic consumption and for artisanal oil production, 10% for seed grain and 75% for oil, processed industrially and destined, for the most part, for export markets. Plant breeders have selected several varieties according to the length of their growing times, adaptation to different agro-ecological zones and final product, i.e. for grain consumption or oil production (Anon. 1982). The principal constraints to peanut stor-

age are attack by insects, especially the peanut bruchid beetle (*Caryedon serratus*), rodents and mould, particularly *Aspergillus flavus* which produces aflatoxin, a carcinogenic mycotoxin causing serious contamination of peanut products (Vol. 1 p. 128, Wareing 2002).

The USA exports about 35% of its production while the bulk is consumed domestically as peanut candy (18%), snack peanuts (20%), peanut butter (39%), roasted in-shell (6%), or as oil (15%) (USDA 2001). Farmer stock peanuts are harvested from late August through to November and are normally stored in flat storage facilities until they are sent for shelling. The warehouse is gradually emptied as the peanuts are transferred to the shelling plants, and some of the peanuts can be stored for as long as 8–10 months. The shelling process removes the shells and the kernels are sized and transferred to bulk containers for use by food processors. In most years, very little farmer stock is carried over to the next harvest. If there is an excess of peanuts from the previous year, then these are usually shelled and kept in cold storage until they can be sold or crushed for oil. Storage of farmer stock peanuts begins at harvest as early as August and extends through to March of the following year. If peanut supplies are excessive then farmers may continue to store until May or June. The main problems are insect infestation, mostly by the Indian meal moth (*Plodia interpunctella*) and mould problems leading to aflatoxin contamination.

Senegal

Historical perspectives

In the 1960s, nearly 50% of arable land in Senegal was used for the cultivation of peanuts. This commodity accounted for 80% of exports and provided the major monetary revenue for rural areas (Anon. 2000). Peanuts are harvested when mature, at the end of the rainy season in October–November; at this time their moisture content is around 40–50%. Peanuts for immediate human consumption, rather than storage, are roasted or boiled, then decorticated and put on the market. Those for storage, which will later be used for seed, oil production or domestic consumption, are first dried in the field. After lifting from the ground they are arranged in windrows and, when sufficiently dry, stacked in the field. At this stage they should have dried to about 10% m.c. Drying is followed by threshing to separate the pods from leaves and stalks (haulm).

The majority of the crop (80%) is sold within a few weeks of harvest at the start of the annual commercialisation campaign. Up to 1980, a state body, ONCAD (National Office for Co-operation and Assistance in Development), intervened through agricultural co-operatives by distributing seed and credit, purchasing the harvest, then selling and delivering it to oil extractors (NDiaye 1981). Since 1984, with the establishment of a new agricultural policy, the state has progressively disengaged from the market to the profit of private operators, with the result that farmers store their own seed grain (NDiaye 1991). Two parastatal organisations, SONACOS (National Society for the Commercialization of Oilseeds in Senegal) and SONAGRAINES (National Society for Grain) together with private enterprises of more limited scope, have assumed the duties of ONCAD. The private organisations have arisen recently and include UNIS (The Inter-professional Union for Seeds), NOVASEN (Body for the Exploitation of Peanuts in Senegal) and CNIA (The Inter-professional Council for Peanuts). The two parastatals and the private bodies are now responsible for regulating the market. However, since 1989 there has been a collapse in national production which now usually only reaches 250 000–300 000 t. This is due to several factors, not least problems with seed grain (Freud *et al.* 1997). However, in the year 2000, rains were exceptionally good and production was estimated to be 1m t.

Physical facilities

Farmers generally store a small quantity of their harvest (about one-fifth) using simple methods such as bulk storage in ventilated granaries (Fig. 9.1) made with local materials such as millet stalks or a basketwork made from the common shrub *Combretum glutinosum.* Now more commonly they use woven polypropylene

Fig. 9.1 A basketwork store for holding groundnuts on farm (Senegal).

sacks, stacked one on the other (NDiaye 1991). Sack storage is useful as it enables easy delivery of the pods to temporary outdoor enclosures referred to as *seccos*, where further drying can take place.

According to DISEM (Division for Seeds), some farmers store their nuts for seed in the pod in metal drums or cans. SONAGRAINES protects peanut seed in ventilated metal stores (Fig. 9.2) of about 250 t capactiy. Technicians, following a manual, are responsible for the management of these stores, ensuring that the sacks are placed in correct and well-identified piles and that the appropriate phytosanitary procedures are applied. This activity continues until the distribution of seeds to the producers, two to four weeks before the start of the rains.

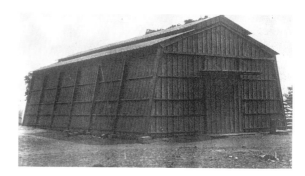

Fig. 9.2 A metal store for preservation of groundnut seed, capacity 250 tonnes (Senegal).

From December onwards, a few weeks after threshing and some conditioning in polypropylene sacks, farmers deliver their pods by cart to small *seccos* of 200–800 t capacity. If the peanuts are destined for oil production then they are resold by SONAGRAINES to SONACOS. The delivery of peanuts to the different processing plants of SONACOS is achieved using large lorries (Fig. 9.3). If processing is not achieved rapidly, the pods enter bulk storage again, this time in large *seccos* in the form of pyramids, with a base and sides formed by sacks of pods to assure stability (Fig. 9.4). These pyramids can hold 10 000–20 000 t. The construction and phytosanitary care of these stocks is under the control of the SONACOS technical teams using specified procedures and products.

Fig. 9.3 Lorry laden with groundnuts being carried to the oil factory (Senegal).

Fig. 9.4 Groundnut collection pyramid, *secco*, managed by SONACOS, capacity 10 000-20 000 tonnes (Senegal).

Objectives of storage

The storage of peanuts assures that they can be consumed throughout the year, that good-quality seed is available for the next agricultural campaign and that there is a constant supply to the oil extraction and peanut confectionery industries for local and export markets.

Farmers' stocks of peanuts are used mainly for domestic consumption, transformed generally into flour or paste for the preparation of cereal-based dishes, which are their main diet. In rural areas, certain quantities are marketed for the production of oil and peanut cake by artisans while in towns the roasted nuts are the subject of dynamic, small-scale commerce. In all these cases, the quality of the grain concerned is affected by how recently the nuts have been harvested and the diligence of producers, traders and buyers.

For the marketing campaign there are official quality standards for seeds, and peanuts for both oil and food. At the time of purchase and assembly of the peanuts by the producers, impurities are removed by passing across a manual screen (Fig. 9.5). The managers check the quality of the grains by a rapid observation and verification of certain key parameters such as the bulk density for seed grain. On arrival at the *seccos* the quality of shelled nuts is judged by OPS (Private Storage Organization) managers who will accept or refuse them after inspection. Those destined to be used as seed grain are controlled by DISEM who are responsible for their certification before storage. Seeds are chosen according to the following quality criteria:

• Bulk density (g L⁻¹ of pods). This is defined officially each year for each variety and grading category.

Fig. 9.5 Groundnuts being passed over a manual screen to remove impurities (Senegal).

- High level of varietal purity, approaching 100%.
- Good sanitary state: the proportion of bruchid-infested grains must be less than 2.5%.
- Impurities not more than 2%.
- Viability more than 80%.

Nuts destined for oil production are subject to systematic quality control but only in cases of doubt, at the time that ownership is transferred, are samples presented to SONACOS for verification. Checks are made of m.c., which should not exceed 5–7%, and oil content that should not be below 38%. In cases of non-conformity in either of these two factors the procurement price is lowered. Nevertheless, impurities are strictly controlled. SONACOS uses a motorised screening at 20 r.p.m. and the impurity levels found in samples are used to make deductions from the price of the whole consignment. Stocks wetted by rainfall are refused or purchased at one-tenth of their value, to be used as furnace fuel. Confectionery peanuts for export are priced according to their size and whether they have been sorted by hand or mechanically.

Major sources of quality decline

Mould, insects and rodents are the principal enemies of stored peanuts. Field drying exposes pods to attack by various rodents such as *Xerus erythropus* and *Cricetomys gambianus* as well as to insect pests. The bug *Elasmolomus* (*Apahanus*) *sordidus* pierces the peanut pods and extracts juices from the nuts when these are drying in the field and when in store. The nuts can become dried and wrinkled as a result of this attack, which lowers the oil content and raises the free fatty acids so causing rancidity and making the nuts unpalatable. The bruchid beetle *Caryedon serratus* is another serious pest of stored peanuts (Vol. 1 p. 100, Hodges 2002a). Adult beetles glue their eggs to peanut pods or on to the nuts after decortication, although most attack starts before storage while the pods are drying in the field. The larvae bore into the seeds (Fig. 9.6), but may leave one seed and attack another. Pupation takes place outside the seed in a paper-like cocoon that is spun by the larva. Besides peanuts, *C. serratus* also develops on several wild hosts (Conway 1974; NDiaye 1991; Sembène 1997). Other insects, such as the beetles *Trogoderma granarium* and *Tribolium castaneum* and the moth *Ephestia cautella* are also associated with stored peanuts.

Early observations on the biology of *C. serratus* concluded that the most important infestation of pea-

Fig. 9.6 Damage to peanuts by the bruchid beetle, *Caryedon serratus*, the predominant peanut pest in West Africa. A peanut (centre) has been cut open to show a larva and a pod cut open (bottom right) to show a pupa between the peanuts in a pod.

nuts by this pest occurred in store (Appert 1957; Green 1959). This led to recommendations for the control of *C. serratus* whereby insecticide was admixed with the outer layers of the stored pods, since infestation was usually restricted to this location (Green 1959). However, subsequent detailed study of the biology of the pest showed that the nuts generally become infested when drying in the field and that there is a sequence of primary host species supporting *C. serratus* throughout the year (Conway 1974). It was shown that elimination of this initial infestation by fumigation of the pods with methyl bromide or phosphine was effective and obviated the need for any insecticide admixture, except in situations where fumigation could not be used. In The Gambia, fumigation became the standard technique for all seed on agricultural stations, commercial farms and the entire confectionery peanut production. Recently in Senegal, morphometric and genetic analyses of different strains of *C. serratus* have shown that the populations of this insect are divided into host races (biotypes) associated with each plant host. There would appear to be a high rate of exchange of *C. serratus* between peanuts and pods of the plant *Piliostigma reticulatum*. During the cooler period of the year there remains a small reproductively active population of *C. serratus* which infest the newly harvested peanuts. This infestation, although feeble during the first two months, later becomes the origin of large quantitative losses since the life cycle of *C. serratus* is rapid and can be completed several times in the course of a year. This analysis of population biology has confirmed the origin of infestation in Senegal of newly harvested peanuts, due to female *C. serratus* issuing from the pods of *P. reticulatum*. Thus, if the peanut pods that become infested in the field could be selected out and if the drying period could be shortened to minimise field exposure, then the occurrence of infestation would be much reduced (Sembène & Delobel 1996, 1998; Sembène *et al.* 1998).

Damage to peanuts varies according to the variety and the length of storage. The subject has been studied for many years on large-scale confectionery and oilseed stocks. For example, after storing an 18 t experimental stock of peanuts for three months, the infestation by *C. serratus* was confined to a few centimetres from the surface although the losses caused by the pest can escalate rapidly from 8% to 40% during six months of storage (Pointel & Yaciuk 1979). However, the losses from stocks of several thousands of tonnes would be expected to be somewhat lower as larger grain masses present a lower surface to volume ratio, and so the proportion of grain on the exterior that is liable to attack is somewhat lower. At least 83% of nuts in pods can be attacked after six weeks' storage in the commercial centres if no phytosanitary treatments are used (NDiaye 1991). Some authors (NDiaye 1991; Sembène 1997) have suggested that the nuts in shell are more vulnerable to *C. serratus* than the decorticated nuts (60–90% attacked compared with 30–70% attacked). This may not be correct since a pod holds two or three nuts and not all these have necessarily been attacked. A realistic estimate of the annual losses due to *C. serratus* in The Gambia amounts to 3% (Friendship 1974).

Commodity and pest management

In the past it was common for farmers to preserve their small stocks of peanuts by placing then in granaries together with the leaves of *Azadirachta indica* (neem), *Boscia senegalensis*, *Datura* spp and *Malpighia* spp (see, for example, Seck *et al.* 1993). However, these botanicals do not seem to be very effective as in real storage conditions they do not prevent the development of infestation by *C. serratus* and they have been more or less supplanted by the use of synthetic pesticides, particularly dilute dust insecticides and the fumigant phosphine. Small-scale farmers sometimes attempt to use these chemicals but are often unaware of the correct application methods. Some recent studies have shown that storage of peanut in pods in woven polypropylene

sacks, with a continuous plastic liner, offers better protection of the harvest than bulk storage in traditional bins (Sembène & Delobel 1998). In this semi-airtight storage, the composition of the interstitial air is modified. There is a decline in oxygen and a rise in carbon dioxide that prevents the development of insects and moulds. An evaluation of such an improved system of storage in tropical conditions has given encouraging results (Sembène 1997). Thus, after threshing and sorting the pods should ideally be put in unholed, lined woven polypropylene sacks and when these are full they should be sealed hermetically.

Seed grains are given priority in the phytosanitary plan of the technical services. In addition, they are given more shelter by being placed in metal stores. However, when the latter are insufficient the surplus is stored in the open-air *seccos*. The nuts for oil, stored in bulk at the *seccos*, are left under the supervision of managers who are entirely responsible for their protection up to their discharge to the oil factory.

The SONACOS–SONAGRAINES group signs contracts with OPS for storage. They are contracted to purchase pods from producers and manage them in *seccos* or metal stores. The stores and pods are subject to various phytosanitary treatments including fumigation, or dusting/spraying with insecticide. The only method used against *C. serratus* is conventional chemical control. Organochlorine pesticides such as DDT and BHC were used but these have been replaced by organophosphorous compounds, first lindane and malathion and more recently by fenitrothion, chlorpiriphos-methyl, pyrimiphos-methyl and phoxim (Appert 1985). Methyl bromide, formerly employed for fumigation, is now often replaced by phosphine, which is somewhat easier to use. These treatments are undertaken according to well-specified protocols, and are of particular importance for the protection of seed. SONAGRAINES allow OPS 0.5% of loss on the quantity of peanut procured, whatever the duration of storage. This generally lasts from December through to March, but can be prolonged until June or later if there has been a good harvest. This risks losses due to rainfall if the stock has not been covered properly with tarpaulins. At the approach of the rainy season the peanuts must be sheeted correctly to avoid wetting that could favour mould growth, of which *A. flavus* is particularly important as the source of aflatoxin. Some sheets are hired for this purpose but the personnel responsible for the work must be trained to ensure that this important job is done well.

If SONAGRAINES' deliveries of peanuts for oil to SONACOS are made late, the risk of infestation can increase, even though during this period they are in the phytosanitary care of OPS. Samples are taken regularly to check for infestation, particularly by *C. serratus*, and regular treatments are applied. At the oil factory peanuts of different origins are mixed when the large-capacity *seccos* (Fig. 9.4) are constructed. This increases infestation problems with *C. serratus* despite the phytosanitary treatments applied.

Good hygiene is also an important element of commodity management. It is essential that any *C. serratus* infesting the harvest of the preceding season are prevented from passing onto the new crop. All stocks should be removed at least two months before the entry of the new harvest and the stores should be cleaned carefully. Particular care should be taken with sacks of peanuts that may be carrying the adults and pupae of *C. serratus*. To minimise infestation problems the period of purchase and collection from farmers should be as short as possible and sacks showing signs of infestation must be isolated and treated.

Other techniques not involving synthetic pesticides are possible. Refrigerated storage for the preservation of first-quality seed grain has been shown to be effective (NDiaye 1981), but this has not been extended on account of its high cost. However, it is used on a small scale by ISRA (the Senegalese Institute for Agricultural Research) for their seed bank. For the same purpose, ISRA has investigated storage under vacuum and under nitrogen. The technique involved packing shelled peanuts into 500 L airtight polythene bags called *Capitainers*, subjecting them to a vacuum of 87 kPa and then closing the bags by heat sealing. The bags were not strong enough to resist either rodent attack or sustained high vacuum (300 mmHg). The latter problem was solved by initial application of a high vacuum followed by reinjection of nitrogen. It was concluded that packing under vacuum or in nitrogen was effective in preventing insect infestation without affecting seed germination (Rouziere 1986a, b). To date the costs of these methods have not been evaluated. Other unconventional methods that have shown promise but have not been adopted include:

- irradiation of nuts or pods (NDiaye 1981);
- biological control using the parasitic wasp *Uscana caryedoni* Viggiani (Hymenoptera: Trichogrammatidae), the egg parasite of *C. serratus* (Delobel 1989);
- admixture of peanut or castor oil to dried nuts (5 cm^3 kg^{-1}) which are then stored in a hermetically sealed container; and

- admixture of an inert dust mixed with the decorticated nuts at 5 g kg^{-1} (NDiaye 1991).

Economics of the operation of the system

Producers prefer to sell the major part of their harvest quickly and keep only a small portion. They do this to avoid losses and to raise cash to pay their debts and cover family expenditure. The cost of storage at this level are not considered to be significant whereas the costs to the large-scale organisations are considerable. For example, for SONAGRAINES the costs include:

- Payment of permanent technical and administrative staff as well as the deputies and mangers of the *seccos* who are recruited specifically for the peanut campaign.
- Maintenance and repair of storage buildings and areas as well as screens and scales. This is estimated to cost between FCFA 10 000 and 25 000 annually for each collection point. Of 1050 points surveyed, the storage capacity varied from 200 t to 800 t during the 2000 season.
- Cost of phytosanitary treatments during the storage period, the cost of purchase of jute or polypropylene sacks as well as the hire of sheets to cover the stock during rainy periods. On average, about 1000 sheets are hired each year at FCFA 45 000 per unit, making FCFA 45m in total.

Taking all these different elements together, SONAGRAINES estimated that its storage costs per tonne would be FCFA 2500 in a campaign where 300 000 t were collected, although economies of scale prevail and this rate is lowered at times of greater procurement. In the year 2000, of a national production of about 1m t, SONAGRAINES procured 550 000 t. The loss allowed to the private storage companies between collection and evacuation of stock is usually set at a maximum of 0.5% of the total initial weight. In the case of heavy infestation by *C. serratus*, some negotiation between the parties concerned can result in a small raise to this maximum.

SONACOS declined to reveal the costs of storage of peanuts for oil purchased by SONAGRAINES, and waiting to be processed, but state that costs increase with the duration of storage.

Future developments

According to Freud *et al.* (1997), the objectives of the State of Senegal for the peanut market up to the year 2010 will be:

- Production of food oil: 190 000 t to meet national requirements and 100 000 t for export, as well as the resulting groundnut cake.
- Production of 100 000–110 000 t of quality seed.
- Production of 200 000 t of peanuts for subsistence consumption and local marketing.

To implement these objectives, storage is an unavoidable activity and therefore requires serious attention. With globalisation, export products must be competitive. SONACOS, for a few years now, has been working to obtain ISO 9002 certification in order to bring about improvements in quality from the time of harvest onwards and ensure that the pesticides employed by the managers of *seccos* conform to the legislation of oil and groundnut cake importers.

Despite the recommendations of the new agricultural policy, farmers do not generally have their own seed grain. They sell their own harvest as soon as the *seccos* are open, from fear that it will deteriorate. Later they will have to purchase seed on credit from SONAGRAINES. These credits are never fully reimbursed (only 76% in 2000). There is insufficient seed of high quality and a good part of the seed that is used actually comes from the better grain set aside for oil production.

International financial institutions have recommended the full privatisation of the peanut market but this has still not been fully implemented. Peanut, being the main source of monetary income for most farmers, constitutes a sensitive sector in which the government will always try to intervene. Nevertheless, the new structure put in place, in particular CNIA, but also UNIS, should take over the organisation of the market. UNIS should in time replace SONAGRAINES for the supply of seeds. The multiplication of seeds selected by plant breeders are generally those adapted for rain-fed conditions. However, this could now be extended to the selection of varieties appropriate to the irrigated conditions of Senegal River valley where production of edible peanut is open to expansion (Dancette 1997).

Cultural practices before storage appear to be important for the protection of stocks against *C. serratus*. In this case, recent studies have shown that the nature of control methods rests on the fact that the principal pest of peanuts is considered as a storage pest when in fact infestation is often already established at the time of drying. The efficacy of control measures appears to be incontestably linked to a knowledge of the origin and evolution of primary infestation in peanuts. The problem of conserving peanuts should be reviewed taking account of pre-infestation in the field.

United States of America

Historical perspectives

During the early 1900s, devastation of the cotton crop in the southeastern USA by the boll weevil prompted farmers to grow peanut as an alternative crop. Peanut production increased during World War I as equipment was developed to improve planting and harvesting (McGill 1973; Hammons 1982). As production rose, the use of off-farm storage facilities increased. The US Government designated peanut for oil as an essential crop in 1941. Research led to improved varieties and production techniques that resulted in significantly higher yields between 1948 and 1950. The vines with peanuts attached, after being dug up, were placed in stacks to cure (dry) for several weeks and sometimes months (Fig. 9.7). When the peanuts were ready for harvest, a stationary thresher was placed near the stacks and peanuts separated from the vine and bagged. The development of the peanut combine in 1948 by W.D. McKinney (USDA) and J.L. Shepherd (University of Georgia) facilitated major increases in production (Hammons 1982). Even with the development of the peanut combine, peanuts were still being bagged and stored in the fields until curing was complete.

Today, a digger-shaker lifts the plant from the ground, gently shakes the soil from the peanuts and places the plant upside down in windrows to cure in the sun for

Fig. 9.7 Peanuts placed on stackpoles for curing and storage (US).

two or three days. Combining is the second phase of the harvest. After drying in the field, a combine separates the peanuts from the vines, placing the peanuts into a hopper on the top of the machine and depositing the vines back in the field. Freshly dug peanuts are then placed into peanut wagons for further curing with forced warm air circulating through the wagon. In this final stage, the m.c. is reduced to 10% for storage. As peanut production increased, storage moved from the farm to commercial facilities owned and operated by the peanut shelling industry. Harvested peanuts that have not been shelled, cleaned or crushed are referred to as the farmers' stock peanuts and over 90% of these stored in the US are kept in commercial facilities. Generally, the only peanuts stored on-farm are those held by the grower for seed.

Physical facilities

Farmer stock peanuts are stored in large, flat, bulk storage facilities. Conventional storage facilities are usually 24 m wide steel buildings (Fig. 9.8) and vary in length from 55 m to 122 m. The height of the eaves is usually 7.3 m. The roof pitch is usually 37° or 45°. Warehouses with a 37° roof usually have an additional structure, called a doghouse, to accommodate a conveyor and ventilation equipment (Fig. 9.9). The steeper roof pitch is easier to build though it increases the amount of free space above the peanuts that must be ventilated. The lower pitched roof (37°) costs less to build but, if the warehouse is overloaded, so that peanuts rest against the roof, it may be damaged. In addition, minimal free space above the peanuts may increase the difficulty of installing ventilation equipment. Another type of storage structure is the Muscogee style building that often has

Fig. 9.8 Typical peanut warehouse with 45° (12 : 12) roof pitch (US).

Fig. 9.9 Typical peanut warehouse with 37° (9 : 12) roof pitch and doghouse that houses conveyors and a ventilation system (US).

a lighter framework than the conventional warehouse, with a roof sloping to all four walls from the centre peak. The eave height is typically lower than that of most conventional warehouses (Smith *et al.* 1995). Desirable characteristics for peanut warehouses are summarised in Table 9.1. Orienting the building to have a north–south roof ridge will give wall and roof surfaces more uniform exposure to solar radiation and so minimise problems with condensation (Smith 1994; Smith *et al.* 1995).

As the warehouse ages, the metal roof may begin to leak. As an alternative to roof replacement, some warehouse owners opt to install a second layer of corrugated metal, separated by an air gap, over the existing roof (Smith *et al.* 1995). The cost of the double roof is about the same as removing and replacing an existing roof. To position the second layer, metal purlins approximately 5 cm tall are installed every 3 m on top of the existing roof, parallel to the ridge. Corrugated metal roofing is then installed on the new purlins. Hardware cloth is fitted at the eaves to prevent birds from nesting in the air space between the roof panels. The double roof reduces the maximum temperature in the free space in the store. In a warehouse with a single roof, the maximum roof temperature often exceeds the free space temperature by as much as 15°C and the minimum temperature may be 4–5°C cooler than the free space. The night-time temperature of the single roof is often below the dew point, causing condensation and drip lines on the peanuts below (Smith *et al.* 1985). The double roof virtually eliminates drip lines (Smith *et al.* 1995) because the roof surface closest to the peanuts is insulated by the air space

Table 9.1 Some desirable design features for farmers stock peanut warehouses (adapted from Smith *et al.* 1995).

Component	Desirable characteristics	Function
Site	Clean, elevated, graded and well-drained	To prevent water entry and remove havens for insect/rodents
Building orientation	North and south	To provide uniform sun exposure and minimise condensation
Approach and exit to dump pit and doors	Concrete or asphalt paving sloped away from dump pit and doors	To prevent water and additional foreign material from entering pit and warehouse
Foundation and floor	Steel-reinforced concrete with vapour barrier underneath floor of 15 cm minimum thickness	To support the weight of the building, peanuts, the loading and unloading vehicles and maintain a dry floor
Exterior walls and roof	Steel or concrete adequately designed for integrity and strength with no cracks or crevices and reflective exterior finish or coating	To withstand loads of peanut, wind, ice, and snow, while preventing leaks, entry by rodents, insects and birds and to reflect solar radiation
Interior	Open floor space with no beams, posts, or other obstructions. No full height partitions if ventilated mechanically	To provide easy removal of peanuts and unrestricted headspace ventilation
Loading equipment	Dump pit, elevator, and overhead conveyor or mobile conveyor loader	To provide rapid but gentle loading at minimum cost
Unloading equipment	Tunnel with floor conveyor, draw ports and front end loader or mobile conveyor unloader	To provide rapid but gentle unloading at minimum cost
Precleaning equipment	Conventional precleaner or belt screen or cleaning equipment on elevator down spouting	To remove foreign materials that restrict air movement during storage and/or that add additional moisture and/or take up storage space
Ventilation system	Headspace ventilation that will provide at least one air change every 3 minutes	To remove excess moisture and heat, to minimise condensation and reduce headspace temperatures
Aeration system	In-floor aeration that will deliver 0.1 $m^3 min^{-1} m^{-3}$ uniformly through the peanuts	To cool peanuts rapidly and remove excess moisture. Maintain moisture content greater than 7%
Insect control system	Designed for fumigation and automated space treatment	To provide good insect control with a minimum risk of adding to moisture problems

between the two roof panels and rarely falls below the dew point of the air in the free space.

The free space in stores must be well ventilated to maintain proper temperature and humidity conditions. Ventilation systems may be natural or mechanical. A warehouse that is naturally ventilated has a continuous vent along the ridge of the roof and inlet vents along the eaves. As the headspace heats up, the ridge vent allows the hot air to escape. As the hot air exits through the ridge vent, it draws in fresh, cooler air at the eaves. However, the warehouse manager has no control over ventilation rates and this may result in over-dried peanuts. Tests have shown that the average headspace temperature in a naturally ventilated warehouse is not significantly different from that in a mechanically ventilated warehouse (Smith *et al.* 1984). Nonetheless, the risk of condensa-

tion is significantly higher in the naturally ventilated warehouse. An adequate mechanical ventilation system includes sufficient fan capacity to completely exchange the air between the top of the peanuts and the warehouse roof every two to three minutes and enough inlet area so that the air velocity entering the warehouse is less than 305 m min^{-1}. Fans are usually installed in the gable end of the warehouse with 45° or 90° exhaust hoods. Hooded inlets are located on the gable opposite the fans and compose approximately 90% of the required inlet area; the remainder of the inlet area is through the eaves. Air is drawn by fans through the inlets, across the top of the pile of peanuts and then passes out through the exhaust hoods (Smith 1994; Smith *et al.* 1985; Anon. 2001).

After farmer stock peanuts are purchased from the grower, they are emptied from the transport container,

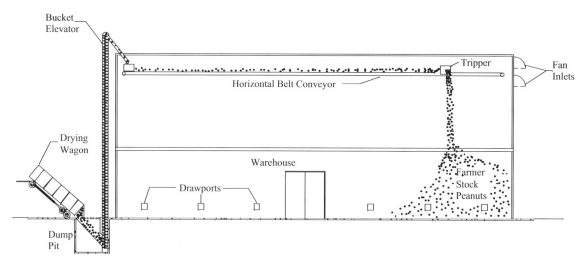

Fig. 9.10 Diagram of a typical peanut warehouse loading system (US).

either a peanut drying trailer or a large truck trailer, into a dump pit (Fig. 9.10). From the dump pit, the peanuts empty into a bucket elevator and into the warehouse. There is usually an in-line cleaner installed in the elevator downspout to remove dirt and other foreign material before the peanuts enter the warehouse. The conveyor belt has a movable device called a tripper (Figs 9.10 and 9.11) that diverts the peanuts off the belt where they fall

up to 18 m onto the floor or onto other peanuts. The tripper is moved back and forth manually over the length of the warehouse, so layering the peanuts until they are within 0.3 m of the eaves. The peak of the pile is approximately 8 m above the eaves.

When it is necessary to transport peanuts from the warehouse to the shelling plant, small doors, or drawports, near the base of the walls are opened. Peanuts flow from the drawport into the hopper of a belt conveyor, up the conveyor and into a waiting hopper-bottom truck. Up to 40% of the peanuts stored in the warehouse may be unloaded through the drawports (Fig. 9.12). Once the

Fig. 9.11 Typical tripper mechanism for distributing peanuts in warehouse (US).

Fig. 9.12 Peanuts damaged by the larvae of the Indian meal moth, *Plodia interpunctella*, the predominant peanut pest in the US. The larvae leave silken trails that build up into a 'webbing' that can be seen connecting the damaged peanuts.

Fig. 9.13 Peanuts flowing from drawport in the warehouse bulkhead (US).

level of peanuts has fallen such that they no longer flow out of the drawport, the bulkhead is removed and peanuts are loaded into the conveyor hopper with a front-end or skid steer loader (Fig. 9.13).

Objectives of storage

The objective of peanut storage is to maintain product quality from the time peanuts are delivered by the farmer until they are ready for shelling and processing. Quality deterioration is minimised by proper maintenance of equipment, proper control of product moisture and temperature, and adequate insect control. Good storage results in peanuts between 7% and 8% m.c., minimal mechanical or insect damage and no detectable increase in aflatoxin contamination. Peanuts at 7.5% m.c. are in equilibrium with air at 70% relative humidity.

Major sources of quality decline

Major sources of quality decline for stored peanuts include improper handling, poor moisture and temperature control, and insect damage. Poor moisture and temperature control increase the risk of aflatoxin contamination during storage (Sanders *et al*. 1981). Improper handling includes poorly maintained elevators, conveyors and cleaners as well as excessive speeds on the elevators and conveyors. Poor handling practices result in increased loose-shelled kernels (LSK), split kernels and foreign material (FM). LSK represent a decrease in value of over $US1 kg^{-1}. Insect populations and insect-damaged kernels increase as LSK percentage rises from 0 to 5% (Arthur & Redlinger 1988). The peanut shell offers some protection to the kernel from insect damage and so a rise

in LSK increases the food available to insects. Above 5% LSK, food is not the limiting resource for insect growth.

The impact of peanuts as they are loaded into a store can be a source of quality decline. When a warehouse is first loaded, peanuts fall 18 m onto a concrete floor. However, the majority of peanuts experience peanut-on-peanut impact during loading with falls ranging from 18 m to 1 m. Foreign material and LSK increase by approximately one percentage point, primarily due to impact damage (Slay 1976; Butts & Smith 1995). Seed germination and milling quality decrease significantly as impact velocity increases (Turner *et al*. 1967). The rate at which FM, LSK and split kernels occur increases as impact velocity exceeds 6 m s^{-1}.

Poorly designed or maintained handling equipment can significantly increase mechanical damage and losses. When bucket elevators operate at belt speeds from 43 to 116 m min^{-1} most damage occurs when belt speeds exceeded 61 m min^{-1} (Slay & Hutchison 1973). Greater conveying rates can be achieved by using large buckets (23 × 14 cm) rather than just increasing conveyor speeds. Using the maximum belt speed of 61 m min^{-1} as a guide, the maximum belt speeds for bucket elevators with various diameter head pulleys have been calculated assuming that the centrifugal force exerted by the 20 cm head pulley would be the same as a larger head pulley (Table 9.2). For instance, the maximum belt speed for an elevator with a 20 cm diameter head pulley is 61 m min^{-1}, while a 90 cm diameter head pulley will have an acceptable belt velocity of 106 m min^{-1} (Smith 1994). The maximum speed for a horizontal conveyor belt should not exceed 61 m min^{-1} (Smith 1994).

In many cases, about half the value loss of peanuts during storage can be attributed to the shrinkage of m.c. below 7% (Butts & Smith 1995). An example of the calculations used to determine the value of peanuts when purchased is shown in Table 9.3. The net weight is the weight of peanuts (LSK and in-shell) at 7% m.c. The gross weight decreases due to handling losses during loading and moisture loss during storage. No adjustment in the price paid to the farmer is made when shrinkage is below 7% m.c. If the percentage foreign material and LSK remain the same as shown in the example, the 0.5% excessive moisture loss leads to a reduction in value of $US10 000. Instead of continuously operating the ventilation fans, ventilating the headspace only when temperature and humidity conditions were unfavourable minimised the economic loss due to excessive moisture loss during storage (Butts & Smith 1994).

Table 9.2 Recommended elevator speeds to minimise mechanical damage to peanut (adapted from Smith *et al.* 1995).

Head pulley diameter (cm)	Effective radius (cm)	Velocity (m min⁻¹)	Head pulley speed (rpm)
20	17.62	61	97
25	20.12	65	83
30	22.62	69	73
35	25.12	73	66
40	27.62	77	61
45	30.12	80	57
50	32.62	83	53
55	35.12	86	50
60	37.62	89	47
65	40.12	92	45
70	42.62	95	43
75	45.12	98	42
80	47.62	100	40
85	50.12	103	39
90	52.62	106	37
95	55.12	108	36
100	57.62	110	35
105	60.12	113	34
110	62.62	115	33
115	65.12	117	32
120	67.62	120	32
125	70.12	122	31
130	72.62	124	30
135	75.12	126	30

Commodity and pest management

Peanut harvest in the southeastern US generally begins during the first week of September and lasts through the second or third week in October. The most southern peanut-producing area in Florida may start as early as the third week in August, while harvest may not begin in the west Texas and Virginia/North Carolina until the third week in September. As long as weather permits, approximately 80–90% of the peanuts are harvested in four to six weeks.

Current marketing regulations require that the moisture content of the peanut kernels be less than 10.5%. Other quality factors measured for incoming peanuts include percentage edible kernels and the presence or absence of the aflatoxin forming mould, *Aspergillus flavus*. The weight of the peanuts in a transport container is calculated based on the gross weight of material in the container, the percentage foreign material, percent m.c. and the percentage of loose shelled kernels measured in a 1.5–1.8 kg sample extracted from the load. The weight of foreign material, more than 7% m.c. and loose-shelled kernels are subtracted from the gross material weight to determine the weight of peanuts for which the farmer is paid (Table 9.3). The farmer currently receives $154 t⁻¹ for the loose-shelled kernels in the load. The price for in-shell peanuts changes according to the international

Table 9.3 Example calculations of peanut value before and after storage, assuming only a change in moisture content.†

	Before storage		After storage	
	Grade factor (%)	Weight (tonnes)	Grade factor (%)	Weight (tonnes)
Gross weight (GWT)		5 000.00		4 858.82
Foreign material (FMWT = FM%/100*GWT)	4.82%	241.00	4.82%	234.20
Gross less foreign material (GLFM=GWT − FMWT)		4 759.00		4 624.63
Moisture content (MC)	9.14%		6.50%	
Excess moisture content (EM = max[O, (MC − 7)/100*GLFM])		101.84		0.00
Net weight (NET = GLFM − EM)		4 657.16		4 624.63
Loose shelled kernels (LSK WT = LSK %/100*GWT)	4.48%	224.00	4.48%	217.68
In-shell peanut weight (ISP = NET − LSK WT)		4 433.16		4 406.95
LSK value‡ US$ 154/tonne*LSK WT		US$34 496		US$33 522
In-shell value§ (US$ 344/tonne*ISP)		US$1 525 006		US$1 515 919
Total value		US$1 559 502		US$1 549 514

†Example grade factors before storage for foreign material, moisture content and loose shelled kernels taken from Butts & Smith (1995).
‡Loose shelled kernel value equivalent to price paid to grower for peanuts crushed for oil.
§In-shell value is the weekly price published by the United States Department of Agriculture for the week of 11 February 2003.

markets, and is published weekly by the United States Department of Agriculture (Farm Service Agency 2003). If the peanuts originated from a non-irrigated field with moderate to severe drought stress, the buyer may assay the sample for aflatoxin. Based on the incoming peanut quality, aflatoxin concentrations and their potential final use, the buyer segregates the peanuts into various warehouses or sends them directly to the shelling plant.

Most peanuts are stored in-shell as this protects the kernel from insect damage and moderates the effects of fluctuations in temperature and humidity. Ventilation fans are operated during warehouse loading to control employee exposure to dust and to begin cooling the peanuts. Most warehouse managers operate the ventilation fans continuously throughout the storage period. Inspections are made weekly or biweekly for proper fan operation, presence of insects and signs of condensation.

A range of pest management strategies is used for the control of storage pests in peanuts, beginning with thorough cleaning of the warehouse following the removal of the previous crop. This prevents any carry-over of insects or rodents between harvests. The warehouse should be inspected and cleaned again immediately prior to loading. After the warehouse and associated equipment are thoroughly cleaned, a general surface, crack and crevice insecticide that is approved for use on peanut storage facilities, such as cyfluthrin, is applied as a barrier to prevent infestation by the predominant post-harvest insect pest of US peanuts, the Indian meal moth (*Plodia interpunctella*) (Fig. 9.14) (Vol. 1 p. 104, Hodges 2002a). In recent years, an alternative to the use of synthetic insecticide has been to dust walls and floors

with diatomaceous earth (DE), an inert dust formed from the skeletal remains of microscopic organisms (Vol. 1 p. 271, Stathers 2002). DE absorbs the lipid layer of the insect exocuticle, leading to death by desiccation.

Plodia interpunctella is controlled during the storage period by a combination of methods. The presence or absence of flying moths is noted during periodic warehouse inspections although very few warehouse managers use insect traps to monitor insect populations to determine when treatments are needed. There are daily or regular automated applications of a non-residual insecticide, dichlorvos (DDVP), by an automated fogging/misting system. Each day near dusk a timer turns off the overhead ventilation fans. Approximately 30 min after the fans have been turned off, the solenoid valves are opened for 12–20 s, emitting an aerosol mist of the insecticide. Fans remain off for approximately two hours, allowing the flying insects to come in contact with the insecticide. The registration of DDVP for insect control in stored agricultural products is currently under scrutiny by the US Environmental Protection Agency. There is also a residual treatment of DE applied as a barrier to the top of the peanut pile at a rate of about 1 kg per 1000 m². Finally, there is fumigation with phosphine in which the goal is to maintain a toxic concentration of this gas long enough to give a complete kill of the pest (Vol. 1 p. 323, Taylor 2002). Warehouses are usually fumigated soon after filling. The success of fumigation depends entirely on whether or not the storage facility is gas-tight enough to maintain sufficient gas concentration. Phosphine is currently the only viable fumigant since the alternative fumigant, methyl bromide, is to be phased out by 2005. For the future, an integrated pest management (IPM) approach in which sanitation, scouting, temperature and moisture control and pesticides are deployed in varying degrees based on information flow, damage thresholds and cost–benefit will optimise profits and reduce environmental and health concerns (Vol. 1 p. 301, Hodges 2002b).

Economics of operation of the system

A typical peanut warehouse has a capacity of 5000–7000 t. Constructing a new warehouse costs on average $110 t⁻¹ (range $90–150 t⁻¹) depending on the environmental control systems, cleaning, loading and unloading equipment. If the average price of peanuts paid to the grower is $344 t⁻¹ then the value of the commodity in

Fig. 9.14 Skid steer loader used to unload a typical peanut warehouse (US).

each warehouse is \$1.7–2.4m. The average storage period in the USA is 150 days.

A warehouse of 5000 t will usually have three or four 3.75 kW fans running continuously. If electricity costs an average of \$0.08 kW h^{-1} then the operating cost for ventilation will be \$22–29 d^{-1} and during 150 days storage amounts to \$3300–4350 (\$0.66–0.87 t^{-1}). Use of automated fan controls, based on temperature and humidity conditions in the free space, can reduce fan operation and hence the associated costs by approximately one-third (Butts & Smith 1994). This can give further savings by preventing moisture shrinkage below the level at which farmers are not compensated for moisture loss (7%).

Insect control costs can be highly variable depending on the size of the storage facility, length of storage period, temperature and humidity during the storage period, and management strategies (Fig. 9.15). Costs of various insect control measures have been determined based on labelled application frequency and amount of chemical used. Using DDVP throughout the 150-day storage period in a 5000 t storage facility costs approximately \$1 t^{-1}. DDVP is purchased and applied in discrete quantities, i.e. whole canisters, and so there is a step-wise increase in DDVP cost as warehouse capacity increases (Fig. 9.15). In contrast, DE costs increase linearly with increased warehouse capacity as it is applied to the surface of the peanut pile at a rate of 110 kg per 1000 m^2. Applying DE to the surface of the peanuts in a 5000 t warehouse costs \$0.56 t^{-1} while sealing the same warehouse and fumigating it with phosphine costs approximately \$0.8 t^{-1}. However, under humid conditions fumigation may be more cost effective than reliance on DE as research indicates that the efficacy of DE on some stored product insect pests declines as relative humidity increases (Arthur 2001, 2002); no data are available for *P. interpunctella*.

Future developments

The peanut market in the US will drive many of the future changes in post-harvest handling and storage. As the US peanut market has been opened up by the North American Free Trade Agreement (NAFTA), the General Agreement on Tariffs and Trade (GATT) and domestic farm policy, prices will change according to the international supply and demand for peanuts. These changes may make it advantageous for the individual grower or grower co-operatives to build and manage peanut storage facilities. If additional storage is needed, peanuts can be stored in facilities initially constructed and used for other agricultural commodities such as cereal grain or cottonseed.

Marketing rules may change to allow the sale of peanuts with more than 10.5% m.c. (Blankenship *et al.* 2000b). If high moisture grading were allowed, then peanuts could be cleaned and sorted prior to curing, resulting in a more homogeneous mixture of peanuts in a single farmer stock warehouse. A more homogeneous blend of peanuts should reduce the risk of condensation or aflatoxin contamination due to pockets of high moisture.

When new warehouse facilities are required, alternative structure and system designs should be considered.

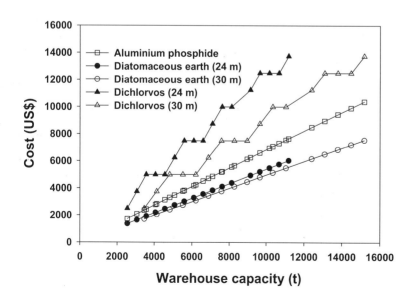

Fig. 9.15 Typical cost of control measures for the Indian meal moth, *Plodia interpunctella*, in 24 m and 30 m wide peanut warehouses using dichlorvos (DDVP), aluminium phosphide or diatomaceous earth.

Building dimensions, aeration/ventilation systems, handling systems and pesticide delivery systems are just a few of factors that could be improved upon. Wright *et al.* (1998) examined the feasibility of using a single container to transport peanuts from the field, cure and store them, as an alternative to the current system. Most of the recently constructed warehouses have been built in west Texas and are at least 100 m long and incorporate in-floor aeration systems to manage the peanut moisture content in storage (Blankenship *et al.* 2000a). Since that time, other storage facilities in the southeastern US have been modified to include in-floor aeration systems, primarily to speed up the equilibration process in storage. Smith and Sanders (1987) showed that storing peanuts in semi-underground storage facilities produced cooler and more uniform temperature conditions and should be considered in the future.

Potato handling equipment could be used to load and unload peanuts from storage facilities and reduce the loose-shelled kernels and split kernels compared to conventional elevator and belt loading systems. These systems may allow handlers to use lower-cost storage facilities (Blankenship & Lamb 1996).

As chemicals for controlling pests become more limited, vigilant sanitation and periodic inspection of warehouses will become ever more important. In addition, diatomaceous earth may see wider usage and insect population monitoring and/or modelling implemented to facilitate IPM for the economic and safe storage of peanuts

References

Adrian, J. & Jacquot, R. (1968) *Valeur Alimentaire de l'Arachide et de ses Dérivés. Collection Techniques Agricoles et Productions Tropicales.* Maisonneuve & Larose, Paris, France.

Anon. (1982) Caractéristiques des variétés d'arachides actuellement recommandées au Sénégal. Centre National de Recherches Agronomiques de Bambey/Institut Sénégalais de Recherches Agricoles, Dakar, Senegal.

Anon. (2000) *Atlas du Sénégal.* Collection les Atlas de l'Afrique. Les Éditions J.A., Paris, France.

Anon. (2001) *Handling and Storage of Farmer Stock Peanuts.* American Peanut Shellers' Association, Albany GA.

Appert, J. (1957) Les insectes nuisibles aux plantes cultivées au Sénégal et au Soudan. Gouvernement Général AOF, Dakar, Sénégal.

Appert, J. (1985) *Le Stockage des Produits Vivriers et Semenciers.* 2 Vols. ACCT/CTA/Maisonneuve & Larose, Paris, France.

Arthur, F.H. (2001) Immediate and delayed mortality of *Oryzaephilus surinamensis* (L.) exposed on wheat treated with diatomaceous earth: effects of temperature, relative humidity, and exposure interval. *Journal of Stored Products Research,* **37**, 13–21.

Arthur, F.H. (2002) Survival of *Sitophilus oryzae* (L.) on wheat treated with diatomaceous earth: impact of biological and environmental parameters on product efficacy. *Journal of Stored Products Research,* **38**, 305–313.

Arthur, F.H. & Redlinger, L.M. (1988) Influence of loose-shelled kernels and foreign material on insect damage in stored peanuts. *Journal of Economic Entomology,* **81**, 387–390.

Blankenship, P.D. & Lamb, M.C. (1996) Handling farmer stock peanuts at warehouses with potato equipment. *Peanut Science,* **23**, 19–23.

Blankenship, P.D., Grice, G.M., Butts, C.L., *et al.* (2000a) Effect of storage environment on farmers stock peanuts grade factors in an aerated warehouse in west Texas. *Peanut Science,* **27**, 56–62.

Blankenship, P.D., Lamb, M.C., Butts, C.L., Whitaker, T.B. & Williams, E.J. (2000b) Summary report on grading high moisture farmer stock peanut lots. Agriculture Research Service, US Department of Agriculture Technical Report, Dawson GA, USA.

Butts, C.L. & Smith, J.S., Jr. (1994) Maintaining peanut quality in storage with automatic ventilation controls. *ASAE Paper No. 94–6516,* American Society of Agricultural Engineers, St. Joseph MI, USA.

Butts, C.L. & Smith, J.S., Jr. (1995) Shrinkage of farmers' stock peanuts during storage. *Peanut Science,* **22**, 33–41.

Conway, J.A. (1974) Investigations into the origin, development and control of *Caryedon serratus* (Col. Bruchidae) attacking stored peanuts in The Gambia. In: *Proceedings of the First International Conference on Stored-Product Entomology,* Savannah, GA, USA.

Dancette, C. (1997) Synthèse des acquis de la recherche/développement, sur l'arachide irriguée dans la vallée et le delta du Fleuve Sénégal de 1993 à 1996. Pôle Systèmes Irrigués (PSI), Travaux et Études No. 4. Institut Sénégalais de Recherches Agricoles, Dakar, Sénégal.

Delobel, A. (1989) *Uscana caryedoni* (Hym. Trichogrammatidae): possibilités d'utilisation en lutte biologique contre la bruche de l'arachide, *Caryedon serratus* (Col. Bruchidae). *Entomophaga,* **34**, 351–363.

Farm Service Agency (2003) United States Department of Agriculture, Washington DC. http://www.fsa.usda.gov/pas/news/releases/index.htm.

Freud, C., Freud, E.H., Richard, J. & Thevenin, P. (1997) La crise de l'arachide au Sénégal, un bilan-diagnostic. Document Ministère de l'Agriculture Sénégal/Commission Européenne/CIRAD, Montpellier, France.

Friendship, R. (1974) A preliminary investigation of field and secco infestation of Gambian groundnut by *Caryedon serratus* (Ol.). Report No. 38. Tropical Products Institute, Slough, UK.

Green, A.A. (1959) The control of insects infesting groundnuts aftre harvest in the Gambia. 1. A study of the groundnut borer *Caryedon gonagra* (F.) under field conditions. *Tropical Science,* **1**, 200–205.

Hammons, R.O. (1982) Origin and early history of the peanut. In: H.E. Pattee & C.T. Young (eds) *Peanut Science and Tech-*

nology. American Peanut Research and Education Society, Stillwater OK, USA.

Hodges, R.J. (2002a) Pests of durable crops – insects and arachnids. In: P. Golob, G. Farrell & J.E. Orchard (eds) *Crop Post-Harvest: Science and Technology. Volume 1 Principles and Practice*. Blackwell Publishing, Oxford, UK.

Hodges, R.J. (2002b) Approaches to pest management in stored grain. In: P. Golob, G. Farrell & J.E. Orchard (eds) *Crop Post-Harvest: Science and Technology. Volume 1 Principles and Practice*. Blackwell Publishing, Oxford, UK.

McGill, J.F. (1973) Economic importance of peanuts. In: C.T. Wilson (ed.) *Peanuts: Cultures and Uses*. American Peanut Research and Education Society, Stillwater OK.

NDiaye, A. (1981) Contribution à l'étude de la biologie de la bruche de l'arachide, I Ol. Effet des rayons X sur la femelle. Thèse de Doctorat de 3ème cycle. Université Paris Sud, Paris, France.

NDiaye, S. (1991) La bruche de l'arachide dans un agrosystème du centre ouest du Sénégal: contribution à l'étude de la contamination en plein champ et dans les stocks de l'arachide (*Arachis hypogaea* L.) par Caryedon serratus (Ol.) – Coleoptera, Bruchidae. Rôle des légumineuses hôtes sauvages dans le cycle de cette bruche. Thèse de Doctorat. Université en Ecologie Expérimentale, Académie de Bordeaux, Bordeaux, France.

Pehaut, Y. (1976) *Les Oléagineux dans les Pays d'Afrique Occidentale Associés au Marché Commun. La producton, le Commerce et la Transformation des Produits*. Champion, Paris, France.

Pointel, J.G. & Yaciuk, G. (1979) Infestation par *Caryedon gonagra* F. de stocks expérimentaux d'arachides en coque au Sénégal et températures observées. *Zeitschrift fur Angewandte Zoologie*, **66**, 185–198.

Rouziere, A. (1986a) Storage of shelled groundnuts in controlled atmospheres, 1. Preliminary trials 1979–1982. *Oléagineux*, **41**, 329–344.

Rouziere, A. (1986b) Storage of shelled groundnuts in controlled atmospheres, 2. Pre-extension trials 1983–1985. *Oléagineux*, **41**, 507–518.

Sanders, T.H., Smith, J.S., Jr., Lansden, J.A. & Davidson, J.I. (1981) Peanut quality changes associated with deficient warehouse storage. *Peanut Science*, **8**, 121–124.

Seck D., Lognay, G., Haubruge, E., *et al.* (1993) Biological activity of *Boscia senegalensis* (Pers.) Lam (Capparacae) on stored grain insects. *Journal of Chemical Ecology*, **19**, 377–389.

Sembène, M. (1997) Modalités d'infestation de l'arachide par la bruche *Caryedon serratus* (Olivier) en zone soudano-sahélienne: identification morphométrique et génétique de populations sauvages et adaptées. Thèse de Doctorat de 3ème cycle. Université Cheikh Anta Diop, Dakar, Sénégal.

Sembène, M. & Delobel, A. (1996) Identification morphométrique de populations soudano-sahéliennes de bruche de l'arachide, *Caryedon serratus* (Olivier) (Coleoptera: Bruchidae). *Journal of African Zoology*, **110**, 357–366.

Sembène, M. & Delobel, A. (1998) Genetic differentiation of groundnut seed-beetle populations in Senegal. *Entomologia Experimentalis et Applicata*, **87**, 171–180.

Sembène, M., Brizard, J. P. & Delobel, A. (1998) Allozyme variation among populations of groundnut seed-beetle *Caryedon serratus* (Ol.) (Coleoptera: Bruchidae) in Senegal. *Insect Science and its Application*, **18**, 77–86.

Slay, W.O. (1976) Damage to peanuts from free-fall impact. ARS-S-173. Agricultural Research Service, United States Department of Agriculture, Washington DC, USA.

Slay, W.O. & Hutchison, R.S. (1973) Handling peanuts with bucket elevators: rates of conveying and mechanical damage. ARS-S-17. Agricultural Research Service, United States Department of Agriculture, Washington DC, USA.

Smith, J.S., Jr. (1994) Considerations for farmers stock peanut warehouses. National Peanut Council, Alexandria VA, USA.

Smith, J.S., Jr. & Sanders, T.H. (1987) Potential for semi-underground storage of farmers stock peanuts. *Peanut Science*, **14**, 34–38.

Smith, J.S., Jr., Davidson, J.I., Jr., Sanders, T.H. & Cole, R.J. (1984) Overspace environment in mechanically and naturally ventilated peanut storages. *Peanut Science*, **11**, 46–49.

Smith, J.S., Jr., Davidson, J.I., Jr., Sanders, T.H. & Cole, R.J. (1985) Storage environment in a mechanically ventilated peanut warehouse. *Transactions of the American Society of Agricultural Engineers*, **28**, 1248–1252.

Smith, J.S., Jr., Blankenship, P.D. & McIntosh, F.P. (1995) Advances in peanut handling, shelling, and storage from farmer stock to processing. In: H.E. Pattee & H.T. Stalker (eds) *Advances in Peanut Science*. American Peanut Research and Education Society, Stillwater OK, USA.

Stathers, T.E. (2002) Pest management – Inert dusts. In: P. Golob, G. Farrell & J.E. Orchard (eds) *Crop Post-Harvest: Science and Technology. Volume 1 Principles and Practice*. Blackwell Publishing, Oxford, UK.

Taylor, R.W.D. (2002) Fumigation. In: P. Golob, G. Farrell & J.E. Orchard (eds) *Crop Post-Harvest: Science and Technology. Volume 1 Principles and Practice*. Blackwell Publishing, Oxford, UK.

Turner, W.K., Suggs, C.W. & Dickens, J.W. (1967) Impact damage to peanuts and its effect on germination, seedling development, and milling quality. *Transactions of the American Society of Agricultural Engineers*, **10**, 248–251.

USDA (2001) Peanut stocks and processing. Agricultural Statistics Board, National Agricultural Statistics Service, United States Department of Agriculture, Washington DC, USA. http://usda.mannlib.cornell.edu/reports/nassr/field/pps-bb/2001/pnut1201.pdf.

Wareing, P.W. (2002) Pest of durable crops – moulds. In: P. Golob, G. Farrell & J.E. Orchard (eds) *Crop Post-Harvest: Science and Technology. Volume 1 Principles and Practice*. Blackwell Publishing, Oxford, UK.

Wright, F.S., Butts, C.L., Lamb, M.C. & Cundiff, J.S. (1998) Quality and economic evaluation of peanut stored in modular containers. *Peanut Science*, **25**, 81–85.

Chapter 10
Copra: The Philippines

D. D. Bawalan

Coconuts (*Cocos nucifera*) are the seeds of the coconut palm and consist of an outer husk and a hard, inner shell to which adheres the coconut kernel or flesh. In many countries in the tropics, the kernel is dried to make copra (Plate 10). Oil for human consumption can be extracted from copra and the solid residue, copra meal (Fig. 10.1), is used in animal feed. Copra making prevents the spoilage of fresh coconut kernel by reducing its moisture content. The kernel at 48–55% moisture content is highly perishable and should be dried to 5–6% for safe storage.

Worldwide, copra is prepared in three forms:

Fig. 10.1 Copra meal, an ingredient in animal feed, loaded on to a bulk carrier (Philippines).

- *Ball copra:* This is a dehydrated whole kernel and is an edible copra found only in India and certain parts of Sri Lanka. It is prepared by placing fully matured or ripe, undehusked coconuts in specially constructed 'ball copra stores' or a wooden platform above a fireplace where the nuts are allowed to dry for 8 to 12 months (Ranasinghe *et al.* 1980).
- *Finger copra:* This is small pieces of dried kernel, normally prepared in the Pacific Islands, where the fresh kernel is removed from split undehusked or dehusked coconuts and then dried by kiln or in the sun.
- *Cup copra:* This is dried kernel in half-cups (Plate 10) and is the most commonly traded copra. It is produced by drying split coconut kernels in shell (husked or unhusked) in the sun, in smoke kilns or a combination of both. The kernel is removed from the shell either when it is partially dried or at the end of the drying operation.

In the Philippines, copra is generally produced by farm workers, tenants and farmers with land holdings of 10 ha and below; the average land holding of coconut farmers is about 1.5 ha. The product is dried either in the sun or using a kiln. Kilns may be direct (*kukum*), where the fire bed is directly below the copra bed, or indirect (*tapahan*) (Fig. 10.2) where the hearth is located to one side of the dryer and connected to the drying bed by a tunnel-like flue. Kilns dry by heat and smoke although there are modified *kukum* dryers where drying is by hot air; this produces a cleaner, whiter product. From the producer, copra passes through a series of traders and is then processed into oil and copra meal. An exception to this is the sale of fresh coconut kernels removed from the shell in the central Philippines (province of Cebu and certain parts of Negros). In these areas, the local trader prepares the copra either by sun-drying or by smoke-drying.

Fig. 10.2 Copra drying by *tapahan*, where hot combustion product is channelled via an underground tunnel (Philippines). Dry coconut husk is used as fuel.

Historical perspectives

The Philippines is one of the biggest coconut producers in the world. Coconut production in the country started in 1642 when the Spanish Governor General Corcuera issued a decree requiring each *indio* (the Spanish colonial term for native Filipinos) to plant 100–200 coconut trees to provide caulking and rigging for Spanish galleons. Copra was first exported to Europe in 1880 (Cornelius 1973) and first milled into coconut oil and copra meal in Manila in 1908 (Boceta 1989). The coconut industry has always been one of the biggest sources of foreign exchange earnings and, among coconut products, oil ranks highest in volume and value.

On average, the Philippines produces about twelve billion coconuts annually, 85% of which are converted into copra and milled into oil and copra meal. There are about 90 registered oil mills scattered all over the archipelago with copra crushing capacities ranging from 5 t d^{-1} to 1000 t d^{-1}. The oil milling industry has grown to the point that there is now substantially over-capacity; the total annual crushing capacity of 5.1m t is now more than double the average annual copra production of about 2.3m t.

Objectives of storage

A distinct copra marketing and storage system evolved in the Philippines from the need to sustain the operation of the oil milling industry. Storage facilities maintained by traders at different layers of the marketing chain are primarily used to serve the following functions:

- *As a means of reducing the m.c. of copra to 5–6% which is preferred by oil millers and processors.* Copra is a hygroscopic material and will take up moisture from, or release it to, the surrounding air until it reaches its equilibrium m.c. This is about 5–6% under the prevailing humidity conditions in the Philippines. Thus, as long as copra is exposed to ambient air while in storage, it can be dried to the level required by processors. Given good ventilation, copra stored at 10–25% m.c. can dry down to 5–7% m.c. in 14 days. It should be noted that copra stored at high m.c. may be attacked by various bacteria and moulds including the mould *Aspergillus flavus*, which produces aflatoxin, a highly toxic mycotoxin (Vol. 1 p. 128, Wareing 2002).
- *As a means by which traders can accumulate copra and can provide larger-scale traders with the flexibility to respond rapidly to copra price fluctuations.* The domestic price of copra is based on the world market price of coconut, which is very variable. Copra storage is essential if traders want to buy copra when prices are low and to sell when prices are high. This also allows the trader to enter into forward contracts with the processors or exporters.
- *As a means by which copra coming from different islands in the archipelago can be accumulated and stored to insure a year-round supply of raw material to processors at the right m.c.* Owing to the large capacities of existing oil mills, not all raw materials can be sourced locally. Likewise, since the Philippines is an archipelago, transport of copra from the coconut-producing islands to processing centres is difficult. It has to be stored and transported in bulk to meet the needs of processors.

Marketing structure

With such a large industry based in an archipelago of 7100 islands, a complex copra marketing and storage system evolved to support the needs of oil mills for a steady, year-round supply of copra. The marketing

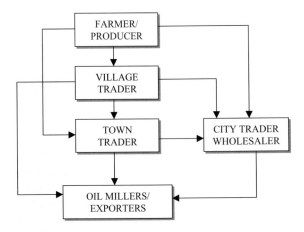

Fig. 10.3 Simplified diagram of copra marketing chain in the Philippines.

Box 10.1 Copra moisture content in the marketing chain

The moisture content of copra that enters the marketing chain is a function of the traditional copra making practices in a particular coconut locality and the prevailing price of copra. For instance, farmers in Marinduque island normally dry only up to the point when the kernel shows shrinkage from the shell, while farmers in the Bicol Region dry up to the point when the kernel can already be easily detached from the shell. Likewise, greater quantities of copra with relatively higher moisture content and *goma* (a lower-quality, rubbery form of copra obtained from drying immature coconuts) enter the chain when prices are high.

---

chain (Fig. 10.3) is composed of several layers of copra traders, from village to town then to city traders and wholesalers who accumulate copra and sell to oil millers and exporters when the price is favourable. A wholesaler may have several village traders or agents and town traders whose operations he finances and who are committed to sell only to him.

Under this system, a farmer/producer either sells to the village trader, the town trader or the wholesaler depending on the volume of copra produced and the geographical area. The coconut farmer normally delivers to a village trader (*suki*) who provides cash advances whenever the need for money arises. In cases where the producers are big landowners, they may enter into a special arrangement with wholesalers or processors/exporters where they receive a more favourable price depending on the volume which they commit to deliver. In certain cases, town traders and wholesalers send their own trucks to villages to buy directly from the producers.

A national survey, involving 4214 farmer respondents from different coconut-producing regions of the Philippines, showed that 92% of the farmers do not store copra (Bawalan 1990). After it is removed from the shell, copra is generally placed in jute or woven propylene sacks and then delivered to a local trader the following day. At this time copra is normally at 10–25% m.c. (see Box 10.1).

The village traders based in the coconut-producing areas then sell to bigger village traders, town traders or to city traders/wholesalers. They also sell to processors in areas where large-capacity mills have established buying stations. Copra stocks are normally sold once 5–10 t has been accumulated. Storage time at this stage is typically

from one to two weeks. In most cases, the village traders' warehouses are not exclusively copra stores. Copra may be stored together with sacks of fertiliser, cement, cases of beer and cola drinks, coconut shell charcoal or with any commodity in which the village traders deal. Copra bought from the producers is either retained in sacks or piled loosely on top of copra already in store. Village traders normally do not re-dry copra bought from producers; however, some occasionally sun-dry if copra deliveries are obviously wet. Copra is returned to sacks before delivery to the next link in the chain, a bigger trader, town buyer, wholesaler or processor.

The town traders and wholesalers generally buy copra from village traders and deliver it to processors and exporters. Trading is done by entering into forward contracts of up to 30 days with oil millers and exporters and they generally receive more than the prevailing mill gate price, based on the volume committed for delivery. Since these traders have more operating capital, they speculate on prices, i.e. they buy and stock copra when it is cheap and sell it when the price is high, and so keep copra in store from two weeks to three months depending on the prevailing prices. Like the village trader, the majority of town traders and wholesalers also do not re-dry copra. Copra deliveries are immediately removed from the sacks and piled loosely in separate heaps or on top of the pile that is already in store. Once the piled copra in store has dried down naturally to an acceptable level and the trader has decided that it is an appropriate time to sell, it is returned to sacks or loaded directly into trucks for delivery to oil millers, exporters or bigger wholesalers.

Over-capacity in the copra milling industry creates competition for the product, so to ensure an adequate supply for their operation, the processors, especially the large mills, buy as much copra as available regardless of quality. Despite the high working capital involved, the majority of oil millers normally stock up inventories and store copra for at least one month. Thus, processors' storage facilities can be one or more warehouses covering an area of as much as 1000 m². Attempts to ensure adequate supplies of copra are made by entering into contracts with big traders, landowners and wholesalers as well as shipping companies for inter-island buying and shipment of copra. The storage practices of millers are usually better than those of traders as there is usually adequate space and ventilation and regular cleaning and fumigation are practiced. Copra is loosely piled in a series of heaps.

Classification standards and pricing scheme

There are two classification standards issued by the Philippine authorities to provide guidelines and quality criteria for the domestic pricing and trading of copra. These are the Moisture Meter Law (Republic Act 1365, 1954) enacted by the Philippines Congress and the New Copra Classification Standard (Administrative Order 01, 1991) of the Philippine Coconut Authority (PCA). The Philippine Coconut Oil Producers Association (PCOPA) also introduced its own classification standards in December 1988 in an attempt to improve the quality of copra entering the mills. However, implementation was stopped within three months after it was realised that this was reducing their copra intake.

To provide a faster method for determining the m.c. of copra during trading than the standard laboratory oven method, the Philippine coconut industry adapted the Brown-Duvel procedure which was developed around 90 years ago in the USA for the analysis of grain of high moisture content (Head & Harris 1995). The Brown-Duvel method involves heating small pieces of copra in an oil bath; the water vapour from the copra is condensed and then collected in a graduated cylinder. This gives a higher m.c. reading than the oven method, with the difference increasing with higher m.c. Consequently, Brown-Duvel readings need to be calibrated to the oven method. However, the Moisture Meter Law requires all copra buyers to use moisture meters for domestic purchases and payments are made with reference to a standard m.c. of 5% (Table 10.1).

The New Copra Classification Standard was developed in response to the European Union setting the maximum aflatoxin limit in copra meal to 20 ppb (see Box 10.2). It was a revision of an existing classification standard to make mould content a more significant criterion of quality. It also recognised the fact that it is a difficult task for the farmer to dry to 5% moisture as prescribed in the Moisture Meter Law. However, to discourage the trading of high moisture-content copra, the New Copra Standard stipulated rejection of copra at a m.c. above 12% (Brown-Duvel method). It is of course essential that the reference method be mentioned because copra at 12% m.c. by oven method may suffer a high degree of mould damage. On the other hand, copra at 12% m.c. as measured by the Brown-Duvel method corresponds to 8.9% m.c. by the oven method, which would suffer limited mould growth. The Standard also stipulated that smoke-dried copra with moisture content of 12–14% could be accepted with corresponding discounts (provided that it will be dried to at least 12% by traders before storage), as the smoke-dried product is less susceptible to mould growth (Head *et al.* 1999).

The price adjustment factors for copra at selected moisture contents and mould contamination rate as stipulated in the New Copra Classification Standards are

Grade	Moisture content (%)	Discount/ premium (%)	Copra price* (Php/kg)
Corriente	25	−22	7.80
Corriente Mejorado	20	−16	8.40
Buen Corriente	15	−10	9.00
Buen Corriente Mejorado	12.5	−7.5	9.25
Semi-resecada	10	−5	9.50
Resecada	5	0	10.00
Resecada Bodega	3	+2	10.20

Table 10.1 Classification grades, discount and premiums according to the Moisture Meter Law.

*Based on a hypothetical copra price of Php 10.00/kg.

Box 10. 2   The New Copra Classification Standard

The scientific basis for the promulgation of the New Copra Classification Standard is the result of a comprehensive copra storage study by the RP-UK Project 'Reduction in Aflatoxin Contamination of Copra in the Philippines'. This was implemented jointly by the Philippine Coconut Authority and the Natural Resources Institute of the United Kingdom between January 1990 and June 1992. It should be noted that, under the EU standard, an aflatoxin limit of 20 ppb was actually stipulated for compound feed. However, EU countries, particularly the Netherlands and Germany, do not distinguish between a feed ingredient like copra meal and compound feed.

ity. This is termed the *pasa* system. The daily *pasa* price is determined by making a fixed percentage deduction on the current mill gate price to account for m.c., weight of dust and weight of copra from immature nuts (*goma*). Deductions can be as high as 20–25% at the discretion of the trader. The prices paid to farmers appear to vary little even when demand is high. Larger traders tend to apply the so-called *resecada* system where price deduction is based on the traders' assessment of the m.c. of copra delivered relative to 5% (Table 10.1). Additional deductions are also made for dust content and amount of *goma*. The resulting weight of copra after all deductions are made is then multiplied by the prevailing mill gate price. In buying copra, the majority of the traders rely on their visual judgement in determining moisture content. It is only the city traders/wholesalers and the processors who use the Brown-Duvel apparatus. Several big oil millers/processors are applying the provisions of New Copra Classification Standard in buying copra but do not use the stipulation on rejection (Table 10.2). It should be noted, however, that the marketing system does not adhere strictly to the New Copra Classification Standard. The traders and millers adopt a pricing system that is more practical and beneficial to their interest, notwith-

shown in Tables 10.2 and 10.3 where the m.c. is determined by the Brown-Duvel method.

Copra pricing at the first point of sale, to small traders, has always been based on the weight regardless of qual-

Table 10.2   Price adjustment for copra at selected moisture content* according to the New Copra Classification Standard.

Moisture content (%)	Discount/ premium (%)	Copra price† (Php/kg)	Moisture content (%)	Discount/ premium (%)	Copra price† (Php/kg)
≤7.0	+5.7	10.57	11.0	+1.1	10.11
7.5	+5.1	10.51	11.5	+0.6	10.06
8.0	+4.6	10.46	12.0	0	10.00
8.5	+4.0	10.40		Rejection level	
9.0	+3.4	10.34	12.5	−1.0	9.90
9.5	+2.8	10.28	13.0	−2.0	9.80
10.0	+2.3	10.23	13.5	−3.0	9.70
10.5	+1.7	10.17	14.0	−4.0	9.60

*As measured by the Brown-Duvel method.
†Based on a hypothetical copra price of Php 10.00/kg.

Table 10.3   Price adjustment for mould content according to the New Copra Classification Standard.

Grade	Mould level (ARM %*)	Discount/ premium (%)	Copra price† (Php/kg)
No. I	≤ 1	+2	10.20
No. II	1.1–10	0	10.00
No. III	10.1–20	−2	9.80
No. IV	> 20	−4	9.60

*% aflatoxin-related mould = % yellow-green mould + % penetrating mould.
†Based on a hypothetical copra price of Php 10.00/kg.

standing the general effect on the quality of copra being traded. The PCA as the government agency mandated to regulate the industry does not have any police power or the resources needed to ensure strict implementation of the Classification Standard. Since 1986 copra marketing has been free from government intervention.

Physical facilities

The national survey of coconut post-harvest activities showed that the majority of traders (62.5% of 1312 trader respondents) maintain warehouses with capacities ranging from 5 t to 20 t (Bawalan 1990) while 12.7% of traders have warehouses with capacities ranging from 30 t to 100 t (Fig. 10.4). They are normally made with concrete walls and floors and roofed with corrugated iron sheet. Air intake ports are provided either through grill openings or holes in moulded decorative concrete on the upper portion of the concrete walls. Ventilation is also provided naturally through the front entrance, which is fully opened during the day. Most of the traders' warehouses are located along roads where trucks can park easily, load and unload. Some warehouses are constructed with wide concrete pavements beside them that are occasionally used for sun-drying.

The warehouses of a village trader, town trader or wholesaler generally differ only in their storage capacities. On the other hand, there is a marked difference between storage facilities in large and small oil mills. Plant buildings are normally made of steel trusses, iron-sheet roofs and concrete flooring. For oil mills with capacities of 50 t d^{-1} and below, copra is piled in vacant spaces around the oil milling equipment. Movement of natural air is much better here than in traders' stores since the

Fig. 10.5 Copra pile in a miller's warehouse (Philippines).

ceiling is high and the enclosure is fully open on two sides. Storage facilities for large-capacity mills are similar apart from the degree of mechanisation in the warehouse (Fig. 10.5). There may be mechanical ventilators, a series of conveyors and an elevator system that moves copra to the top of the warehouse as it is discharged from the truck at ground level. From the top, copra is released to form piles on the warehouse floor. A study in 1987 of the m.c. of copra at four mills of 400–900 t d^{-1} capacity showed little variation during the year (Fig. 10.6). The fairly uniform m.c. of copra in these mills can be attributed to good ventilation in the millers' warehouses and the fact that the majority of the copra entering the mills had already dried down to equilibrium m.c. while in the traders' warehouses. Three of the oil mills were located in Mindanao where rainfall is fairly even throughout the year and fourth was in Luzon where there is a distinct dry and wet season (Lozada *et al.* 1988).

Major sources of quality decline

Since 1989, the PCA has implemented various copra quality improvement programmes, both locally funded and foreign-assisted, but to this day copra quality remains a big problem. The low quality of copra traded in the country can be attributed to socioeconomic factors, discovered by the PCA.

Lack of quality-based pricing

Drying the copra down to the required moisture content for safe storage entails additional time and extra effort for the farmer. With the *pasa* system of trading, the

Fig. 10.4 A copra trader's warehouse (Philippines).

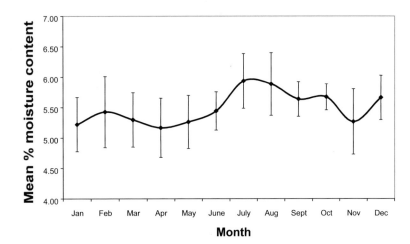

Fig. 10.6 Monthly average moisture content (± s.e.) of copra for 1987 in four mills of 400–900 tonnes per day capacity (data from Lozada *et al.* 1988).

higher the moisture content of copra, the more money the farmer tends to get from it because the copra is heavier. This is the major reason why copra with high moisture content enters the marketing chain.

Dependence of the coconut farmer on the copra trader

To meet their cash needs for their family's expenses, coconut farmers receive informal credit from copra traders using their future harvest as collateral. Since the crop is already committed to the traders there is no incentive for farmers to deliver at anything above minimum quality and the trader has to accept the copra delivered, regardless of quality, because payment has already been made.

Over-capacity in the oil milling sector

At present, there is a gross over-capacity in the oil milling sector that is compounded by the concentration of mills in certain areas. Competition for copra is very keen where clusters of mills are located, while supply of copra from areas without mills is hampered by poor or inadequate transport facilities. In this situation, processors have to buy whatever quality of copra is available to sustain the operation of their plant. Since any quality for copra is accepted, the natural tendency for farmers is to produce copra that will entail the least effort on their part.

Deterioration during storage

Deterioration of copra in storage is generally due to the attack of fungi, bacteria, moulds and insects, and the severity of attack by all these is correlated with m.c. The actual quantity of mouldy, degenerated copra present in a particular location normally determines the intensity of insect infestation (Nathanael 1965). The red-legged copra beetle, *Necrobia rufipes* (Vol. 1 p. 106, Hodges 2002), is apparently the most numerous insect in copra piles and in mixed storage of copra and copra meal or cake, though a survey of oil mills, copra stores and warehouses on the island of Luzon and sampling at a number of other locations lists a total of fifteen insect species that thrive on copra during storage (Table 10.4) (Zipagan & Pacumbaba 1996). The same study failed to record any mites infesting copra despite the high moisture content

Table 10.4 Insects found on copra in the Philippines. Source: Zipagan & Pacumbaba (1996).

Order	Family	Species
Lepidoptera	Phycitidae	*Ephestia cautella*
Dermaptera	Furficulidae	*Forficula* sp (near *auricularia*)
Dictyoptera	Blattidae	*Periplaneta americana*
Coleoptera	Anthribidae	*Araecerus fasciculatus*
	Cleridae	*Necrobia rufipes*
	Cucujidae	*Cryptolestes ferrugineus*
	Dermestidae	*Dermestes ater*
	Nitidulidae	*Carpophilus dimidiatus*
		Carpophilus pilosellus
		Carpophilus mutilatus
		Carpophilus obsoletus
		Carpophilus maculatus
	Silvanidae	*Ahasverus advena*
		Oryzaephilus mercator
	Tenebrionidae	*Tribolium castaneum*

and nutritional value of the product; it is possible they may have been overlooked. One mite species, *Tyrophagus putrescentiae*, is frequently associated with copra in other countries and is sometimes termed the 'copra mite'.

Another factor contributing to the decline in the quality of stored copra is inadequate ventilation and lack of storage hygiene, especially where deteriorated and moist residues of copra are left on the floor and mixed with fresh stocks (Nathanael 1965). Mould and subsequent bacterial attack causes heating in poorly ventilated stores; this can lead to charring of copra and even spontaneous combustion. There is a very real risk of warehouse fires and fire on board ships transporting copra and copra by-products.

Storage losses associated with the deterioration of copra are as follows:

Loss in dry matter weight and oil

When copra is stored at relatively high moisture contents there are losses in both dry matter and oil (Table 10.5).

Loss in oil quality due to increased colour and free fatty acid content

Elevated fatty acid levels are an indication of rancidity and additional processing is needed to reduce them. The fatty acid content of oil from stored copra is proportional to its initial moisture content at the start of storage. The relationship between the FFA increase in oil and the initial moisture content of copra, when stored after eight weeks, was found (Nagler *et al.* 1992) to be:

$$\text{FFA increase} = 0.2 \times \% \text{ m.c.} - 1.26$$

Table 10.5 Estimates of mean material losses from copra stored for 8 weeks with different initial moisture content. Source: Head *et al.* (1999).

Initial moisture (%)	Loss in dry matter (%)	Loss of oil (%)
6	0.2	−0.6*
8	0.7	−0.1*
10	1.3	0.4
12	1.8	1.0
14	2.4	1.5
16	2.9	2.0
18	3.5	2.5
20	4.0	3.0

*Apparent increases in oil content.

The deterioration of copra during storage also affects the colour of oil; the greater the degree of deterioration, the deeper is the hue of the oil. This results in increased processing costs as dark-coloured oil needs to be bleached to be commercially acceptable.

Loss in copra meal quality due to aflatoxin contamination

Aflatoxin contamination occurs within two weeks of opening the nuts. This corresponds to the period of on-farm drying and storage in a primary or village trader's store. Smoke-dried copra is more resistant to aflatoxin contamination than sun-dried or hot-air-dried copra (Head *et al.* 1999). When copra is processed using small-scale extraction equipment, aflatoxin is distributed without loss as 72% in cake, 21% in oil and 7% in the foots (fine particles of seed debris). However, under the more severe conditions of commercial expellers, up to 45% of the aflatoxin is lost; in addition any aflatoxin present in crude coconut oil is removed during alkali refining (Head *et al.* 1993).

Commodity and pest management regimes

Processors state that they carry out regular fumigation of their warehouses to control insects that thrive on moist and mouldy copra, but there is no information available on how this is performed. There is also no information on what action traders take to prevent infestation of their copra stocks.

Economics of operation of the system

Traders and processors are not willing to provide details of the costs of their operation. Warehousing costs have been estimated at $0.01 kg^{-1} per month based on a building cost of $54.54 m^{-2}, packing density of 375 kg m^{-3}, packing height of 2 m and interest rates of 16% (Lozada *et al.* 1988). The cost did not take storage losses into account.

With regard to capital investment for storage facilities, it is quite obvious that it is much higher for the processor than for the trader, due to the higher storage capacity, handling system and mechanical ventilators.

Considerable economic losses result from copra deterioration in storage. Based on the Philippine coconut production of 2.3m t with 85% actually converted to copra, the material loss to the industry due to deterioration while in storage amounts to about 43 000 t which includes 21 000 t of oil valued at $21.5m. Further, there is an additional processing cost due to increase in FFA and colour in oil that must be reduced to meet the quality standard stipulated for commercial oil. Estimated annual losses to the industry due to aflatoxin content in copra meal are about $6.45m, considering the $10–12 t^{-1} price discount being imposed on Philippine copra meal vis-à-vis Indonesian copra meal. It should be noted that when the first EU standard on aflatoxin B_1 level in copra and copra meal (200 ppb maximum for feed ingredient; 50 ppb for compound feed) was first imposed on December 1989, it was estimated that the potential loss to the industry would be about $67.5m, which was equivalent to the total Philippine copra meal exports to Europe in 1988. However, so far no copra meal shipment from the Philippines has been rejected due to aflatoxin contamination. The maximum limit for aflatoxin B_1 content of compound feed was lowered to 20ppb in December 1991 although the limit for a feed ingredient remained at 200 ppb.

Future developments

Recent developments in the world market such as the implementation of the General Agreement for Tariff and Trade (GATT), trade liberalisation and the demand to standardise product quality through the application of quality management systems such as ISO 9000 are forcing industries to re-evaluate their existing production and manufacturing methods. With the demands and standards for good-quality products getting higher, the Philippine coconut industry cannot remain complacent and continue with its traditional ways. Since the export earnings of the industry are mostly derived from the oil milling sector, change is required soon to ensure that the Philippines does not lose its competitive edge in the world market for coconut oil.

In the last ten years, there has been renewed interest in developing oil extraction methods that obviate copra production. Mini-oil mills have been established in different parts of the country that use fresh coconuts as raw material instead of copra. It has already been demonstrated that a very high-quality oil (water white in colour with FFA content of 0.06–0.08% without refining) and an aflatoxin-free meal can be produced if the starting material is fresh coconut. Likewise, there is now growing interest in evaluating the economic feasibility of integrated coconut processing plants where all parts of the coconut fruit (husks, kernel, shell and water) will be processed into various products. There will definitely be a major change in the type and capacity of storage systems for coconut if the coconut industry finally decides to adopt processing technologies that bypass copra.

References

Bawalan, D.D. (1990) Coconut Postharvest Practices in the Philippines. Unpublished Report, Philippine Coconut Authority, Quezon City, Philippines.

Boceta, N.M. (1989) Coconut processing in the Philippines. In: *Proceedings of the Working Group Meeting on Coconut Processing.* UNDP-FAO Project on Improved Coconut Production in Asia and the Pacific (RAS/80/032). December 1989.

Cornelius, J.A. (1973) Coconuts: a Review. *Tropical Science,* **15**, 15–37.

Head, S.W. & Harris, R.V. (1995) Comparison of the Brown-Duvel and oven methods for the determination of moisture in copra and the development of a simplified Brown-Duvel technique. *Coconuts Today,* **12**, 58–62, 73–76.

Head, S.W., Nagler, M.J. & Harris, R.V. (1993) An assessment of aflatoxin transfer from copra to products during the manufacture of crude coconut oil. In: *Proceedings of the Seminar on European Research Working for Coconut,* 8–10 September 1993, CIRAD, Montpellier, France.

Head, S.W., Swetman, T.A. & Nagler, M.J. (1999) Studies in deterioration and aflatoxin contamination in copra during storage. *Oleagineux Corps Gras Lipides,* **6**, 349–359.

Hodges, R.J. (2002) Pests of durable crops – insects and arachnids. In: P. Golob, G. Farrell & J.E. Orchard (eds) *Crop Post-Harvest: Science and Technology. Volume 1 Principles and Practice.* Blackwell Publishing, Oxford, UK.

Lozada, E.P., Benico, J.B. & Hao Chin, V.R., Jr. (1988) *The aflatoxin problem: a driving force to improve the productivity of the coconut industry.* PCRDF Report, Philippine Coconut Research and Development Foundation, Pasig City, Philippines.

Nagler, M.J., Head, S.W., Bennet, C.J., *et al.* (1992) *Copra storage trials, Final Project Report: RP-UK Reduction in Aflatoxin Contamination of Copra in the Philippines,* Chapter 6, Volume 3: Technical Studies. Natural Resources Institute, Chatham, UK.

Nathanael, W.R.N. (1965) Some aspects of copra deterioration. *Ceylon Coconut Quarterly,* **16**, 111–120.

Ranasinghe, T.K.G., Catanaoan, P.C., Patterson, H.B.W. & Abaca, P.M. (1980) Coconut Harvesting and Copra Production, Part 1 of 7: Coconut Processing Technology Information Documents, United Nations Industrial Development Organization (UNIDO) – Asian and Pacific Coconut Community (APCC). Project on Establishment of a Coconut Processing Technology Consultancy Service UF/RAS/78/049.

Wareing, P.W. (2002) Pest of durable crops – moulds. In: P. Golob, G. Farrell & J.E. Orchard (eds) *Crop Post-Harvest: Science and Technology. Volume 1 Principles and Practice.* Blackwell Publishing, Oxford, UK.

Zipagan, M.B. & Pacumbaba, E.P. (1996) Insect pests of copra and copra meal in the Philippines: their occurrence, biology of the key species, natural enemies and contributory factors to their proliferation. Paper presented during the 1996 PCA Research and Development, Extension and Training, In-house Review and Planning Meting, 18–19 June 1996. Philippine Coconut Authority, Zamboanga Research Centre, Zamboanga City, Philippines.

Chapter 11
Coffee

P. Bucheli

Coffee beans are produced from the berries of a small, tropical evergreen tree, and are the source of one of the world's most popular beverages. There are two common species of coffee, *arabica* (*Coffea arabica*) that came originally from Ethiopia, and was first cultivated in the Arabian peninsula, and *robusta* (*Coffea canephora*) which came from the Congo. *Arabica* is restricted to highland areas while *robusta* flourishes at altitudes below 1000 m. The two species are now widely distrubted in the tropics; *arabica* accounts for about 75% of the coffee produced, the remainder is mostly *robusta*.

After coffee berries have been picked, the outer gummy fruit and various other layers are removed. This is done by either a wet process of scraping and fermentation or a dry process of just drying and hulling. These processes reveal the coffee bean; however, if the wet process is used then a thin parchment layer remains over the 'parchment bean' that will be removed later by hulling. Both processes finally end with green coffee beans (Plate 11), which are roasted and then ground before consumption.

Successful storage of green or parchment coffee beans can be achieved provided the initial quality of beans is good, they are well dried to 10–12% moisture content and the relative humidity in storage is low enough (50–70%) to prevent excessive moisture uptake. Maintaining the temperature below 20°C will further reduce coffee deterioration. Constant temperature and humidity conditions are believed to be more beneficial than widely fluctuating ones. Parchment coffee stores better than green coffee, and *robusta* beans store better than *arabica* beans.

In the non-humid tropics, good storage in bag and bulk can be achieved by using fan ventilation. In the humid tropics, air conditioning is cost effective provided there is sufficient thermal insulation of the warehouse or silo.

Other cost-effective storage options for maintaining bean quality under difficult climatic conditions involve sealing bagged beans into polyethylene envelopes or holding beans in bulk in moulded plastic containers under vacuum. Based on a storage period of one year, costs for vacuum storage are up to 40% cheaper than those for warehouse storage, while ventilated storage is better than sealed storage, especially at higher temperature.

There have been efforts to find ways of assessing the potential beverage quality characteristics of consignments of stored coffee beans. To date the most reliable indicator is the glucose concentration; higher glucose concentrations are associated with poorer cup quality.

Historical perspectives

One of the first records of coffee storage dates from 1683 when 500 sacks of coffee were discovered in a warehouse after the retreat of the Turkish army from Vienna. The interpreter for the Turkish army had discovered the coffee and, because he had learned how to roast the beans and prepare them for drinking, he was later able to open a famous Viennese coffee house.

Since World War II, considerable increases in coffee consumption, especially in Europe and the United States, have led to a continuous increase in coffee production. As coffee is a seasonal crop, some of the production has inevitably to go through a period of storage. This may be either in the warehouses of the producing country or at its overseas destination, and on-board ship during transit. For economic reasons, coffee is increasingly handled in bulk by consuming countries, and so silo storage is now an integral part of the coffee industry. Storage also plays a crucial role in periods of over-production, serving as a buffer for the stabilisation of coffee export prices (Sassen 1994).

Physical facilities

Since the early days of coffee trading, beans have been transported and stored in bags. However, for economic and logistical reasons, transportation and storage in bulk are increasingly common. Storage in silos is gaining importance and sealed and vacuum storage methods are more recent, cost-effective innovations in some situations.

Bag storage

The majority of modern coffee warehouses are designed to accommodate bags in which green coffee and sometimes parchment coffee are stored. The bags are usually made from jute and conform to the standards of the International Coffee Organization (ICO) which specifies a bag that is large enough to hold 60 kg of clean coffee and still leave a space at the top. The storage of green coffee occupies significantly less storage space than parchment coffee.

Typically, bagged coffee is palletised in storage. Normally each pallet holds 20 to 25 bags in four or five layers, arranged in alternate directions. Pallets are usually stacked three to five units high, but not higher because they are relatively unstable and can fall. Compared with large bag stacks, this arrangement provides accessibility to small lots, good air circulation as well as easy inspection and sampling. The coffee bags must be kept at least 15 cm off the floor and stacked so that they are not too close to the roof, where temperatures rise rapidly, and at least 20 cm away from the walls for the same reason (Mabbett 1990).

The warehouse building should be completely clean and have smooth walls, and all openings should be screened to prevent insects from entering. Coffee should not be stored alongside other farm products or agrochemicals, as it takes on foreign odours easily. Any contact of the bags with metal can be a cause of moisture condensation on coffee and must therefore be avoided.

Bulk storage

In producing countries, bulk storage is uncommon except where companies handle large quantities of relatively uniform beans, for example, at ports or points of grading and export. In general, bulk storage lasting more than a few months requires several precautions to avoid compaction and deterioration. Air circulation through the beans to control r.h. as well as bean m.c. and temperature will be necessary. If good air circulation cannot be assured, then the beans may have to be turned by movement between silos. The incorporation of a temperature monitoring system, consisting of regularly distributed temperature sensors throughout a silo, will allow the early detection of hot spots before any serious caking of the silo contents can occur. In addition, specific devices permitting small samples to be taken several metres deep within a coffee bulk can provide additional quality assessment during storage.

Bulk handling at ports in the importing countries is common. Some coffee arrives from the producing country in bulk but most is discharged from bags arriving by container at the port. After cleaning, coffee is allocated to various silos according to grade (based on bean size and defects), and can be sold to the roaster in bulk in any quantity desired, ready to roast and blend with beans of other origin. Very large storage facilities of this kind exist in receiving ports; for example, the Silocaf Company (New Orleans, USA) has a 40 000 t silo complex (John 1995), while Nestlé have a 5400 t silo complex in Thailand (Fig. 11.1). It is important that metal and concrete storage silos are equipped with anti-breakage devices such as chutes, to prevent bean damage.

Bulk handling and storage of coffee in importing countries reduces the labour costs associated with bag handling but also has other benefits as greater quality control and product differentiation are possible. However, bulk transport and storage require considerable capital investment, which only a few speciality roasters and coffee manufacturers can afford.

Fig. 11.1 Silo complex (nine silos of 600 tonnes), Nestlé, Thailand, with a secondary wall to provide shielding against temperature extremes.

Air-conditioned storage

Storage losses, under warm and humid conditions, can be reduced considerably by the use of cold preservation. This is accomplished by passing conditioned air, cold and relatively dry, through silos and warehouses of any size (Stirling 1980). Usually, cooling is required only once every few months of storage, and will assure the maintainance of green coffee within the range of 10–12% m.c. However, before installing air-conditioning equipment, it is worthwhile establishing that air-conditioned storage is more cost effective than any alternatives. Air-conditioned and aeration storage for the maintenance of green coffee at a safe m.c. has been compared in silo and bag storage in Thailand (Bucheli *et al.* 1998). Air conditioning was very efficient, but an equally good coffee cup and product quality was also obtained using aeration which kept coffee m.c. at about 13% throughout storage (Fig. 11.2). Cost-effective air conditioning requires adequate insulation. Galvanised steel and aluminium are suitable silo materials, neither causing any tainting. The silos should be completely protected by a light external wall construction (Fig. 11.1) in order to minimise temperature fluctuations and direct solar heating.

Sealed and vacuum storage

The sealed storage method is refered to as the ballooning technique. Stacks of coffee bags are sealed into low-density polyethylene envelopes (Narasimhan *et al.* 1994). A similar technology has been developed for milled rice storage (Nataredja & Hodges 1989) and a video is available on how the technique can be implemented (Natural Resources Institute 1996). Once sealed, the stored coffee can be fumigated with phosphine or carbon dioxide. The envelope prevents contact with high ambient humidity so that well-dried coffee can be stored under wet season conditions without appreciable quality change.

Vacuum storage is also possible. This involves packing pre-cleaned and well-dried coffee beans into purpose-built polyethylene containers (200–1200 L) that are impervious to moisture and highly airtight. A vacuum of about 0.4 bar is created. The coffee beans continue to absorb the remaining oxygen inside the container, and this is replaced by the carbon dioxide resulting from coffee respiration. Thus, the airtightness of the container permits the storage of coffee for several years under harsh ambient conditions and avoids the need to apply chemicals to prevent spoilage (Sassen 1994).

Objectives of storage

As coffee supplies often exceed demand, the world's producers frequently have to store their coffee for long periods to stabilise prices. Storage can be an important instrument to improve export prices but the cost efficiency of storage is key to the success of this strategy (Sassen 1994). The dilemma is that high ambient temperatures and humidities in the tropics make storage for long periods virtually impossible without substantial losses of quantity and quality. Beyond a certain storage period

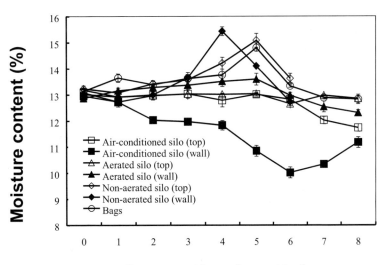

Fig. 11.2 Moisture content of green *robusta* coffee stored in Thailand under different silo and bag conditions (bean sampling positions in parentheses).

the cumulative storage costs can increase to above the remaining economic value of the coffee stored.

To enable a continuous supply of coffee, manufacturers often need storage between harvests, a period of normally about one year. Some coffee is processed in the producing countries and in certain developing countries the direct procurement of green coffee from farmers has strengthened the link between farmers and processors and has led to an improvement in coffee quality.

Major sources of quality decline

Good drying of parchment or green coffee is recognised as the key step in which cup quality is won or lost (Mabbett 1990). This process can fail in several ways. There is evidence that when drying is done too rapidly using high temperatures, the potential cup flavour is largely lost from coffees that would otherwise have been excellent. Drying coffee cherries too slowly when ambient humidities are high can lead to mould growth (Bucheli *et al.* 2000). If beans are over-dried, so that m.c. is reduced to 8–10%, chemical oxidation can occur that will result in reduced extraction yield for soluble coffee manufacture (Rodríguez *et al.* 1976). This damage cannot be reversed by returning the coffee to a higher equilibrium moisture content. Another dangerous situation arises when the incoming crop is much larger than the drying facilities. In this case, cherries and beans are sometimes stored in plastic bags or unventilated bins, respectively, before being properly dried. This can lead to mould development and quality deterioration. The upper m.c. limit is of much greater importance than the lower limit in defining the conditions required for quality preservation in storage. This is because high bean m.c. can lead to insect infestation, mould growth and biochemical processes that are responsible for the degradation of coffee components. Respiration of stored green coffee, which generates water and carbon dioxide, is dependent on both bean m.c. and temperature. A high respiration rate contributes to coffee bean heating, loss of dry matter and the induction of biochemical processes responsible for colour and flavour degradation.

Effect of post-harvest processing on the initial state of coffee

Various agronomic factors (species, variety, origin, etc.) and processing and handling factors (fermentation, dehulling, drying, transport) can determine how difficult it will be to store a particular coffee (De Carvalho & Chalfoun 1989). In the case of wet processing, pulping and fermentation, which are affected by temperature, pectin-degrading enzyme activities, pH and other factors, can affect green coffee and beverage quality upon storage (López *et al.* 1989).

Storage moisture

The biochemical modifications occuring during storage are likely to be a function of m.c., temperature, storage period, origin and initial state of the beans, post-harvest fermentation, drying procedure and other variables. Thus it is difficult to define an exact upper threshold for bean m.c. The difficulty lies also in the method of determination of bean m.c., which can give results differing by 1%. A bean of 12% m.c. is generally considered to be safe for storage; however, it is important to bear in mind that *arabica* beans pick up moisture more rapidly than *robusta* beans. This was shown in an experiment at 80% r.h. where after only two months of storage the m.c. of *arabica* and *robusta* beans was 14.5% and 12%, respectively (Narasimhan *et al.* 1972). Under such conditions, green *arabica* beans may have a storage life of only six months (longer if stored as parchment) whereas green *robusta* beans will remain in a marketable condition for much longer. For example, the coffee cup and product quality of *robusta* beans was unaffected by eight months' storage in jute bags at an average 11.5% m.c. (Bucheli *et al.* 1998).

Storage temperature

Temperature can also have an important effect on coffee quality during storage. For example, *arabica* parchment coffee kept at 10.8% m.c. in a storage trial in Kenya was unacceptable after eight months' storage at 35°C. On the other hand, the same coffee beans stored at 15.5% m.c. and 10°C managed to retain their initial quality (Stirling 1980). The effects of temperature should be kept in mind for coffee storage in producing countries where average day/night temperatures can easily range between 28° and 30°C (Bucheli *et al.* 1998). In fact, several storage studies in Kenya with sensitive *arabica* beans, which had undergone wet processing, indicated that maintaining the storage temperature below 20°C will make a major contribution towards coffee quality (Muriithi 1977).

Moulds

Mould infestation of green coffee can result from inappropriate storage conditions and lead to quality loss and the production of the mycotoxin ochratoxin A (OTA) where there is growth of *Aspergillus carbonarius*. In a detailed study of the microflora of stored coffee the fungus species *Aspergillus ochraceus* and *Penicillium* spp were found to affect bean colour and beverage quality (López Garay *et al.* 1987). The presence of *Aspergillus fumigatus*, *Aspergillus niger*, *Aspergillus* series *restrictus*, *Penicillium* sp and *Wallemia sebi* was reported on green *robusta* beans under industrial storage conditions (Bucheli *et al.* 1998).

In the south of India it is necessary to keep the m.c. of coffee beans during the monsoon rains below 14.5% to prevent mould growth during storage (Gopalakrishna Rao *et al.* 1971) and 14% m.c. and 75% r.h. appear to be the maximum limits if spoilage of green coffee due to fungal growth is to be avoided (Betancourt & Frank 1983). During eight months of bag storage (r.h. = 81%, m.c. = 13.5%), there was neither any mould biomass increase, nor any indication of OTA production even though coffee beans can be a good substrate for *A. ochraceus*. However, at least in the case of *arabica* coffee, high amounts of OTA are only produced when beans have an m.c. of 20% or higher (Taniwaki *et al.* 2001). Beans in storage do not normally become so moist; if they did this would also lead to the multiplication of yeasts and bacteria. Another potential mycotoxin contaminant of coffee is aflatoxin, produced by *Aspergillus flavus*. However, the incidence of this appears to be infrequent, with apparently only one case being reported, from Guatemala (De Campos *et al.* 1980).

Insects

Insects commonly infest green coffee beans. This was demonstrated in a three-year survey of green coffee offered for import to the USA. Beans were found to have insect damage in 71% of the cases studied, with an average insect damage of 1.7% in the 1020 lots tested (Gecan *et al.* 1988). Two insect pests are particularly associated with the spoilage of green coffee and both are beetles: the coffee berry borer *Hypothenemus hampei* and coffee bean weevil *Araecerus fasciculatus*. *Hypothenemus hampei* mostly attacks coffee cherries in plantations. *Araecerus fasciculatus* may infest coffee cherries in the field but is mostly found on dry beans during processing, storage and transport. In an experimental study at 80% r.h. and 25°C, *arabica* beans were much more quickly infested with *A. fasciculatus* than were *robusta* beans (Lavabre & Decazy 1968).

Commodity and pest management regimes

Essential storage parameters

Good raw material quality and storage conditions are necessary to minimise the risk of coffee deterioration. The following factors are of particular significance:

(1) Only well-dried parchment or green coffee of good quality should be stored. In fact, in coffee-producing countries with high ambient humidities, coffee should not be put in storage above 13% m.c. and should be rejected for purchase at above 14% m.c. It is essential that coffee is sampled at intake to ensure that it conforms to quality standards (Fig. 11.3). Liquor quality is the most important criterion on which coffee is judged, but size, colour and defects are also considered (Australian Coffee Research and Development Team 1995). Defects are more common in dry-processed coffee.

(2) Storage under high humidity and temperature can lead to mould infection. A knowledge of the relationship between prevailing humidity and the associated bean moisture content, termed the equilibrium moisture content (e.m.c.), is a reliable means of defining safe storage conditions. Mould growth may be initiated at moisture contents in equilibrium with 70% r.h. and so this defines the safe storage limit (Fig. 11.4). The exact relationship between r.h. and m.c. depends upon the type of coffee, including origin and processing (Fig. 11.4), as well as storage form (parchment or green coffee) and to some

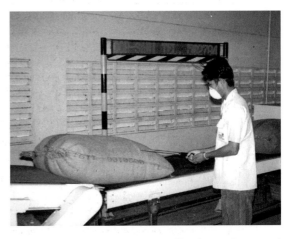

Fig. 11.3 Sampling incoming coffee for quality control in Thailand.

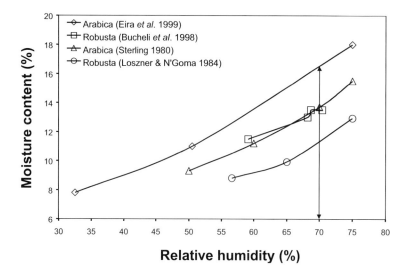

Fig. 11.4 Relationship between relative humidity and bean moisture content for *arabica* and *robusta* coffee (data from various sources).

extent the prevailing temperature. For coffee stored under industrial conditions in Kenya and Thailand moisture contents in equilibrium with 70% r.h. were 13.5% and 14%, respectively (Stirling 1980; Bucheli *et al.* 1998). Such values are also typical of cereal grains like wheat or maize. Some laboratory data suggest a different relationship. At 70% r.h. the e.m.c. has been observed to vary between 11% (Loszner & N'Goma 1984) and 16% (Eira *et al.* 1999) at 20° and 25°C, respectively (Fig. 11.4). The different methods used for m.c. determination as well as differences in the beans themselves might explain these widely ranging results.

(3) The proportion of defective coffee beans (black, partly black, broken, infested, husks, cherries, foreign matter) should be kept as low as possible, to reduce OTA contamination by *A. carbonarius* on drying coffee cherries (Joosten *et al.* 2001), and risks of coffee spoilage during storage. OTA contamination can be particularly high on coffee husks (Bucheli *et al.* 2000).

(4) Several studies indicated that storage of dry parchment beans, rather than green beans, reduces the development of off-flavour attributes such as woodiness, although even parchment coffee must be handled with care (Wootton 1970; Stirling 1980). The parchment, which is the seed coat of the coffee seed, is in fact an efficient natural barrier against moisture transfer and provides a good protection against damage from insects and micro-organisms, and the potential contamination of coffee by pesticides.

Continuous monitoring of temperature and r.h. at any given storage site is highly recommended, and should be combined with regular (e.g. monthly) determination of bean m.c. This monitoring is inexpensive and provides valuable information about the critical periods when ventilation or cooling would be beneficial. The accumulation of monitoring data also provides a basis on which decisions can be made about future investment in storage technology such as the adoption of ventilation or air conditioning.

Insects infesting coffee and other stored products can easliy be destroyed by fumigation with the gas methyl bromide. However, this gas is listed as a depleter of stratospheric ozone under the terms of the Montreal Protocol, and so it is being phased out. The most convenient alternative to methyl bromide is the gas phosphine (PH_3), which is generated from solid formulations of aluminium or magnesium phosphide. The application and fate of phosphine was investigated in India in several studies using *arabica* and *robusta* beans. Surprisingly, the phosphine-holding capacity of *robusta* beans was found to be higher than that of *arabica* beans, and this behaviour was explained by differences in wax characteristics (Rangaswamy & Sasikala 1991). The authors concluded that coffee can be fumigated safely with aluminium phosphide preparations generating 3 g of gas/tonne of coffee without having any perceptible free or bound phosphine residues in the product. Phosphine is widely used throughout the world and when correctly applied will kill all infesting insects.

Monitoring coffee quality during storage

Green coffee can be stored under well-controlled conditions for three or more years. Nevertheless, even under the best storage conditions, colour and flavour characteristics gradually decline and good storage is more difficult to achieve when the ambient conditions fluctuate widely. Frequent m.c. determinations are a simple and efficient tool to monitor whether stored coffee beans have picked up moisture and therefore are at risk of deterioration. However, determining m.c. does not provide any direct information about the gradual decline in important characteristics such as flavour and colour. Other measures are needed to detect at an early stage any quality decline that will be perceived later at cup tasting.

During the last 30 years, research into the biochemical processes that are likely to be responsible for coffee quality changes in storage has been very limited. In theory, an ideal indicator of coffee quality would be seed viability, assuming that germination rate would correlate with raw material quality. In practice, however, seed germination is often poor and slow. It is therefore not a practical marker for monitoring deterioration in storage. Nevertheless, *arabica* seeds can survive desiccation to about 10% m.c. for 12 months of hermetic storage at 15°C (Ellis *et al.* 1990). At a lower temperature or lower m.c., deterioration occurred more rapidly, indicating that storage at a low m.c. and temperature does not necessarily maintain seed quality.

Potential chemical and physical markers that correlate with coffee deterioration during storage are scarce, and those that have been studied have often proved inconclusive. However, the determination of endogenous coffee enzyme activities is one potential tool for monitoring coffee deterioration. Polyphenol oxidase (PPO), an enzyme often associated with browning of plant raw materials, was measured and reported to be an indicator for coffee beverage quality (Melo *et al.* 1980). In fact, it decreased during storage, and was lower in spoiled coffee. Its variability and instability during storage suggest, however, that there is no direct relationship between PPO activity and cup quality (Bucheli, unpublished). It seems possible that it is acting as a proxy for other endogenous enzymes that preferentially modify proteins, lipids, carbohydrates or precursors of coffee flavour during storage, and that thereby affect final beverage quality. Such biochemical changes are likely to be induced at bean m.c. of 12–16%, and average temperatures of 25–30°C that are frequently encountered in tropical climates.

A somewhat more reliable marker appears to be glucose content. Accelerated storage conditions in the laboratory (37°C, 82% r.h.) generated glucose in green coffee as a result of sucrose hydrolysis. After three months, glucose concentrations in *arabica* and *robusta* coffee were found to have increased 6- to 12-fold (Bucheli *et al.* 1996). This suggested that glucose could be a reliable marker for coffee quality since mature *robusta* and *arabica* beans usually contain only very small amounts

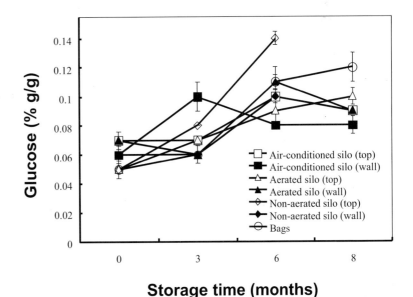

Fig. 11.5 Glucose content of green *robusta* coffee stored in Thailand under different silo and bag conditions (bean sampling positions in parentheses).

of glucose, 0.03 to 0.06% g g^{-1} (Rogers *et al.* 1999). This hypothesis was verified in an industrial storage trial (Bucheli *et al.* 1998) where it was found that the increase of glucose was linked to the appearance of a woody or rubbery taste in the coffee cup, and that a glucose increase of about 50% took place during the rainy period when coffee m.c. increased the most (Fig. 11.5). These results demonstrate that increases in glucose, for example, from 0.05% to 0.1%, can indicate whether coffee was subjected to conditions of increased moisture.

Economics of storage

The economic role of coffee storage in producing countries during coffee overproduction has been discussed in detail by Sassen (1994). Comparing the costs of warehouse and vacuum storage (Fig. 11.6), it was shown that capital investment for vacuum storage was about 40% lower than for warehouse storage. The total storage costs per tonne of coffee, based upon one year of storage, were \$US206 for warehouse and \$121 for vacuum storage. Several factors explained these differences. First, traditional storage has fixed costs of 70–80%, whereas vacuum storage has less than 60% because it is essentially a packaging activity. Vacuum storage is more flexible, allowing the selection of a storage site based entirely on efficiency of location and transport logistics. The benefits of vacuum storage are particularly apparent for long-term storage, since warehouse storage for more than 18 months risks significant deterioration which is not apparent if vacuum storage has been used. For short-term storage, such as in port facilities, vacuum storage is not competitive; it becomes structurally cheaper beyond the breakeven period of around three to four months.

The costs in using the ballooning technique were estimated over a two-year period in a study in India (Narasimhan *et al.* 1994). It was established that a substantial financial advantage could be obtained if the beans were stored in sealed enclosures for at least one year.

The question of whether silo storage is more economic than bag storage is not easily answered. Certainly, bulk handling reduces transportation costs, and labour for the unloading of the coffee at the receiving factory. Nevertheless, investments in a silo structure are very high, and the time to payback can easily be in the range of eight to ten years, which might not be cost-effective. Investments in silos will be more justified when the factory is lacking space that could be used for other purposes, or when bag storage is extremely expensive. As an example, storage in 150 t silos can save more than 80% of the space used for bag storage on pallets (25 bags of 60 kg stored four pallets high).

Future developments

Bulk handling is anticipated to increase further, not only in the consuming countries, but also in producing countries. It is possible that large roasters will increasingly manufacture coffee in producing countries, and ship products to consuming countries only for conditioning. Continuous coffee over-production will keep prices depressed and increase the pressure to reduce warehouse

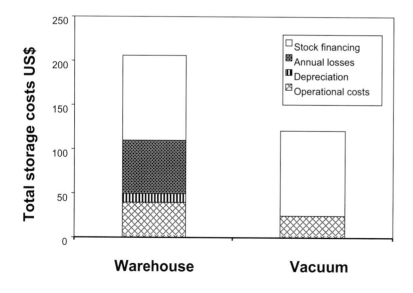

Fig. 11.6 Comparison of total storage costs of coffee per tonne based upon one year's storage (adapted from Sassen 1994).

and silo storage costs. This will act as a strong incentive to adopt cheaper and more efficient storage methods such as vacuum storage.

References

Australian Coffee Research and Development Team (1995) *Coffee growing in Australia: a machine-harvesting perspective*. Rural Industries Research and Development Corporation, Barton ACT, Australia.

Betancourt, L.E. & Frank, H.K. (1983) Bedingungen des mikrobiellen Verderbs von grünem Kaffee. *Deutsche Lebensmittel Rundschau*, **79**, 366–369.

Bucheli, P., Meyer, I., Pasquier, M. & Locher, R. (1996) Determination of soluble sugars by high performance anion exchange chromatography (HPAE) and pulsed electrochemical detection (PED) in coffee beans upon accelerated storage. 10th FESPP, Florence, Italy. *Plant Physiology and Biochemistry*, Special Issue, L-12.

Bucheli, P., Meyer, I., Pittet, A., Vuataz, G. & Viani, R. (1998) Industrial storage of green robusta coffee under tropical conditions and its impact on raw material quality and ochratoxin A content. *Journal of Agricultural and Food Chemistry*, **46**, 4507–4511.

Bucheli, P., Kanchanomai, C., Meyer, I. & Pittet, A. (2000) Development of ochratoxin A during Robusta (*Coffea canephora*) coffee cherry drying. *Journal of Agricultural and Food Chemistry*, **48**, 1358–1362.

De Campos, M., Crespo Santos, J. & Olszyna-Marzys, A.E. (1980) Aflatoxin contamination in grains from the pacific coast in Guatemala and the effect of storage upon contamination. *Bulletin of Environmental Contamination and Toxicology*, **24**, 789–795.

De Carvalho, V.D. & Chalfoun, S.M. (1989) Quality of coffee. Influence of post-harvest factors. *Indian Coffee*, **August**, 5–13.

Eira, M.T.S., Walters, C. & Caldas, L.S. (1999) Water sorption properties in *Coffea* spp. seeds and embryos. *Seed Science Research*, **9**, 321–330.

Ellis, R.H., Hong, T.D. & Roberts, E.H. (1990) An intermediate category of seed storage behaviour? I. Coffee. *Journal of Experimental Botany*, **41**, 1167–1174.

Gecan, J.S., Bandler, R. & Atkinson, J.C. (1988) Microanalytical quality of imported green coffee beans. *Journal of Food Protection*, **51**, 569–570.

Gopalakrishna Rao, N., Balachandran, A., Natarajan, C.P. & Sankaran, A.N. (1971) Variations in moisture and colour in monsooned coffee. *Journal of Food Science and Technology*, **8**, 174–176.

John, G.A. (1995) Handling coffee its own. *Tea & Coffee Trade Journal*, **March**, 12–18.

Joosten, H.M.L.J., Goetz, J., Pittet, A., Schellenberg, M. & Bucheli, P. (2001) Production of ochratoxin A by *Aspergillus carbonarius* on coffee cherries. *International Journal of Food Microbiology*, **65**, 39–44.

Lavabre, E.M. & Decazy, B. (1968) Contribution à l'étude des problèmes posés par le stockage des cafés dans les pays de production. *Café, Cacao, Thé*, **12**, 321–342.

López, C.I., Bautista, E., Moreno, E. & Dentan, E. (1989) Factors related to the formation of 'overfermented coffee beans' during the wet processing method and storage of coffee. In: *Proceedings of the 13th ASIC Coffee Conference*, Paipa, Columbia. Association Scientifique Internationale du Café, Paris, France.

López Garay, C., Bautista Romero, E. & González, M. (1987) Microflora of the stored coffee and its influence on quality. In: *Proceedings of the 12th ASIC Coffee Conference*, Montreux, France. Association Scientifique Internationale du Café, Paris, France.

Loszner, G. & N'Goma, A. (1984) Bestimmung der Sorptionsisothermen von Rohkaffee nach der statischen Methode. *Beiträge zur tropischen Landwirtschaft und Veterinärmedizin*, **22**, 269–274.

Mabbett, T. (1990) Rules for origin storage. *Coffee and Cocoa International*, **5**, 29.

Melo, M., Fazuoli, L.C., Teixeira, A.A. & Amorim, H.V. (1980) Alterações físicas, químicas e organolépticas em grãos de café armazenados. *Ciência e cultura*, **32**, 468–471.

Muriithi, G.K. (1977) Coffee storage research. Bulk storage in an autocool silo. *Kenya Coffee*, **July**, 245–252.

Narasimhan, K.S., Majumder, S.K. & Natarajan, C.P. (1972) Studies on the storage of coffee beans in the interior parts of South India. *Indian Coffee*, **October**, 327–330.

Narasimhan, K.S., Rajendran, S., Jayaram, M. & Muralidharan, N. (1994) Technologies for storage and preservation of coffee beans in India. In: *Proceedings of the 6th International Working Conference on Stored-product Protection*, 17–23 April 1994, Canberra ACT, Australia. CAB International, Wallingford, UK.

Nataredja, Y. & Hodges, R.J. (1989) Commercial experience of sealed storage of stacks in Indonesia. In: *Fumigation and Controlled Atmosphere Storage of Grain*. ACIAR Proceedings No 25, Canberra ACT, Australia.

Natural Resources Institute (1996) Sealed bag stacks for better grain storage: a guide to long-term storage of bagged grain in sealed stacks. Video. Natural Resources Institute, University of Greenwich, Chatham, UK.

Rangaswamy, J.R. & Sasikala, V.B. (1991) Comparative responses of coffee species to phosphine fumigation. *Journal of Food Science and Technology*, **28**, 222–225.

Rodríguez, J., Fritsch, G. & Díaz, J. (1976) Estudio por resonancia magnetica de algunos efectos del estado del agua contenida en el cafe verde. In: *Proceedings of the 7th ASIC Coffee Conference*, Hamburg, West Germany. Association Scientifique Internationale du Café, Paris, France.

Rogers, W.J., Michaux, S., Bastin, M. & Bucheli, P. (1999) Changes to the content of sugars, sugar alcohols, myo-inositol, carboxylic acids and inorganic anions in developing grains from different varieties of Robusta (*Coffea canephora*) and Arabica (*C. arabica*) coffees. *Plant Science*, **149**, 115–123.

Sassen, J. (1994) Can the coffee pact be successful? (The crucial role of storage). *F.O. Licht International Coffee Yearbook 1994*, C19–C24.

Stirling, H. (1980) Storage research on Kenya arabica coffee. In: *Proceedings of the 9th ASIC Coffee Conference*, London. Association Scientifique Internationale du Café, Paris, France.

Taniwaki, M.H., Urbano, G.R., Cabrera Palacios, H., Leitão, M.F.F., Menezes, H.C., Vicentini, M.C., Iamanaka, B.T. & Taniwaki, N.N. (2001) Influence of water activity on mould growth and ochratoxin A production in coffee. In: *Proceedings of the 19th ASIC Coffee Conference*, 14–18 May 2001, Trieste, Italy. Association Scientifique Internationale du Café, Paris, France.

Wootton, A.E. (1970) The storage of parchment coffee. *Kenya Coffee*, **May**, 144–147.

Chapter 12
Cocoa: West Africa (Ghana)

W. A. Jonfia-Essien

Cocoa (*Theobroma cacao*) is a small tree that originally occurred in the wild in the Amazon basin but is now cultivated widely in the tropics. The tree bears pods that each contain about fifty cocoa beans (Plate 12). These are fermented and on processing yield cocoa butter used in the cosmetics industry and cocoa solids that are made into chocolate confectionery and drinks.

Cocoa is an international commodity worth an estimated $US5bn in world trade in 1993 (ICCO 2002). Approximately 60% of the world's supply of cocoa originates in West Africa, where Côte d'Ivoire is the largest exporter of cocoa in the world, supplying about 39% of the market. Crop yields show wide year-to-year variation, so prices fluctuate considerably on futures markets. The producing countries consume a small proportion of this cocoa; the remainder is exported (Fig. 12.1).

After harvest, cocoa beans are fermented, sun dried and then stored briefly on-farm. Cocoa yields are very low in Ghana and production has always been through small-scale operations using low technology. Nevertheless Ghana supplies a product with a high fat content and good flavour, for which confectionery manufacturers are prepared to pay a premium. Until 1992–93, the state-owned Ghana Cocoa Board held a monopoly on cocoa purchases from farmers through its subsidiary, the Produce Buying Company (PBC). Subsequently private companies have also been licensed as buyers and the PBC privatised, with the Cocoa Board remaining as majority shareholder and the bulk of remaining shares held by staff. The licensed buying companies (LBCs including the PBC) were obliged to sell their beans at a fixed price to the Cocoa Marketing Company (a subsidiary of the Ghana Cocoa Board) so that the Board had a monopoly over exports. However, from the 2000–01 cocoa season the export of cocoa beans has been partially liberalised, with five buying companies being qualified to export

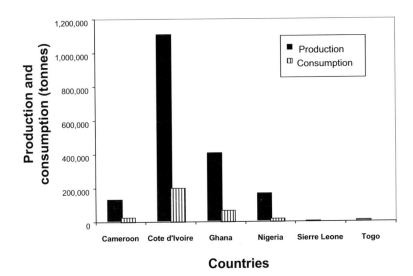

Fig. 12.1 Production and consumption of cocoa beans by West Africa countries in 1997/98 (data from ICCO 2002).

30% of their bean purchases, the remainder being sold to the Cocoa Marketing Company (CMC). However, until at least 2003 these companies were not in a position to be able to make exports.

For safe storage, the moisture content of cocoa beans should be between 6% and 7%. Above 8% there is danger of moulds developing on the beans and below 5% the beans will be very brittle. After drying the beans are placed in jute sacks. Cocoa is purchased from the farm and is then moved as quickly as possible to stores where it is held in sack. Here, the cocoa may be reconditioned to meet quality standards, graded and sealed before dispatch to the port. In Ghana, cocoa deliveries from Brong Ahafo and Ashanti Regions and parts of Western, Central and Eastern Regions are first moved to one of two inland ports at Kaase or Abuakwa, where they are briefly in transit to the main exit sea ports at Tema and Takoradi. Within the port area, cocoa is held either in short-term transit sheds awaiting shipping or in long-term storage in warehouses. In other countries the export market has been fully liberalised, however this is believed to be responsible for some decline in quality.

Historical perspectives

Cocoa was first planted in the State of Bahia in Brazil in 1746, and cocoa of the Amelonado type was subsequently taken from Bahia to the West African islands of São Tomé and Bioko. The cocoa industries developed initially with liberated slaves and later with migrant labour from the West African mainland. The returning labourers took cocoa pods to their various homelands and attempted to grow the crop. This was how planting started in Ghana and Nigeria. Cocoa developed later in Côte d'Ivoire and Togo and was also of the typical Amelonado variety (Wood & Lass 1985).

In Ghana, the size of farmers' holdings and the manner in which the farms are held vary enormously (Hill 1962). In Nigeria, individual farmers cultivate small plots of cocoa (Galletti *et al.* 1956) and the same picture is probably true of the other main cocoa-growing countries in West Africa – Côte d'Ivoire, Cameroon and Togo.

Côte d'Ivoire fully liberalised its external marketing system in October 1999 and several other cocoa producers have followed suit. However, based on recommendations by Landell Mills Commodities, a decision was made in Ghana in 1996 to keep external marketing of cocoa as the preserve of the CMC while cocoa can be purchased from the farmer by LBCs, co-operatives or the PBC. The experiences of countries such as Nigeria, Togo

and Cameroon, which have liberalised their external marketing systems, are that the farmers' share of the free on board (FOB) price is increased (Ministry of Finance 1998). For example, in 1998, the FOB price in Ghana was 97.7%, 75.4% and 66.7% of that in Côte d'Ivoire, Togo and Cameroon, respectively. It may, however, be argued that the FOB price in some of these countries is lower in absolute terms than in Ghana. Ghana spends roughly 3% of the FOB price of cocoa directly on research and extension whereas in some other countries of the sub-region, research and extension are funded through the government budget (Ministry of Finance 1998). It is believed that Ghana must liberalise its external trade to remain competitive with its neighbours, especially Côte d'Ivoire. Liberalisation is in line with the Ghanaian Government's policy of privatisation, and private sector participation has initially permitted LBCs to export 30% of their purchases, which leaves majority control in the hands of CMC. This cautious approach is a test of whether complete liberalisation of the external market would be in Ghana's best interest.

In the last 25 years in Ghana, the main centre of cocoa production has shifted from the Eastern to the Western Region. Private individuals own all the storage systems up-country, from the farm storage to the LBC at the society, district and regional levels. The LBCs rent these warehouses for temporary storage of cocoa before dispatch to central warehousing and, while PBC rents warehouses occasionally, it currently owns and manages most of the existing warehouses that were originally Cocoa Board property. As a result of liberalisation of the cocoa market, the Cocoa Board owns only the storage systems at the inland ports. All the storage systems at the sea ports and the surroundings areas are rented terminal warehouses. Before liberalisation, the Ghana Cocoa Board used about 75 000 t of storage capacity, most of which was in the Eastern Region and it is now surplus to requirements. At the same time, many of the LBCs do not have adequate warehouse space. At least in the Eastern Region there is no need to encourage the building of new stores since surplus stores can be rehabilitated and arrangements are being made to lease or sell them to the LBCs.

Physical facilities

Farmer storage systems

After harvesting, the pods are broken to extract the beans, which are then fermented for about six days. The fermentation procedure varies from country to country (Wood & Lass 1985). After fermentation, the beans are sun dried on mats raised off the ground in Ghana or on concrete floors in Côte d'Ivoire. Solar dryers have been designed for small-scale farmers in Côte d'Ivoire and Cameroon (Richard 1969, 1971) and a large number have been built in Cameroon for small cooperative societies.

In Ghana, the Cocoa Board supplies farmers with jute sacks for packing their beans and the sacks are assembled in stores typically made of mud, bamboo, wood and thatch, but in some instances iron sheets are used for the roofing (Fig. 12.2). The produce is stored only for a short period in such stores, often on a raised platform that aids drying; no pesticides are applied. The cocoa may be placed together with other produce in such stores and this can produce off-flavours and allows exposure to insect infestation.

Cocoa warehousing

Cocoa purchased from the farm is moved as quickly as possible from the initial assembly points to larger stores of the LBCs, PBC and co-operatives. Shed-gangs build the bags of cocoa into stacks by hand or sometimes mechanically, using high-mast battery-operated forklift trucks. Every effort is made by dealers to avoid storing cocoa in the open, but if this is unavoidable then pallets are doubled, to prevent problems from ground moisture, and the produce covered immediately with a durable tarpaulin.

Ideally, cocoa warehouses are easily accessible by road or railway. About 56% of the 24 000 km of feeder roads in Ghana are in the cocoa-growing areas. Even though the network of feeder roads appears adequate, their condition is very poor (Ministry of Finance 1998). The same picture is probably true of the other main cocoa-growing countries in West Africa. The feeder roads are not uniformly distributed in the cocoa-growing areas in all of the sub-region. Both road and rail transport are used to evacuate cocoa from up-country to the ports. Fixed telecommunication facilities are available in only a few cocoa-growing areas but communications are improving with the introduction of mobile telephones.

The stores must be rain-proof, the floor must be dry and properly constructed of cement or concrete, and the warehouse must be provided with sufficient doors to allow adequate ventilation. The inside walls of the warehouse must be whitewashed annually. The ideal warehouse must have sufficient pallets made with seasoned wood and provide at least an 8 cm air space between the cocoa and floor (Fig. 12.3) to give good ventilation between floor and stack. The warehouse construction must facilitate pest control and maintenance of shed hygiene by having adequate gas-proof sheets, sand snakes (sand-filled tubes) and a wooden shovel.

During the colonial era in Ghana, up to 1958, expatriate LBCs administered the cocoa trade. For successful operation the system relied on cocoa packed in jute sacks stacked on seasoned pallets, with good hygiene and the use of insecticides. There were three basic designs of store:

Fig. 12.2 Traditional cocoa warehouse built in 1920 with mud, bamboo, wood and iron sheet roof in the Eastern Region (Ghana). It is privately owned and rented by the Produce Buying Company.

Fig. 12.3 A stack of cocoa bags built on seasoned wood pallets, carefully arranged to prevent collapse of the stack (Ghana).

- Rectangular mud-walled warehouses that were water-proof and with ventilation achieved by opening or closing the warehouse doors.
- Rectangular cement block warehouses equipped with hoppers. Prior to bagging beans would be graded, quality improved by removing flat and broken beans, and mixed to improve homogeneity. Ventilation was achieved by opening or closing the warehouse doors.
- Rectangular sheds clad with iron sheets, painted black to reduce heat gain and with some limited, gable ventilation (Fig. 12.4).

After 1958 a wider range of more modern warehousing became available. The most up-to-date are cylindrical concrete warehouses with automated filling and emptying. A good example can be found at the Cocoa Processing Company (CPC) in Tema. At these stores the beans are aerated with ambient air to stabilise temperatures, residual insecticides are applied and insecticidal space treatments reduce infestation. Also in use are rectangular concrete warehouses with iron sheet roofs (Fig. 12.5a–c) and concrete floors with doors variously arranged and from six to nine eave ventilators on each long wall. These warehouses are larger and offer much better ventilation than earlier designs, and enable easier pest control operations and movement of produce. They also facilitate modern storage operations such as palletisation and the use of forklift trucks. The stacking of cocoa is carefully controlled to optimise the use of space but also to allow easy access for store management and pest control operations and to aid better ventilation.

A recent innovation by a private warehousing company has been the bulk storage of beans in rectangu-

(a)

(b)

(c)

Fig. 12.5(a,b,c) Some cocoa warehouses at the port of Tema, Greater Accra Region (Ghana).

Fig. 12.4 An iron-sheet clad cocoa shed, painted black and built by Cadbury & Fry (Eastern Region, Ghana). Currently in use by the Produce Buying Company.

lar, concrete-walled warehouses with bitumen floors. Adequate ventilation is assured by way of large doors. Conveyor belts are used to build heaps of cocoa on the floors and to load vessels. This type of storage offers ease of shed hygiene and insecticide application, good space utilisation and the use of vehicles to scoop beans from heaps onto conveyor belts during discharge. The only disadvantage is the high level of cocoa beans that are crushed during operations. Storage space in these warehouses can be rented and the system is well established in Côte d'Ivoire, Cameroon and Ghana.

Objectives of storage

The objectives of farm storage are mainly to collect the harvested cocoa, facilitate packaging and to organise disposal. In LBC, co-operative and PBC stores, the cocoa may be reconditioned or graded to meet quality standards, bagged and weighed (new jute sacks must be used when cocoa is exported, in line with ISO 1973), then sealed before evacuation to the port. In many countries the net weight of a bag of cocoa is 62.5 kg, i.e. there are 16 bags to a tonne. This weight is accepted internationally and anything less is considered as short weight. Either the cocoa is rejected or the equivalent in amount is deducted from the purchase price. Some countries have adopted other net weights. For example, in Ghana, the approved net weight of a cocoa bag is 63.8 kg. This has been adopted to ensure that, even if sampling removes some of the produce, or delay in export results in some moisture shrinkage, the bags will not be underweight. Most of the bags of cocoa beans from Ghana on the external market weigh between 62.5 kg and 63.0 kg.

Within the port area are short-term storage facilities (transit sheds) and long-term facilities (warehouses). Long-term storage provides additional sources of revenue if managed efficiently, and can improve the port's reputation and encourage trade.

Short-term storage

Short-term storage in ports is particularly important for break-bulk cargoes like cocoa that are normally carried in general cargo vessels. The stocks are normally held in transit sheds. This facilitates quality control, administrative formalities and cocoa consolidation, and evens the imbalance between the quantities of cocoa carried by each inland transport vehicle and the large capacity of the ship.

Long-term storage

Cocoa is normally only stored long-term at the port. This may be to balance differences between supply and demand in the market or to permit economies of scale in purchasing and transport.

Market imbalance

Long-term storage is often needed because of the differences in timing between supply and demand for cocoa. This is particularly the case with agricultural produce such as cocoa where the crop ripens, is harvested and processed over a comparatively short period but is consumed steadily throughout the year. Climatic conditions can reduce or suspend the transportation of produce. Thus, stocks of export cocoa are often maintained by CMC in a port's long-term storage area as a buffer to avoid the risk of delay due to bad weather. The long-term storage also provides a safety stock to protect shippers' interests. This cocoa, referred to as policy stock, is generally the largest and held to protect shippers against political and economic developments and to allow speculative purchases. The long-term stocks are the property of Ghana Cocoa Board, managed by its subsidiaries CMC and Quality Control Division (QCD). QCD has unrestricted access to all stocks of cocoa in the country, whether they are in the hands of the LBCs, PBC or CMC.

Economies of scale

Shippers often like to purchase large quantities of commodities because they can obtain good discounts, and charter their own vessels at attractive rates. Cocoa is then put into long-term storage; the economies of scale can often provide discounts that more than compensate for the extra cost of storage. These benefits may be lost, however, if the cocoa is discharged slowly from storage.

Major sources of quality decline

The storage of cocoa beans in West Africa suffers three potential sources of quality decline – the development of mould, infestation by stored product pests and, if there is prolonged storage in hot conditions, a rise in free fatty acids (Wood & Lass 1985). It is safe to store well-dried cocoa beans for two or three months, but for longer periods special precautions are needed to ensure against quality decline.

Humidity

Cocoa beans are hygroscopic and will absorb moisture under very humid conditions. Beans with moisture contents of at least 8% will become mouldy and such beans are in equilibrium at a relative humidity of 75% or more (Gough & Lippiatt 1978). To help to maintain suitable conditions, especially where relative humidity is exceptionally high, as in Cameroon, the store should be open during normal working hours but closed at other times. In Nigeria, it has been found that in large bag stacks of up to 1000 t, the temperature rises several degrees above ambient (Riley 1964). This leads to a fall in moisture content in the inner bags, and a slight fall in moisture content in the outer bags. In Côte D'Ivoire, storage in 1 t vacuum packs of high-density polypropylene has been tested successfully; unfortunately, the Capitainer company responsible for the system ceased trading and the method has not been adopted despite the promising results.

Insect infestation

In West Africa all cocoa is liable to insect infestation, especially by the tropical warehouse moth (*Ephestia cautella*). At the farm, drying mats are an important source of infestation. At the end of the season these are usually rolled up and stored under the eaves; however, they often carry pupae from which moths may emerge to infest the new crop. Similarly, the area around mechanical dryers can provide a breeding ground for pests (Wood & Lass 1985). In stores infestation can be minimised by good hygiene measures and, when necessary, the application of pest control. Where the crop is seasonal the store will be empty for some of the year and this enables thorough cleaning, but in countries where there is no slack season emptying the store may present a problem. Nevertheless, cleaning should be done regularly.

Apart from the tropical warehouse moth there are two beetles, *Lasioderma serricorne* (the tobacco beetle) and *Araecerus fasciculatus* (the coffee bean weevil), which may infest cocoa beans. Both can pierce the shell of the bean, thereby providing an entrance for *E. cautella* and for moulds. In Ghana, dry cocoa beans were monitored for insect pests associated with the cocoa in storage from 1995 to 2000 and eleven species identified (Table 12.1).

Free fatty acids

Both mould and insect infestation results in an increase

Table 12.1 Insect pests associated with dry cocoa beans in storage in Ghana. Source: Jonfia-Essien (2001).

Order	Family	Species
Lepidoptera	Pyralidae	*Ephestia cautella*
		Corcyra cephalonica
Coleoptera	Anthribidae	*Lasioderma serricorne*
		Araecerus fasciculatus
	Tenebrionidae	*Tribolium castaneum*
	Nitidulidae	*Carpophilus hemipterus*
		Carpophilus dimidiatus
	Cucujidae	*Cryptolestes ferrugineus*
		Cryptolestes pusillus
	Silvanidae	*Oryzaephilus mercator*
	Bostrichidae	*Rhyzopertha dominica*

in free fatty acids (FFA) in cocoa beans. High levels of FFA can render the beans unsuitable for chocolate and cocoa butter production (Anon. 1970). Good healthy beans contain low concentrations of FFA (about 0.5%) but generally the level of free fatty acids in the beans must be less than 1% to meet the acceptable level of 1.75% in cocoa butter extracted from the beans (Anon. 1996).

International cocoa standards

The various cocoa standards that apply relate to flavour and purity or wholesomeness. The most important of these standards are the ISO standards (ISO 1973) and those in the contracts of various trade associations such as Cocoa Association of London, the Association Française du Commerce des Cacao, and the US Cocoa Merchants Association. The grade standards are based on the cut test that allows certain gross flavour defects to be identified by cutting open beans to reveal the colour of the dried cotyledons (Anon. 1996). Defective beans may either be mouldy or slaty. In the latter case the bean has been dried before fermentation and is unsuitable for chocolate production.

Quality standards are also set by the authorities in the producer countries; these are their practical internal quality standard and against which the quality is assessed throughout the internal marketing channels in the country. An example is the grade standards based on the cut test in Ghana (Table 12.2). It is only in Ghana that quality control measures are strictly enforced to ensure the marketing of high-quality cocoa, and this probably

Table 12.2 Grade standards based on the cut test in Ghana (Ghana Standards).

Grade	Maximum percentage defect		
	Mouldy	Slaty	Insect damaged, germinated, flat
I	3	3	3
II	4	8	6

explains why Ghana has been, and is still, the leading producer of good-quality cocoa worldwide.

Effect of liberalisation on quality

In those countries that have liberalised their cocoa market, the effects on quality have been mixed. In Cameroon and Nigeria, where liberalisation took place without putting in place adequate institutional measures for quality control checks, the quality deteriorated initially, but is now recovering (Ministry of Finance 1998). In the case of Togo, liberalisation has been successful without loss in quality because the country put in place adequate quality control measures with the participation of industry stakeholders. Forward selling of the crop ended with liberalisation and cocoa is sold on a spot basis. This situation continues today, with many sales only concluded once the shipment has been inspected in a western European or North American port warehouse (LMC 2000). The evidence goes to support the view that it is not liberalisation *per se* that leads to decline in quality but the institutional mechanisms put in place for quality control and certification (Ministry of Finance 1998).

Cameroon

Liberalisation in Cameroon was implemented in stages beginning in 1991. Prior to liberalisation, quality controls were enforced by the state both up-country and at the port. Following liberalisation, up-country checks were stopped and quality control at the port was privatised (LMC 2000). Quality deteriorated at both farm and export levels. There was increased demand, which meant that farmers effectively had a guaranteed buyer for their crop whatever the quality, and so attention to quality declined (LMC 2000).

Cote d'Ivoire

External marketing of cocoa was liberalised in October

1998 and in June 1999 quality control was liberalised. The latter became the responsibility of two private companies, SGS and Corlender, which have since merged (LMC 2000). Licensed agents buy the cocoa through traders from farmers and deliver the crop to licensed exporters who have their own warehouses. Exporters can also buy cocoa directly from farmers. Quality control has been reduced with a negative effect on the quality of the cocoa beans exported (Ministry of Finance 1998).

Nigeria

Until liberalisation, the Nigerian Cocoa Board (NCB) had responsibilities very similar to those of the Ghana Cocoa Board. In 1986 the Government of Nigeria, as part of a general move towards liberalising pricing system and the marketing of commodities, disbanded the NCB and completely deregulated the internal and external marketing of cocoa. The licensing of internal buyers and exporters was abolished and quality control procedures abandoned (LMC 2000). These changes took place almost overnight with no effective alternative put in place. Liberalisation initially boosted production and exports but the quality of cocoa delivered to end-users plummeted and became notoriously unreliable. The problem originated at both the farm level and at the ports. Operators had no concept of what they were buying and, consequently, often sold anything, even moist unfermented beans, simply because many unscrupulous exporters wanted to take delivery as quickly as possible. They then mixed the higher quality cocoa with the poorer grades and shipped the resulting uneven blend. Cocoa was purchased at high prices with little regard to quality, which declined along with the premium that Nigerian cocoa used to fetch on the export market (LMC 2000).

Commodity and pest management regimes

Cocoa is stored on-farm only short-term and this does not involve any chemical application. Disinfestation of cocoa begins in the up-country stores and continues through the inland ports to the sea ports. Cocoa which has become infested may need fumigation using methyl bromide or phosphine under gas-tight conditions.

Methyl bromide is applied to the commodity as a gas at the rate of 50 g t^{-1} of cocoa through piping terminating at suitable points on top of bag stacks. The fumigation can be completed in one day. Those countries using methyl bromide would normally give a single treatment just prior to export. Multiple fumigations would run the risk of the accumulation of unacceptable bromide residues. Phosphine is much simpler to use, as it is available in the form of sachets, tablets or pellets of aluminium or magnesium phosphide. The gas is applied at a rate of $7 \, g \, t^{-1}$. The phosphide preparations absorb moisture from the air and release phosphine. It takes a little time for the concentration of the gas to build up, so the use of phosphide preparations requires longer fumigation periods than methyl bromide, not less than five days. Cocoa stocks can be fumigated with phosphine without the risk of accumulation of any significant residues.

In Ghana, the QCD of the Ghana Cocoa Board undertakes disinfestations of cocoa in storage. QCD has 18 disinfestation zones within which a trained pest control staff operate from strategic locations including the ports, regional capitals and a few district offices (COCOBOD 2000). Operational staff inspect storage sheds and disinfest produce to ensure that only insect-free cocoa is exported. The disinfestation exercise involves spraying, fogging and fumigation, although only cocoa in large warehouses is fumigated. In Ghana, water-based insecticide is used for spraying (e.g. fenitrothion EC) while oil-based insecticide is used for the fogging (e.g. pyrethrin).

At the ports, every parcel of cocoa is fumigated with phosphine to completely disinfest the produce before shipment. After every fumigation, bags of cocoa are selected at random and sieved to ascertain whether fumigation was successful. In the case of failure, stocks are re-fumigated. In addition to insect control activities, QCD also applies rodent control in all storage premises (COCOBOD 2000). Both liquid and solid poison (e.g. Racumen, an anticoagulant coumarin) are used by placing them at baiting points regularly.

Not much information is available on the pest management regimes of stored cocoa in other West Africa countries. In Ghana, phosphine is the only fumigant currently in use whereas all others use methyl bromide, although there is also extensive use of phosphine in Côte d'Ivoire, Nigeria and Cameroon. Methyl bromide is a depleter of stratospheric ozone and for this reason is due to be phased out under the terms of the Montreal Protocol. The timetable for phase-out in developed countries is by 2005 but developing countries have use of the chemical until 2015. Nevertheless, it may be increasingly difficult to sustain methyl bromide fumigation as supplies will become more difficult to obtain and/or more expensive, and consumers in developed countries may reject produce treated with it.

Economics of operation of the system

The costs of cocoa storage are not readily available as they are the subject of confidential arrangements between the Ghana Cocoa Board and the management of storage facilities. The costs include store rental as well as the price of disinfestation, equipment and labour. The cost implications of long-term storage are dependent on storage management and the external market index. With good storage management, long-term storage may yield a good dividend if the world market price is rising but a poor dividend if price is falling. With poor storage management, long-term storage may yield poor dividends whether the world market price is rising or falling.

During the post-independence era of the 1960s, the governments of the countries that derived a large proportion of their foreign income earnings from cocoa found that the volatility of the world cocoa market adversely affected the management of their balance of payments, national reserves and budgets. It was difficult for them to prepare and implement sound fiscal, monetary and trade policies. Similarly, in the importing countries, this instability made business planning and budgeting difficult for the processors and manufacturers (ICCO 2002). To balance differences between supply and demand in order to meet market conditions and to provide a safety stock for export, substantial quantities of cocoa are held to protect the cocoa price. In addition, to allow speculative purchases large storage systems were established in the main ports. These storage systems are mostly state owned to permit stockpiling by governments for strategic and economic development.

Unfortunately, liberalisation of the cocoa market created undesirable effects such as a fall in cocoa quality and an increase in counterpart risks for banks and international traders. This in turn has led to the withdrawal of finance for small operators and the disappearance of a forward market. To address and correct these effects, a project for the improvement of cocoa marketing and trade in liberalising cocoa producing countries was implemented in Cameroon, Côte d'Ivoire and Nigeria in 1999 (ICCO 2002). One of the primary objectives of the project was the promotion of a privately run system of field warehouses and the issue of warehouse receipts. This allows farmers to borrow against their crop once this has been harvested, rather than having to sell immediately after harvest at a low price, and so increases the benefit that they receive from the market (ICCO 2002).

Future developments

A major task for cocoa-producing countries is to adjust to the effects of liberalisation. Stakeholders must come together to develop good and effective private sector storage systems with good quality control schemes. Coherent systems must be put in place for quality assurance, finance of trade (minimising trade and price risk) and dissemination of marketing information, which is efficiently operated by the private sector, within a legal and policy framework provided by the public sector. The further development of co-operatives will help to expand the cocoa industry and bring more benefits to cocoa farmers.

Bulk transportation of cocoa beans to overseas destinations in containers is already commonplace. To complement this there will be an increasing move towards bulk storage, and the most up-to-date storage facilities already use bulk methods. Adding value to the crop is an important long-term objective and in 2002, a new cocoa processing factory was opened at Tema to produce semi-processed cocoa for export.

References

Anon. (1970) International Cocoa Standards. *Cocoa Growers' Bulletin*, **4**, 28.

Anon. (1996) *Chocolate Manufacturers' Quality Requirement*. The Biscuit, Cake, Chocolate and Confectionery Alliance, London, UK.

COCOBOD (2000) *Ghana Cocoa Board Handbook*. 8th edition. Ghana Cocoa Board, Accra, Ghana.

Galletti, R., Baldwin, K.D.S. & Dina, I.O. (1956) *Nigerian Cocoa Farmers: an Economic Survey of Yoruba Cocoa Farming Families*. Oxford University Press, Oxford, UK.

Gough, M. & Lippiatt, G.A. (1978) Moisture humidity equilibria of tropical stored produce Part III – Legumes, spices and beverages. *Tropical Stored Products Information*, **35**, 15–29.

Hill, P. (1962) Social factors in cocoa farming. In: J.B. Wills (ed.) *Agriculture and Land Use in Ghana*. Oxford University Press, Oxford, UK.

ICCO (2002) International Cocoa Organization. http://www.icco.org.

ISO (1973) Cocoa beans – Specification. ISO 2451:1973. International Organization for Standardization, Geneva, Switzerland.

Jonfia-Essien, W.A. (2001) The effect of storage on the quality of cocoa beans. Internal Report, Ghana Cocoa Board, Accra, Ghana.

LMC (2000) *Liberalisation of External Marketing of Cocoa in Ghana*. LMC International Ltd & School of Administration, University of Ghana. Prepared for the Government of Ghana, through the Coordinator, Cocoa Sector Reform Secretariat, Accra, Ghana.

Ministry of Finance (1998) *Ghana Cocoa Sector Development Strategy. A Task Force Report*. Ministry of Finance, Accra, Ghana.

Richard, M. (1969) Un nouveau type de séchoir solaire expérimenté en Côte d'Ivoire. *Café, Cacao, Thé*, **13**, 57–64.

Richard, M. (1971) Le séchage natural du cacao: un nouveau type de séchoir solaire expérimenté en Côte d'Ivoire. In: *Proceedings 3rd International Cocoa Research Conference*, 23–29 November 1969, Accra, Ghana.

Riley, J. (1964) Temperature fluctuations and moisture content changes in a large stack of cocoa beans stored for three months. Annual Report 1964. *Nigerian Stored Product Research Institute, Technical Report* **1**, 17–24.

Wood, G.A.R. & Lass, R.A. (1985) *Cocoa*. 4th edition. Longman Science & Technical, London, UK.

Chapter 13
Dried Fruit and Nuts: United States of America

J. Johnson

The dried fruit and tree nut industry in the USA is centred in the Central Valley of California. It is a large and diverse mixture of co-operatives, individual processors of various sizes and marketing orders. Marketing orders are government programmes voluntarily adopted by producers to facilitate orderly marketing of agricultural commodities through unified action. Federal marketing orders are administered by the United States Department for Agriculture (USDA).

More than a million tonnes of almonds, hazelnuts, macadamias, pecans, pistachios, walnuts, dates, figs, prunes and raisins (Plate 13), worth about $2.5bn, is produced each year. Dried fruits and nuts are relatively high-value products that are used primarily for snack foods or as confectionery ingredients, and their successful marketing requires strict attention to quality control. Of primary concern is preventing the development of aflatoxins, disinfestation of incoming product of field pests and maintaining quality during storage. Marketing orders play a large part in quality control within the industry, and may institute volume restrictions to minimise price fluctuations. Storage facilities and methods depend upon the product and individual processor, and range from outdoor bin stacks or silos to refrigerated warehouse storage. Insect pest management depends largely on scheduled fumigation, both to disinfest raw product of field pests and to control storage pests in processed product. Due to increasing regulatory restrictions on available fumigants, the industry is actively searching for alternative treatment methods.

Historical perspectives

The United States produces significant amounts of ten different dried fruit and tree nut crops, producing about 1.2m t each year (Table 13.1). For this account, empha-

Table 13.1 Annual dried fruit and nut production and value in the USA (5-year average).

Product	Production (tonnes)	Value (US$ '000)	Principal producing states
Nuts			
Almonds	271 234	888 155	California
Hazelnuts	28 766	28 538	Oregon, Washington
Macadamias	25 045	39 639	Hawaii
Pecans	117 976	225 378	Georgia, New Mexico, Texas
Pistachios	67 514	168 258	California
Walnuts	221 597	303 644	California
Total	732 132	1 653 612	
Dried fruits			
Dates	20 690	24 806	California
Figs	14 610	13 724	California
Prunes	164 065	169 083	California
Raisins	298 725	677 678	California
Total	498 090	885 290	
Grand total	1 230 222	2 538 902	

sis will be placed on those commodities grown in the Central Valley of California, which is responsible for about 80% of the total US crop. In particular, California produces all or nearly all of the almonds (*Prunus dulcis*), Persian walnuts (*Juglans regia*), pistachios (*Pistacia vera*), dates (*Phoenix dactylifera*), figs (*Ficus carica*), prunes (*Prunus* x *domestica*) and raisins (*Vitis vinifera*) in the US (USDA 2002).

Although most of these crops were introduced to coastal California by Franciscan monks during the period of the Spanish missions, the modern dried fruit and tree nut industry began in the late 1800s when production moved toward the interior valleys. The flat Central Valley is blessed with deep, rich alluvial soil, underground aquifers and nearby rivers that provide irrigation water, mild winters and hot, dry summers that are perfect for dried fruit and nut production. By World War II, the almond, walnut, fig, prune and raisin industries were well established in the Central Valley, while the southern Coachella Valley was home to a growing date industry. Pistachio nut is a relative newcomer to the Central Valley, with the first commercial crop harvested in 1976. Now, California is second only to Iran in pistachio production.

The success of agriculture in the Central Valley is largely due to its diversity, a characteristic that may also be attributed to the dried fruit and nut industry. The size of processors can vary considerably, as can the processing and storage methods that are used. Grower co-operatives and marketing orders play an important role in these industries, with federal marketing orders operating for California almonds, walnuts, raisins, prunes and dates, and hazelnuts from Oregon and Washington.

Dried fruits and nuts are relatively high-value commodities when marketed as snack foods or confectionery ingredients. Although they are often considered luxury foods and rarely serve as substantial nutritional staples unlike grain, legume or root crops, they are excellent additions to a well-balanced diet. With an annual value of over $2bn, dried fruits and nuts produced in California provide considerable revenue to the economy of the state. Export sales are of particular importance in the marketing of these products, with 30–80% of the production of almonds, walnuts, pistachios, raisins and prunes being sold for foreign export (Table 13.2). In the intense competition that exists for world markets, processors who succeed in maintaining product quality have a trading advantage.

Table 13.2 Annual exports of US dried fruits and nuts (5-year average).

Product	Volume (tonnes)	Value (US$ '000)
Almonds	223 762	781 562
Hazelnuts	12 433	23 700
Macadamias	944	10 104
Pecans	11 105	51 548
Pistachios	24 967	100 834
Walnuts	71 580	168 123
Prunes	67 027	136 148
Raisins	109 603	192 050

Physical facilities

A wide variety of storage facilities are used for dried fruits and nuts, depending upon the specific commodity and individual processor. Tree nuts in-shell are stored in bulk in variously sized silos (Fig. 13.1), flat storages, or in wooden, cardboard or plastic bins in warehouses. Dried fruits are most commonly bulk-stored in wooden bins, although plastic bins are becoming more popular.

Fig. 13.1 Silos for walnut (top) and almond (bottom) (California, US).

Fig. 13.2   Prunes stored in wooden bins under an open-sided pole barn (California, US).

Dried fruit bins may be arranged in stacks within enclosed or open-sided warehouse structures (Fig. 13.2). Processors often store large volumes of raisins in outdoor stacks of bins. These are covered with a double layer of plastic-coated, tar-laminated paper, held in place by vertical wooden slats nailed to a wooden framework built over the stack.

Although refrigerated storage is recommended to maintain product quality for tree nuts, conventional processors do not often use refrigeration for bulk storage of raw product. Pecan (*Carya illinoinensis*) processors are the most likely to use cold storage of bulk product. Large walnut processors often have some refrigerated warehouse storage available, and may have silos fitted with chilled aeration systems. Organic processors of both tree nuts and dried fruits are more likely than conventional processors to use refrigerated storage to prevent infestation by storage insects.

Objectives of storage

Generally, harvested product is processed as quickly as possible to a state in which it may be stored for several months or even years with minimal quality decline. Processing usually involves a certain amount of dehydration, either mechanical or sun-drying, and, in the case of tree nuts, removal of the hull. Such raw product may then be stored in bulk until final processing and eventual marketing. Usually, any required finished processing (sorting, cleaning, washing, shelling) is done shortly before delivery to customers, but individual processors may handle and store a variety of different product lines, from bulk raw product to finished consumer packs, and storage regimes may vary depending upon the product

type. Processors may market directly to consumers, retail outlets, food manufacturers or to exporters. In order to minimise price fluctuations, marketing orders may authorise volume regulations, keeping some product in reserve, or diverting to other markets. However, such restrictions are often unnecessary.

Dried fruit and nut processors must comply with food safety standards enforced by the US Food and Drug Administration. In addition, voluntary quality grades are issued by the US Agricultural Marketing Service under the authority of the Agricultural Marketing Act of 1946, which provides for the development of official US grades to distinguish between produce of different quality. These grades are available for use by producers, suppliers, buyers and consumers, and are designed to facilitate orderly marketing by providing a convenient basis for buying and selling, and for determining loan values. In most cases, marketing orders use these grades for their own quality control programmes. In-house laboratories, governmental agencies or independent inspection services such as the Dried Fruit Association of California may perform quality and grade inspections.

Grade determination is highly specific for the product type and is usually based on size, colour, texture and level of defects such as mould, insect infestation, foreign material, dirt, injury or surface imperfections. The end-use of products is generally determined by its grade. Product with cosmetic imperfections may go for chopping, blanching or the production of pastes, butters or oils. Smaller sized product may be more suitable for confectionery uses. Ultimately, customer requirements drive the industry, and processors are highly sensitive to complaints or returns by consumers.

Major sources of quality decline

Much of the dried fruit and nut crop is marketed as whole fruit or nutmeats; consequently any factor that affects the appearance or taste of the product will reduce quality. Quality issues at the time of commodity intake depend largely on the specific commodity, and are strongly affected by cultural, harvesting and processing methods. Pre-harvest conditions such as weather, irrigation levels, insect infestation and disease may affect overall quality by influencing size, cosmetic damage, physiological defects or mould growth. Shell coloration, particularly in walnuts and pistachios, is also an important factor. Tree nuts and sun-dried fruits such as raisins or figs typically have one or more pre-harvest insect pests that feed directly on the product and are capable of causing considerable damage and quality loss (Simmons & Nelson 1975). Although many of these may be present at the time of harvest and are often brought into storage, they generally do not reproduce under storage conditions. However, because they may continue to feed and cause additional damage, and often present phytosanitary problems for exporters, they are considered post-harvest pests. Feeding damage by these insects in tree nuts may also provide entry to aflatoxin-producing moulds (*Aspergillus* spp) (Vol. 1 p. 128, Wareing 2002).

Quality decline in storage is manifest in several ways. Tree nuts, particularly those with high oil contents such as walnuts and pecans, may become rancid and develop off-flavours. Aflatoxin contamination due to mould growth in stored tree nuts is a concern. Stored dried fruits are susceptible to sugaring – the appearance of noticeable external or internal sugar crystals – which may affect the texture and taste of the product. Once in storage dried fruits and tree nuts may be attacked by rodents and are susceptible to infestation by a number of common stored product moths and beetles (Table 13.3), the most serious being the Indian meal moth, *Plodia interpunctella* (Fig. 13.3) (Simmons & Nelson 1975). While it is occasionally found in the field, *P. interpunctella* is typical of food processing and storage facilities. Larvae feed on a wide variety of products, including all dried fruits and nuts, although populations do not develop as rapidly on dried fruits as they do on nuts (Johnson *et al.* 1995). Indian meal moths overwinter as diapausing fifth instar larvae, pupating and emerging as adults in late April or early May. There are several generations each year, and adult activity is often high during autumn harvest and processing operations. Storage facilities are infested when adults deposit eggs on or near the product. Newly hatched larvae are unable to penetrate packag-

Fig. 13.3 Dried apricot infested by the Indian meal moth, *Plodia interpunctella*. Adult moths and larvae can be seen.

ing material, but are able to find and enter through tiny cracks or holes. Adult activity ceases in late autumn as larvae enter diapause. Diapausing larvae readily survive the relatively mild California winters, even in unheated warehouses or outdoor storage.

Commodity and pest management regimes

Pre-harvest factors

Product quality at the time of commodity intake is highly dependent on both orchard management practices and processing methods. Strict control of insect pests that feed directly on the product not only reduce damage levels but may also reduce associated aflatoxin development. Pre-harvest pest management practices are specific to particular commodities, and may in-

Table 13.3 Post-harvest insect pests of California dried fruits and nuts.

Common name	Scientific name	Pest status
	Lepidoptera	
Codling moth	*Cydia pomonella*	Quarantine pest in walnuts, does not reproduce in storage
Navel orange worm	*Amyelois transitella*	Field pest in tree nuts and figs, may cause phytosanitary problems in exported product, does not normally reproduce in storage
Raisin moth	*Ephestia figulilella*	Field pest in dried fruits, does not normally reproduce in storage
Indian meal moth	*Plodia interpunctella*	Most serious storage pest, common on all dried fruits and nuts
	Coleoptera	
Dried fruit beetles	*Carpophilus* spp	Field pests in dried fruits, does not normally reproduce in storage
Saw-toothed grain beetle	*Oryzaephilus surinamensis*	Common in stored dried fruits and nuts
Red flour beetle	*Tribolium castaneum*	Common in stored dried fruits and nuts

clude pest monitoring, timed pesticide applications, biological control and orchard sanitation. Recently, an increased awareness of food safety issues has resulted in the adoption of practices throughout the production and marketing chain that minimise contamination by potentially dangerous bacteria such as *Salmonella*. The best example of this is the Food Quality and Safety Program of the Almond Board of California, which includes Good Agricultural Practices within its Hazard Analysis and Critical Control Point (HACCP) programme (Vol. 1 p. 502, Nicolaides 2002). The recommended practices seek to reduce potential contamination sources by using composted manure, vertebrate pest control, worker sanitation, cleaning of harvesting equipment and storage bins, careful attention to keeping irrigation water free from contamination and keeping harvested almonds as dry as possible (Almond Board of California 2001).

Dehydration methods

The type of dehydration used is largely dependent on the commodity involved, and strongly affects the degree of insect contamination at commodity intake. High-temperature mechanical dehydration methods are sufficient to kill all or nearly all field pests. Low-temperature mechanical dehydration and sun drying are not completely lethal to insects, and relatively high numbers of insects may subsequently attack the raw product. Initial dehydration may be done on the farm, at remote dehydrators or drying yards, or at the processing plant. Processors that take in product from numerous dehydrators may have less control of pest management practices, and may risk receiving infested material.

Pistachios must be rapidly hulled to avoid unsightly stains on the shell that can reduce marketability. Dehydration reduces the moisture content from about 40% to about 5%. The most common dehydrators in use are large rotary drum dryers. These use air temperatures of 66–82°C, and typically take about 8–10 h to dry a batch of nuts to 5% moisture content (Crane & Maranto 1988). A newer, more energy-efficient two-stage process uses modified grain dryers with air temperatures of 70–90°C to dry the nuts to 9–12% moisture content, then completes the drying process with forced warm air while the nuts are in storage silos (Vol. 1 p. 157, Boxall 2002). The air temperatures and drying times used in both methods are sufficient to kill all or nearly all insects that may be present in the nuts (Johnson *et al.* 1996).

Owing to the relatively high oil contents common to walnuts, pecans and hazelnuts, lower dryer air tempera-tures are used to prevent the development of rancidity. Drying that is too rapid may also cause splitting and cracking of shells. Forced-air dryers operating at air temperatures of 37–43°C can reduce harvested nuts to safe storage moisture contents in less than 24 h (Thompson *et al.* 1998). Almonds are normally allowed to sun-dry naturally in the orchard, usually taking about one week to reduce hull moisture to levels suitable for hulling (15%). Wet weather may prompt almond handlers to use mechanical dryers before hulling and subsequent processing (Thompson *et al.* 1996).

Prunes were originally sun-dried, a process that took about 10–14 days to complete. At present, all California prunes are dried in heated, forced-air dehydrators, using air temperatures of about 75°C and drying times of 16–20 h (Thompson 1981). As with pistachios, these temperatures are sufficient to kill any insects that may be present. After the fruit leaves the dehydrators at about 18–22% moisture content, it is placed in storage bins and held for a conditioning period of several weeks to allow overdried fruit to gain moisture from underdried fruit, resulting in moisture contents of 16–18%.

Most raisins produced in California are sun-dried and are known as natural raisins. Grape bunches are cut by hand and laid on paper trays placed between the vine rows (Fig. 13.4) (Christensen & Peacock 2000a). Complete drying takes about three weeks, during which time the raisins are vulnerable to attack by many different insect species (Lindegren *et al.* 1992). Raisins are normally dried in September, when occasional early rains may prove disastrous. A few producers, particularly organic growers, dehydrate their raisins mechanically, avoiding problems with insects and rain, but at consid-

Fig. 13.4 Natural raisins being sun-dried on paper trays (California, US).

erable expense (Thompson 2000). Other producers are turning to the dried-on-the-vine (DOV) method, which allows for mechanical harvesting of the raisins, so reducing labour costs (Christensen 2000; Christensen & Peacock 2000b). Since DOV raisins take longer to dry than conventional sun-dried raisins, insect infestation may be more prevalent.

Dried figs are allowed to ripen and partially dry on the tree. They finish drying on the orchard floor or on trays in drying yards, or they may be dried mechanically (Obenauf *et al.* 1978). Dates are placed on wooden or wire trays and dried to desired moisture contents with warm forced air.

Storage methods

Storage methods are highly specific to the commodity and processor. While all dried fruits and nuts may benefit from refrigerated storage, cold storage of raw, bulk product is often too costly. However, because most dried fruits and nuts are harvested in the autumn, processors take advantage of low ambient winter temperatures in outdoor storages or unheated warehouses during the first few months of storage. Raw, dried fruits are typically stored at moisture contents below those that allow microbial growth, thus avoiding the need for refrigeration. Unwashed raisins, still with capstems, may be stored efficiently for months in paper-covered yard stacks before undergoing final processing. Processors are more likely to move dried fruits to cold storage after final processing or packaging, particularly if processing increases the moisture content of the product. Tree nuts are often refrigerated at some time during storage, primarily to slow the development of rancidity, but also to reduce insect infestations. Care must be taken with the refrigeration systems used for nut storage, as ammonia leaks can cause serious damage to the product. Consequently, ammonia-cooled refrigeration systems are not recommended for nut storage. Generally, products in long-term storage should be kept at temperatures below 7°C, with 0–2°C being optimal. Optimal relative humidity levels in storage are 55–70% for tree nuts and 55–75% for dried fruits.

Disinfestation of raw product

Post-harvest pest management of dried fruits and nuts begins with the control of field pests in the raw product. Although most field pests do not reproduce well enough under storage conditions to become established

in stores, they may continue to cause problems if no control measures are employed. Harvest activities in California almonds disturb navel orange worm larvae, causing them to move to previously undamaged nuts to feed. Raisin moth larvae move about as they feed, and each may damage 10–20 raisins in the course of their development. Walnut processors must subject product destined for certain countries to a rigorous quarantine treatment to control codling moth. Processors also have a limited amount of time in which to disinfest, inspect and certify as insect-free in-shell walnuts destined for Europe at Christmas time, a market that is vital to the survival of the walnut industry.

The treatments used to control field pests in raw product depend upon the type of commodity. Products that are dried with high temperatures, such as prunes, pistachios, and some raisins, often have no immediate need for treatment. For sun-dried products, or those that use low-temperature dehydration, other treatments are needed. Typically, some type of fumigant is applied, usually phosphine or methyl bromide although, in accordance with the Montreal Protocol, the use of methyl bromide in developed countries will nearly be eliminated by January 2005 (Vol. 1 p. 328, Taylor 2002). In some cases the grower may fumigate on-farm. For almonds, early fumigation of product on the farm reduces damage by navel orange worm and other insects. Some growers may store product in the hope of selling at a better price, and may need to fumigate to maintain quality. Growers normally fumigate stacks of bins under plastic cover, and tarpaulin-covered almond piles can be fumigated efficiently with phosphine.

Many processors fumigate at least once to disinfest incoming product. Raisin processors accomplish this most often by fumigating their yard stacks, covered with laminated paper, with phosphine. To meet the demands of the European Christmas market, walnut processors must process and fumigate the early season in-shell product rapidly. The fumigant of choice for this application is methyl bromide, which allows complete treatment of walnuts within 24 h. Large processors may use methyl bromide under vacuum to further decrease treatment time. For walnuts exported to countries with quarantine restrictions against codling moth, a more stringent methyl bromide vacuum fumigation is used. Currently, while quarantine treatments for codling moth are considered exempt from the ongoing phase-out of methyl bromide, disinfestation treatments for the European Christmas market are not. There is also concern that the phase-out will continue to increase the price and decrease the supply of methyl bromide, making the fumigant unavailable even for exempted uses. Consequently, walnut processors are actively searching for a substitute treatment.

Processors of organic dried fruits and nuts must use non-chemical methods to disinfest raw product. Most organic processors use below or near-freezing temperatures to ensure that product is insect-free, although processors that use high temperature dehydration for prunes, raisins or pistachios may rely on lethal temperatures to provide clean product.

Pest management practices in storage

The diversity within the dried fruit and nut industry results in the use of numerous pest management strategies. As in other food processing facilities, the most successful strategies begin with careful attention to plant design and sanitation, and may also use cold storage to reduce insect population growth. However, most conventional (non-organic) processors will at some point use fumigation to ensure clean product.

Raisin processors usually keep product in paper-covered yard stacks until just before final processing. The paper laminate that covers the yard stacks helps prevent insect infestation, but stacks are fumigated with phosphine on a regular basis, beginning just after receipt of the product and usually every 90 days thereafter. Should the paper covering be disturbed between the scheduled treatments, stacks are recovered and fumigated again. Processors may delay scheduled treatments during the winter months, when insect activity is low. To prevent any contamination of the processing lines, most processors will fumigate raisins before they enter the processing area. Methyl bromide is preferred for these final fumigations because treatment times for this compound are relatively short. However, the restrictions on methyl bromide have prompted some processors to switch to phosphine for all fumigations.

Prunes are dehydrated at high temperatures and since *P. interpunctella* population growth is slow on dried fruits, processors have relatively few pest problems as storage begins. Most processors fumigate the entire facility, including bulk product in storage, once all product has been received, usually in late November. Some processors will then fumigate small lots as needed, and will often fumigate product as it leaves the facility.

Nut processors typically fumigate storage silos once they have been filled. In California, this occurs in the autumn, shortly after harvest, when *P. interpunctella*

adult activity may be very noticeable. Such treatments may be to kill field pests or developing *P. interpunctella* populations. Pistachio processors often fumigate storage silos shortly after harvest, even though dehydration kills all or most field pests. Such treatments serve to reduce overwintering *P. interpunctella* populations, thereby minimising adult emergence in the spring. Although many nut processors use cold storage to maintain product quality and increase shelf life, silos are rarely refrigerated. Some large walnut facilities use chilling silos, but such storage is costly and seldom employed. Cold storage facilities, when available, may be limited and used only for finished product. Generally, processors attempt to market product as quickly as possible to reduce the number of fumigations, but may schedule a final pre-shipment treatment to insure infestation-free outgoing product.

In addition to fumigants, other pesticides may be used to manage pest populations. Residual insecticides or desiccants applied as crack and crevice treatments to areas surrounding storage structures help to reduce infestations. Aerosol fogs may be used as space treatments to kill flying insects (Vol. 1 p. 251, Birkinshaw 2002). Some prune processors apply such fogs during the spring and summer when adult moth activity becomes noticeable.

Organic dried fruit and nut processors are limited to non-chemical means of post-harvest insect control. This is largely accomplished through temperature treatments. Product is frozen if it must be disinfested quickly, and refrigerated to prevent reinfestation. Another option for non-chemical disinfestation is modified atmospheres, but this is rarely used.

Monitoring and quality control

Management programmes for stored dried fruits and nuts rarely rely on monitoring of pest populations to make treatment decisions. Monitoring for insects with pheromone traps is recommended but few processors use them. This is largely due to the difficulty in interpreting trap catch and the lack of clear action thresholds. At present, pheromone traps are best used to identify the presence of insect populations and to locate problem areas. Some prune processors use *P. interpunctella* pheromone traps to identify moth activity and schedule aerosol fogging treatments.

The quality of harvested product is maintained throughout the processing and marketing chain by direct inspections of product. A number of parameters are considered in these inspections, including insect infestation. To encourage growers to manage post-harvest insects efficiently, marketing orders and individual processors may use price incentives or penalties, based on the results of product inspections. Kawaga (2000) outlined the inspection process for raisins. Processors may also sample product at certain points during processing and storage, and may make management decisions based on the results.

Economics of operation of the system

Due to the size and diversity of the dried fruit and nut industry within the United States, and recent volatility of energy and fumigant prices, it is difficult to obtain accurate estimates for the cost of storage or pest management programmes. Many of the processing and storage facilities used in the industry have been in existence for decades, so most expenditure is on maintenance, administration and energy. Because extensive storage facilities are often too expensive for small producers, most are provided by large commercial processors or co-operatives. Individual growers make use of temporary yard storage when needed, or, in the case of some nut crops and organic growers, may contract commercial cold storage facilities for long-term storage.

Fumigation accounts for the bulk of the cost for pest management in stored dried fruits and nuts. Recent estimates give fumigation costs for dried fruits and nuts as $3.60–4.55 t^{-1} for methyl bromide, based on a price of $0.60 kg^{-1} for the fumigant. The estimated cost for phosphine fumigation is $5.15–6.90 t^{-1} based on a price of $0.86 per 100 pellets. It can be seen that labour and other operating expenses are responsible for most of the costs of fumigation. Most of the treatments proposed as alternatives for methyl bromide are either more expensive or have longer treatment times. For some alternatives, processors will need to increase storage space to accommodate product waiting for treatment, and risk missing time-sensitive markets such as Europe at Christmas. Thus, the phase-out of methyl bromide is expected to increase processing and storage costs for products dependent on this fumigant for insect disinfestation.

The market conditions for dried fruits and nuts have a strong influence on the length of time that they are stored. Processors draw down walnut and fig stores rapidly because large portions of these products are destined for Christmas markets shortly after harvest. The demand for other products is more constant throughout the year. With all dried fruit and nut products, oversupply may oc-

casionally cause a delay in marketing. When this occurs, products may be diverted to other markets, such as oil or butter production, distilleries, or animal feed, or stored until the market improves. This may result in added storage costs due to additional fumigation treatments or cold storage. Storage costs for organic produce are generally higher than for the conventionally processed crop, as it must be frozen and later refrigerated.

Future developments

Successful marketing of dried fruits and nuts requires high-quality product free of insects or other contaminants. Management programmes for dried fruits and nuts have relied on remedial treatments with the fumigants methyl bromide or phosphine to ensure uninfested product. With the nearly complete phase-out of methyl bromide in 2005, phosphine can be adopted as a suitable substitute in many dried fruit and nut applications but it is relatively slow and occasionally causes quality problems. In addition, regulatory agencies have considered additional restrictions on phosphine use, which would increase the cost of this fumigant. Restrictions on these fumigants present a serious challenge to the industry. For this reason, the industry strongly supports research into alternative treatments, both chemical and non-chemical, for the control of post-harvest pests infesting dried fruits and nuts.

Of particular interest are alternative fumigants, which may be used in much the same way as methyl bromide or phosphine. Of the new fumigants being considered, sulphuryl fluoride shows the most promise (Zettler & Gill 1999), and the California dried fruit and nut industry has been assisting in the research needed to register this product. Other fumigants under consideration are carbonyl sulphide, methyl iodide, proponyl oxide and ozone. Also of interest are methods that reduce atmospheric emissions of methyl bromide, which may allow its continued use.

Although conventional processors already take advantage of high temperature dehydration to disinfest incoming product and cold storage to prevent insect infestation, temperature treatments are rarely used solely for pest control. In an attempt to reduce the need to fumigate, some processors are considering adding or enlarging their cold storage facilities, and are looking at methods to manipulate storage temperatures as efficiently as possible. There is also interest in rapid high-temperature treatments using radio frequency or microwave energy (Wang *et al.* 2002).

Research has shown that controlled atmospheres (low oxygen or high carbon dioxide) may be used to disinfest or protect product from insect infestation (Soderstrom & Brandl 1984), but the process requires long treatment times and is usually more costly than fumigation. With the development of inexpensive portable storage units made of flexible polyvinyl chloride, it may be possible to reduce the cost of controlled atmosphere treatments. Flexible storage units may also be used to apply vacuum treatments to bulk product, thereby shortening the treatment times for controlled atmospheres.

Biological control agents may also be incorporated into pest management programmes for dried fruits and nuts. A granulosis virus applied as a dust or aqueous spray has been demonstrated to be effective in protecting product from *P. interpunctella* infestation for months at a time. Pest population levels in stored dried fruits and nuts are often reduced by the activity of a number of naturally occurring parasites and predators. At present, US regulations do not allow the intentional release of beneficial insects into bulk dried fruits and nuts, but an exemption, similar to that given for stored grains, may eventually be granted.

There has also been success in combining two or more non-chemical treatment methods. Relatively short-term very low oxygen (0.4% O_2) disinfestation treatments were combined with long-term protective treatments such as sustained low oxygen (5.0% O_2), cold storage or *P. interpunctella* granulosis virus. These methods have been successfully demonstrated for walnuts, almonds and raisins (Johnson *et al.* 1998, 2002).

References

Almond Board of California (2001) *Good Agricultural Practices*. Almond Board of California, Modesto CA, USA.

Birkinshaw, L.A. (2002) Insect control. In: P. Golob, G. Farrell & J.E. Orchard (eds) *Crop Post-Harvest: Science and Technology. Volume 1 Principles and Practice.* Blackwell Publishing, Oxford, UK.

Boxall, R.A. (2002) Storage losses. In: P. Golob, G. Farrell & J.E. Orchard (eds) *Crop Post-Harvest: Science and Technology. Volume 1 Principles and Practice.* Blackwell Publishing, Oxford, UK.

Christensen, L.P. (2000) Current developments in harvest mechanization and DOV. In: L.P. Christensen (ed.) *Raisin Production Manual.* University of California, Agriculture and Natural Resources, Oakland CA, USA.

Christensen, L.P. & Peacock, W.L. (2000a) Harvesting and handling. In: L.P. Christensen (ed.) *Raisin Production Manual.* University of California, Agriculture and Natural Resources, Oakland CA, USA.

Christensen, L.P. & Peacock, W.L. (2000b) The raisin drying process. In: L.P. Christensen (ed.) *Raisin Production Manual.* University of California, Agriculture and Natural Resources, Oakland CA, USA.

Crane, J.C. & Maranto, J. (1988) *Pistachio Production.* University of California, Agriculture and Natural Resources, Oakland CA, USA.

Johnson, J.A., Wofford, P.L. & Gill, R.F. (1995) Developmental thresholds and degree-day accumulations of Indianmeal moth (Lepidoptera: Pyralidae) on dried fruits and nuts. *Journal of Economic Entomology*, **88**, 734–742.

Johnson, J.A., Gill, R.F., Valero, K.A. & May, S.A. (1996) Survival of navel orangeworm (Lepidoptera: Pyralidae) during pistachio processing. *Journal of Economic Entomology*, **89**, 197–203.

Johnson, J.A., Vail, P.V., Soderstrom, E.L., *et al.* (1998) Integration of nonchemical, postharvest treatments for control of navel orangeworm (Lepidoptera: Pyralidae) and Indianmeal moth (Lepidoptera: Pyralidae) in walnuts. *Journal of Economic Entomology*, **91**, 1437–1444.

Johnson, J.A., Vail, P.V., Brandl, D.G., Tebbets, J.S. & Valero, K.A. (2002) Integration of nonchemical, postharvest treatments for control of postharvest pyralid moths (Lepidoptera: Pyralidae) in almonds and raisins. *Journal of Economic Entomology*, **95**, 190–199.

Kawaga, Y. (2000) Quality standards and inspection. In: L.P. Christensen (ed.) *Raisin Production Manual.* University of California, Agriculture and Natural Resources, Oakland CA, USA.

Lindegren, J.A., Curtis, C.E. & Johnson, J.A. (1992) Stored raisin products pests. In: D.L. Flaherty (ed.) *Grape Pest Management.* University of California, Agriculture and Natural Resources, Oakland CA, USA.

Nicolaides, L. (2002) Food safety and HACCP. In: P. Golob, G. Farrell & J.E. Orchard (eds) *Crop Post-Harvest: Science and Technology. Volume 1 Principles and Practice.* Blackwell Publishing, Oxford, UK.

Obenauf, G., Gerdts, M., Leavitt, G. & Crane, J. (1978) *Commercial Dried Fig Production.* University of California, Agriculture and Natural Resources, Oakland CA, USA.

Simmons, P. & Nelson, H.D. (1975) *Insects on Dried Fruit.* USDA Agricultural Handbook **464**. United States Department of Agriculture, Washington DC, USA.

Soderstrom, E.L. & Brandl, D.G. (1984) Low-oxygen atmosphere for postharvest insect control in bulk-stored raisins. *Journal of Economic Entomology*, **77**, 440–445.

Taylor, R.W.D. (2002) Fumigation. In: P. Golob, G. Farrell & J.E. Orchard (eds) *Crop Post-Harvest: Science and Technology. Volume 1 Principles and Practice.* Blackwell Publishing, Oxford, UK.

Thompson, J.F. (1981) Methods of prune dehydration. In: D.E. Ramos (ed.) *Prune Orchard Management.* University of California, Agriculture and Natural Resources, Oakland CA, USA.

Thompson, J.F. (2000) Tunnel dehydration. In: L.P. Christensen (ed.) *Raisin Production Manual.* University of California, Agriculture and Natural Resources, Oakland CA, USA.

Thompson, J.F., Rumsey, T.R. & Connell, J.H. (1996) Drying, hulling and shelling. In: W.C. Micke (ed.) *Almond Production Manual.* University of California, Agriculture and

Natural Resources, Oakland CA, USA.

Thompson, J.F., Rumsey, T.R. & Grant J.A. (1998) Dehydration. In: D.E. Ramos (ed.) *Walnut Production Manual*. University of California, Agriculture and Natural Resources, Oakland CA, USA.

USDA (2002) Statistics of fruits, tree nuts, and horticultural specialties. *Agricultural Statistics*, V1–64 United States Department of Agriculture, Washington DC, USA.

Wang, S., Tang, J., Johnson, J.A., *et al.* (2002) Process protocols based on radio frequency energy to control field and storage pests in in-shell walnuts. *Postharvest Biology and Technology*, **26**, 265–273.

Wareing, P.W. (2002) Pest of durable crops – moulds. In: P. Golob, G. Farrell & J.E. Orchard (eds) *Crop Post-Harvest: Science and Technology. Volume 1 Principles and Practice.* Blackwell Publishing, Oxford, UK.

Zettler, J.L. & Gill, R.F. (1999) Sulfuryl fluoride: a disinfestation treatment for walnuts and almonds. *Annual International Research Conference on Methyl Bromide Alternatives and Emissions Reductions*, San Diego CA, **108**, 1–3.

Chapter 14
Cured Fish

A. Guèye-NDiaye and P. Golob

Fish represents a relatively cheap source of human dietary protein, particularly for coastal and riverine populations. Of the total world catch about 8% is cured to improve its keeping qualities. Curing greatly extends the shelf life of fish and fish products, and allows communities remote from the sea and major rivers to have access to good-quality fish (Plate 14). Almost two-thirds of the world production of cured fish is produced in the developing world, and of that 12–14% is produced in Africa (Table 14.1).

Landed fish is either sold immediately, iced or refrigerated, or cured. Cured or dried fish, the focus of this chapter, can be regarded as having a moisture content below 50%. Fish that has been cured in the tropics by artisanal processors is considered here; fish cured by industrial process is generally subject to stringent quality control and is expected to be of good quality, unless the processing plant malfunctions.

Fish in the tropics are cured within a day or two of being caught. They may be laid out in the sun each day (Fig. 14.1) until sufficiently dry to package and store or, if natural fuel is available, may be dried by smoke curing in small kilns or over racks. In hot dry climates, sun

Fig. 14.1 Fish drying in the sun on platforms (Senegal).

curing is common, whereas smoke curing is practised in places where the humidity is high or where periods of exposure to the sun are interrupted by frequent and prolonged periods of rain.

Immediately after landing, when fish start to lose moisture, they are susceptible to infestation by blowflies (Diptera), particularly of the families Calliphoridae and

Table 14.1 Fish production statistics, 1991–98.

	1992	1993	1994	1995	1996	1997	1998
Cured fish production ('000 tonnes)							
World	3 647	3 697	3 909	4 136	4 280	3 668	4 136
Developing countries	2 066	2 166	2 435	2 654	2 843	2 292	2 836
Africa	344	329	342	353	362	386	381
Total fish catch ('000 tonnes)							
World	2 066	2 166	2 435	2 654	2 843	2 292	2 836
Developing countries	29 437	32 541	35 667	40 241	43 040	47 119	48 561
Africa	2 940	2 877	2 713	2 952	2 946	3 054	3 247

Sarcophagidae. Eighteen species are known to cause post-harvest deterioration in the tropics and a comprehensive account of their biology and control is given by Johnson and Esser (2000). Losses during this initial drying phase, which may take several days, can be as high as 40% by weight; it is not uncommon for fish to be unfit for human consumption as a result of blowfly damage. However, once the fish falls below 50% m.c. it becomes increasingly resistant to blowfly damage but instead becomes prone to damage by beetles. It can remain under threat from beetles for the entire storage period, which may last for many months, even when the m.c. is below 10%. This chapter is therefore mainly concerned with the damage caused by beetles to fish once it is fully cured, beyond the time it is likely to be attacked by Diptera.

Beetle pests are generally regarded as causing significant losses, both in quantity and quality, to cured fish and fish products. These include *Dermestes* spp and *Necrobia rufipes* (Vol. 1 p. 106, Hodges 2002). Although the biology of the main insect species is well known, quantitative information on losses is sparse. The damage caused by beetles has been recorded from situations where the storage period of the dried product is relatively short, such as at fish landing sites while it awaits transportation to urban communities. Losses have not been assessed in fish stocks kept at schools, hospitals, prisons and other such institutions where storage periods may be as long as six months. The general nutritional requirements of the beetles are known but their effect on the nutritional composition of fish that has been attacked is not. In contrast, the effects of bacterial and fungal invasion on fish biochemistry are better documented.

Despite the paucity of valid loss data it is generally accepted that beetle infestation is of sufficient magnitude to require the application of loss reduction strategies. In some circumstances insecticides and fumigants are used; however, good storage practice is key to maintaining the quality of cured fish and depends on the same principles and practices that are necessary for quality maintenance of all food commodities. Hygiene is of prime importance: stores should be clean and cleaned regularly, the physical condition of the store must be good so that rain cannot leak in and rodents and other pests can be kept out, and the fish should be stored in such a way to allow easy regular inspection and the introduction of pest management practices if these be required. Very frequently these conditions are not met, dried fish is stored poorly (Fig. 14.2), frequently in jute or hessian sacks, and becomes heavily infested with beetles (Fig. 14.3). Where

Fig. 14.2 Dried fish bundles in poor storage conditions (Senegal).

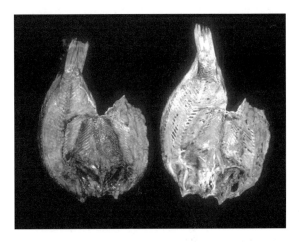

Fig. 14.3 Dried fish in good condition (left) and showing typical damage due to beetle infestation (right).

storage occurs in more humid conditions the fish may also become contaminated with mould and be unfit to consume; the high oil content of fish predisposes it to fungal damage.

This chapter first considers the problems of maintaining the quality of cured fish in Senegal and then reviews the chemical, physical and biological options available for preventing beetle infestation of cured fish in tropical countries.

Senegal

Fishing is an essential sector of the Senegalese economy and fish provide the main source of animal protein for the population (Anon. 2000a). Artisanal fishing accounts for an annual catch of about 400 000 t of mostly marine fish but also molluscs and crustaceans (Durand 1981; Anon. 1997). About 30–40% of this catch is cured by various traditional methods such as fermenting, braising, salting, smoking and drying. The three most popular fish specialities are *gejj* (Fig. 14.4), where the fish is fermented then split and dried, *tambajen* where fish is fermented then dried unsplit (Fig. 14.5) and *keccax* (Fig. 14.6)

Fig. 14.6　Braised dried sardines, *keccax* (Senegal).

where fresh *Sardinella* or *Ethmalosa* spp are braised entire and then dried after removing skin, head and gut (Guèye-NDiaye & Gningue 1995).

In 1999, national production was 34 785 t of cured product of which 2217 t of smoked fish and salted dried fish were exported for a commercial value estimated to be over FCFA5bn (Anon. 2000b). Fishing is a seasonal activity with peaks in catch and so the processed products are stored to provide a steady supply throughout the year, particularly during the rainy season, to the rural zones and for export markets in Africa and elsewhere.

Fish are generally prepared and preserved in conditions of poor hygiene, leading to attack by bacteria, mould and especially insects. They may suffer serious losses, according to the length of storage and the prevailing conditions (FAO 1981; Guèye-NDiaye 1991). The insects that do significant damage to dried fish are mostly beetles of the family Dermestidae including *Dermestes maculatus*, especially *D. frischii* and sometimes *D. ater.* Storage can be undertaken in the house for the benefit of a family, at production sites, or in large or small markets. Dried fish destined for both local and international markets are subjected to quality regulations and a certificate of origin and of hygiene is delivered following a satisfactory sanitary inspection. Despite these arrangements the quality of cured artisanal products always creates problems when they are exported to more developed countries where standards are very strict.

Fig. 14.4　Fermented split dried fish, *gejj* (Senegal).

Fig. 14.5　Fermented whole dried fish, *tambajen* (Senegal).

Historical perspectives

Artisanal preparation of cured products was originally

encouraged so that the wives of fishermen could barter with farmers for cereal and vegetables. As fish catches have risen due to the modernisation of fishing practice so production of artisanal products has increased (Kébé 1994). An exodus of people from rural areas due to drought has gradually increased the manpower available to work in this sector, which has become commercially more significant, reaching from the coastal zones into the interior of the country. More recently exports have increased to places where there are large Asian and African communities, e.g. Europe and the USA.

During and after World War II, there was a demand from France for products from its colonies, which led to the development of fish processing plants, especially for salted, dried fish. These were preferred in export markets, over products prepared using more traditional methods, due to better quality control and pest management by fumigation (Blanc 1955; Mallamaire 1955, 1957). Subsequently, fish processing was forbidden except at the authorised landing sites. By 1957 quality regulations had been established to regulate produce for export while produce to be consumed locally was not the subject of regulation until 1969 and even then these regulations only defined chemical and non-microbiological standards (Seydi 1991).

Since 1970 female processors have been organised by the authorities into local groups linked to development partners. Improvements have been made to processing sites by the installation of modern drying racks, the construction of storage sheds on hardstanding and access to micro-credit. Despite these improvements, production remains at the level of the individual. The processors do not store for long, on average 2 months, thus avoiding the losses and deterioration that would reduce their profit margin.

In 1994, the devaluation of the currency of francophone West Africa, the Franc CFA, made the purchase of fresh fish, especially of the better types, very expensive for local people. This increased the demand for dried fish, but at the same time the supply in primary product was reduced since many fishermen preferred to sell their catch for export to the European Union as fresh or frozen fish. Consequently there is a continuing need to reduce losses during storage and to increase the value of products to improve the livelihood of the producers. To this end projects, financed by UNIDO, the Japanese Government and the European Union, are developing landing stages with infrastructure to facilitate healthier handling of fresh fish and better pre-treatment of fish intended for processing.

Physical facilities

The means of storage and equipment used vary according to the type of product, the quantities and the destination – for sale or for family consumption. The cooked products such as *keccax* and smoked fish are generally stored in woven baskets prepared with twigs or leaves of *Borassus flabellifer* (Palmae), while fermented and dried products, *gejj* and *tambajen*, are generally packed in bundles, or in jute or polypropylene bags.

Family storage

This generally involves small quantities and much care is taken with preparation and conservation. Fish are stored in baskets or fired pots. Some examples are as follows:

- At the fishermen's neighbourhood of Saint Louis, during the peak of the fishing season, fresh fish are hung and cleaned, heavily salted and then dried. The product is consumed during the lean season after desalinating with several washes of fresh water. The strong salt content prevents the proliferation of insects.
- On the Petite Côte in south Dakar, *keccax* are braised and dried, then filleted to remove the vertebral column and preserved as mildly salted strips. This type of conservation offers fewer crevices in which dermestid beetles can hide and breed.
- In the Zone of the Bandial in Casamance fish are hung above the hearth where the heat and the smoke drives away insects.

Commercial storage

The processing sites are at the water's edge and storage is usually in the open air, beneath or above the drying racks (Fig. 14.7). The piles of product are covered with canvas sheets to protect them against moisture. If need be the product may be redried from time to time. After purchase fish are packed at the buyer's expense in baskets, or in bags of jute or polypropylene (Fig. 14.8). This storage is only short-term.

Storage sheds are of modest size and do not have shelves or racks. Construction is difficult by the water's edge although these stores have a paved floor and their roof is generally made from asbestos cement. The fish products are just piled up or packed in bags or baskets (Fig. 14.9). Any infestation that has occurred in one of these batches spreads quickly to the others and by being at the water's edge the products may be dampened by sea spray.

Fig. 14.7 Storage of braised dried fish above a drying rack (Senegal).

Fig. 14.8 Cured fish packed in bags of jute or polypropylene (Senegal).

Fig 14.9 Cured fish in bags or baskets in a storage shed (Senegal).

Objectives of storage

The families of fishermen store only small quantities of cured products. However, much larger quantities are involved in commerce and there may be several middlemen between producers and consumers (NDiaye 1997). The producers and tradesmen often check the product they wish to sell for infestation, rehydration or crumbling, and take what corrective measures may be needed to avoid any reduction of their profit. Fermented dried fish is sold approximately by batch according to size, or per unit of volume, such as the bag or the basket, or more rarely per unit of weight. *Keccax* is sold by weight or, when it is crumbly, in small heaps.

It is important to keep the products attractive for the consumer who is sensitive to colour, texture and flexibility of the flesh, odour and taste. In all cases a heavy infestation by insects, especially dermestid beetles, is a particular worry as this leads to a substantial price reduction.

Legally enforced quality standards exist for fresh fishery products and salted dried fish (Table 14.2), while the other fish specialities, fermented, smoked or braised have chemical and moisture content limits only (Table 14.3). However, cured fish, like other fishery products, is subject to sanitary inspection and certified by the issue of a CCOS (Certificate of Origin and Hygiene) before removal from the processing site. This inspection is visual and samples will only be taken for closer examination in the

Large or medium-sized traders and market middlemen from urban centres store temporarily in baskets or bags or keep the product in bundles under their stalls or in small, poorly ventilated stores rented to several traders. The packages are piled up on one another on the floor. They are generally transported by road but sometimes by boat or train.

Certain large-scale operators, authorised to export to Europe, store on clean racks placed in enclosures surrounded by a wire fence while waiting for sufficient stock to accumulate for a batch to be transported, generally by air. Their products are packed in cartons of double thickness kraft paper placed in canvas sacks. They have facilities for fumigation, unlike the small producers who have to rely on contact insecticides.

event that a problem is suspected. The presence of pests is not tolerated, consequently there is no specific limit for insects or their damage. For export, in particular to the European Union, quality control is stricter although this only applies to salted dried or smoked fish as fermented products are not accepted by the EU. The exporter must obtain the approval of the BCPH (Bureau for the Control of Cured Products) by providing the CCOS as well as the results of chemical and microbiological analyses in conformity with standards (Tables 14.2 and 14.3). Two approved laboratories can undertake these analyses. The BCPH also has responsibility for technical inspection of the production facilities.

Control is very strict for exports to the EU because of the risks of rejection or withdrawal of approval. However, controls are more flexible for national and African markets which, apart from sensory aspects, respond to the state of infestation by *Dermestes*, the principal factor in price-cutting and loss.

Major sources of quality decline

Open-air drying and storage on racks exposes dried fish to losses from birds and other animals such as dogs, cats or rodents, but more especially from micro-organisms and insects. In the period of peak production drying racks may be in short supply and the time available for

drying shortened; the result is that finished products are less well dried and may become infested by insects originating in waste and stocks present at the processing sites. During the rainy season ambient humidity is very high (90%) and impedes thorough drying and so in prolonged storage these products become rotten or mouldy; producers avoid storing them. The average length of storage by producers is 2–3 months.

When products are salted the result is not a homogeneous treatment (Guèye-NDiaye & Gningue 1995) and the sea salt used is raw salt, containing many impurities and halophilic micro-organisms. Under these conditions, if measures are not taken, insects will multiply and cause much damage due to the holes made by larvae, pupal cases and the insect frass that soils the product. In the same way, high moisture content in the presence of halophilic microbes causes a significant development of moulds, leading to changes of colour, odour and taste that can in the long term make fish unsuitable for consumption. In stocks of moist and poorly ventilated *keccax* the high temperatures and moistures are the origin of greenish coloured moulds that are accompanied by a strong odour, making the fish rancid and inedible. This problem occurs mostly in tradesmen's warehouses in the markets.

At production sites sea spray can rehydrate the products, resulting in a proliferation of moulds, particularly

Table 14.2 Microbiological standards (NS-03-016 of 1989) and heavy metal limits for salted dried fish in Senegal.

Factor	Maximum limit
Micro-organisms aerobic at 30°C	5×10^3 g^{-1}
Faecal coliform bacteria at 44°C	25 g^{-1}
Faecal *Staphylococcus* at 37°C	10×10^2 g^{-1}
Spores of sulphate-reducing anaerobes at 46°C	10×10^2 g^{-1}
Faecal *Streptoccocus* at 37°C	40 g^{-1}
Yeast and moulds at 25°C	50 g^{-1}
Cadmium (standard NS 03-045)	2 ppm
Mercury (Ministry of Fishing and Aquaculture 2001)	1 ppm

Table 14.3 Standards for cured fish products in Senegal (Decree 69-132 of 1969 and NS-03-016 of 1989).

	Salted dried fish	Smoked dried fish	Braised dried fish	Fermented dried fish
Moisture content*	≤ 35%	≤ 30%	≤ 28%	
ABVT†	≤ 350 mg/100 g	≤ 2%	≤ 2%	4–8%
Salt content*	10–20%	—	—	—
Histamine*	≤ 20 mg/100 g	—	—	—

*Moisture content, salt content and histamine are expressed relative to the weight of the crude product.
†Expressed ABVT related to the total nitrogen content.
Note: Pests, mould and pesticide residues must be absent.

during the rainy season. Heavy infestation by flies is only really a problem during the winter when the fish are slow to dry and infestation can start in the gills; mould damage is also most prevalent at this time due to the higher moisture levels. The operators combat moulds by cleaning the surface of fish, brushing them in water and then redrying. On the other hand, dermestids are ever present in fish residues, stored fish, the recesses of the traditional storage racks and the rough ground at unmade-up sites (Guèye-NDiaye & Gningue 1995). In a dry atmosphere long storage causes significant dehydration which, combined with the insect infestation, leads to crumbling of the product. This is severe in the places where fish are braised. The storage of catch residues to be used as manure or for poultry food constitutes a permanent source of reinfestation. The larvae and adults of the clerid beetle *Necrobia rufipes* are also present but are not important pests in Senegal, unlike in Nigeria (Osuji 1975a). Mites are present, particularly *Lardoglyphus konoi* on *keccax* and other types of cured fish and *Suidasia pontifica* on *keccax* only (Guèye-NDiaye & Fain 1987; Guèye-NDiaye & Marchand 1989), but fish operators ignore them; they consider beetles as the principal pests of cured fish.

Commodity and pest management regimes

The average length of storage with the producers is 2 months (Diouf 1987); they sell as quickly as possible to avoid losses due to mould and insects. The producers use a lot of salt, especially during hot and humid periods when drying is slow and so infestation by flies is a potential problem. Salting also limits infestation by dermestids when salt contents are at least 15% (Guèye-NDiaye 1991; Sembène 1994). This is in contrast to the conclusions of previous studies (Mushi & Chiang 1974; Osuji 1975b; Wood *et al.* 1987) where salt contents of 9–10% were insufficient to control *D. maculatus* and *D. frischii* in Senegal.

Pesticides may be secretly applied to the products by being sprinkled on with salt as the fish is dried when in store. Mbour in 1991 (unpublished data) discovered the use of chlorpyriphos-methyl, deltamethrin, fenitrothion, lindane, malathion, pirimiphos-methyl and propoxur. The formulations used are generally intended for public health or agriculture and their use on fish is officially prohibited by DOPM (Directorate of Oceanography and Marine Fish). Some packs of insecticide are unlabelled and it is possible they contain DDT although this has not been confirmed. Nevertheless, treatment with insecticide is permitted provided it is done according to recommended practice. The producers' federation in Ziguinchor (south Senegal) have received hands-on training in the correct use of pirimiphos-methyl (Actellic), following FAO guidelines for the protection of dried fish from insect infestation (FAO 1981; Guèye-NDiaye *et al.* 1996). This involved steeping fermented dried fish for 5 s in an aqueous emulsion of pirimiphos-methyl, before drying on racks. The treatment gave effective protection against insects and residues did not exceed 2.4 ppm. The method is now used, particularly during the rainy season, under the supervision of a Fishing Agent. Traditional methods of fighting infestation have largely fallen into disuse although processors may place jute bags containing neem or *Boscia senegalensis* (Capparidaceae) leaves beneath dried fish during storage (see below).

Dried fish has been fumigated with methyl bromide in Senegal (Mallamaire 1955, 1957). Fumigation plants were proposed for countries in the region but one at Dakar was very much under-utilised because the fish was stored for too short a time before export to justify disinfestation, and so the suggestion was not taken up.

The large-scale operators approved for export, in particular to the EU, are better able to manage the product and to fulfil the quality requirements of recipient countries. The raw material is refrigerated in cold rooms before treatment in hygienic enclosures that use clean running water to wash the fish, are cleaned and regularly disinfected, and where produce is smoked and dried in solar, gas or electric furnaces, or sun-dried under mosquito netting. The finished products are exported continuously, by air in some cases, so storage periods are short (2–5 weeks). Other exporters dispatch full containers that are fumigated with phosphine.

Economics of operation of the system

Expenditure during the first 2 months is used for the purchase of packing such as jute or polypropylene bags and baskets, for canvas sheets to provide protection from rain and dew, and for salt and sachets of insecticidal powder. This expenditure does not exceed 5% to 10% of the selling price. The store rooms, where they exist, are in general built by the authorities or development partners although maintenance is at the expense of the users. The tradesmen and urban wholesalers working in the markets rent space in a store while waiting to bring fish to market; they also keep part of their stock in packages under their stalls.

Exporters generally do not store dried fish: they often work to order and sometimes subcontract to small producers, although they may reprocess or recondition the products before dispatching them. Stock may be stored in a cold room or treated by fumigation before storage in containers. The storage of dried fish is actually inexpensive; storage is often unnecessary or limited to only very short periods as the small producers fear heavy losses from insect infestation and large-scale producers fear being unable to fulfil quality standards for export.

Future developments

With much greater export of fresh fish to Europe, and a local fall in purchasing power following the devaluation of the FCFA, local consumption of dried fish has increased. The intra-regional trade in dried fish is also developing through the market of Diaobé to eastern Senegal. A regional workshop held in Dakar in May 2001 listed the constraints to the development of the West African market for processed fish and recommended harmonisation of tariff barriers in the Community of West African States (CDEAO). Quality standards must be set up for the different types of cured seafood products and a scale of prices subsequently established.

The large African and Asian immigrant communities in developed countries provide a demand for cured products but market penetration has been weak due to frequent rejections on quality grounds. Of 99 companies approved in Senegal for the export of cured products in 2003, only 22 make processed products, the majority of these produce salt-dried products; seven have approval for export to Europe, the remainder only to Africa and other continents.

There is a need for better quality products for local and continental markets and to increase exports to more developed countries which have stronger purchasing power. There is a will on the part of the authorities and development partners to improve this branch of industry for food self-sufficiency and for an improvement in livelihoods, particularly of women. To this end, action is being taken to install better unloading and processing sites, construct drying platforms on hardstanding, install latrines and running water and provide access to microcredit. But there is still much to be done including the training of producers in good handling practice, preparation and storage of dried fish.

For better on-site control the processors should be gathered in covered, laid-out workshops arranged to avoid any cross-contamination by having separate processing surfaces, zones for drying and zones for handling the finished products. Organic waste and refuse must be regularly removed so as not to constitute permanent sources for insect reinfestation. Infrastructure should be installed so that products can be fumigated at the principal sites.

Research on more suitable packing as well as the use of drying tents and solar furnaces should be continued and the effectiveness of traditional methods against infestation evaluated and improved as required. An inventory of traditional processing methods is required to establish standards of quality for each type of product. The production of cured seafood products must pass from being an artisanal enterprise to being an industrial or semi-industrial activity undertaken in well-designed facilities with good management.

Beetle infestation and control in the developing tropics

Occurrence of beetle pests

Several different species of *Dermestes* have been recorded infesting cured fish or fish meal, of which *D. maculatus*, *D. frischii* and *D. ater* are common and cosmopolitan (Haines & Rees 1989). *Necrobia rufipes* is also commonly found on cured fish in many regions around the world, and a related species, *N. ruficollis*, has been found in Bangladesh (Walker & Greeley 1991) but, like a third species, *N. violaceae*, is only an occasional pest. Various predatory histerid beetles, especially species of *Saprinus* and its relatives, may be found in association with *Dermestes* on which they prey. Other beetle adults are sometimes found on cured fish as a result of incidental cross-infestation from other commodities.

In Africa, *D. maculatus* is dominant in sub-Saharan countries while *D. frischii* is more abundant in the north and the Arabian peninsula (Green 1967); *D. maculatus* occurs in Egypt (Azab *et al.* 1972a). In Zambia, *D. maculatus* and *D. ater* are the most common (Proctor 1972) but *D. frischii* is also found (Blatchford 1962). In Sahelian Francophone Africa, *D. frischii* is important (Mallamaire 1955). *Dermestes* are able to attack fish of both freshwater and marine origin though *D. maculatus* tend to be more common on the former and *D. frischii* on the latter. *Dermestes ater* is the third most common species in warm climates. Although *D. maculatus* is predominant in Asia, *D. carnivorus* is not uncommon and has been recorded in Indonesia, the Philippines and Pakistan. *Dermestes carnivorus* appears to be more tolerant of fish with high salt content than *D. maculatus* (Madden *et al.* 1995). Both *D. ater* and *D. maculatus* have been recorded in Bangladesh (Walker & Greeley 1991).

Much of the work on insect pests of cured fish has concentrated upon blowfly control at landing sites, during the initial drying stages. As a consequence problems of beetles have been researched predominantly at these locations and in nearby retail markets where turnover is rather fast. For example, in Indonesia at processing sites, Indriati *et al.* (1986) found dried unsalted fish, such as anchovy (*Stolephorus* sp), were commonly infested with *Dermestes* spp and mites; fatty fish, catfish (*Pangasius* sp), were infested with *N. rufipes*. However, these fish and other species, which were all sun-dried, were particularly heavily infested by blowflies and moulds. As the activity at landing sites is transient, storage beetle infestation is rarely a problem at these locations. For example, in Bangladesh Walker and Greeley (1991) found residual populations of beetles were common in retail and wholesale stores, in both commodity and the structural fabric of the premises. Beetles were also found on carrier boats. In Malawi and Zambia, beetle infestations were most significant when fish was stored for prolonged periods in rural institutions such as boarding schools, prisons, hospitals and military establishments (P. Golob, unpublished).

Losses

Apart from data from experiments in which untreated fish have been subjected to infestation by *Dermestes* and *N. rufipes* (e.g. Golob *et al.* 1987, 1995), where losses were found to be up to 19% dry weight and 12% wet weight after 6 months' storage, only Awoyemi (1991) has assessed loss directly attributable to beetle pests. In the Kainji Lake area of Nigeria, he examined losses in dried samples of the three most common aquatic species, *Sarotherodon galilaeus*, *Tilapia nilotica* and *Alestes*. Within 45 days, most of the fish samples showed visible signs of damage and by 60 days they were reduced to frass and bones. The fish, which were stored at $25 \pm 2.5°C$ and had salt contents of less than 1%, all showed similar levels of damage; wet weight losses after 120 days were in excess of 16% and increased to 26–34% thereafter. The insect population, exclusively *D. maculatus,* increased rapidly between 30 days and 90 days, during which time the fish decreased in quality and quantity and then became mere powder mixed with dead larvae and insect excreta. However, despite the conclusions that strongly supported the need to develop remedial procedures to reduce losses, Awoyemi (1991)

made no attempt to obtain views on losses from producers themselves or from consumers.

There has been no direct assessment of losses due to beetle infestation in processors' or traders' stores, or under conditions of long-term storage which might occur in rural schools, hospitals and prisons. However, Ward (1996) examined physical and economic losses in the Lake Victoria and Mafia Island fisheries in Tanzania by participatory rural appraisal methods to obtain people's perceptions. Very little loss was attributable to insects but this was not distinguished between loss caused by blowflies and loss resulting from beetle damage. As the survey concentrated on fish processors it is almost certain that the insect damage would have been almost exclusively the result of blowfly attack as the storage period of the dried product close to the landing areas is relatively short.

Other estimates of loss either do not distinguish between causes of loss or may have wide margins of error. The following are examples.

During 6 months' storage in Mali, *Hyperopisus* sp lost 55% dry weight matter (Aref *et al.* 1964). Although it was inferred that this loss was due to insect infestation there was no evidence from the methodology described that this was the case, though heavily brined *A. dentex* only lost 10%. Weight loss of dried fish in markets in Sudan, as a result of beetle infestation, was estimated at 50% (Cachan 1957) and nutritive loss may have been much greater. Experts have estimated that losses due to insects (beetles) may reach 25–30% of the total tonnage of dried and smoked fish marketed in the regions of the Middle Niger and Chad Basin in West Africa (Daget 1966). This represents 6000–9000 t y^{-1} or 2500–3800 t of edible protein.

Despite the paucity of hard evidence it is clear that government fishery agencies and other organisations seeking to improve the livelihoods of artisanal fisherfolk are concerned about the level of loss caused by beetles. Even casual observation, such as that by Green (1967) who reported that fish reaching stores in Aden was commonly unfit for human consumption, has highlighted the need to take measures to prevent damage.

Few studies have been undertaken to assess the value losses caused by beetle infestation. In Bangladesh, the presence of beetle-damaged fish would result in price loss of 10–15% over and above the value of the weight loss sustained by the trader (Walker & Greeley 1991). On the shores of Lake Chad in Nigeria, annual losses of dried fish as a result of beetle damage was thought to be worth about £500 000 y^{-1} (Rollings & Hayward 1962).

An important approach to limiting losses in cured fish is to prevent the development of beetle pests. To do this, a knowledge of the development of these insects is essential and so the following section considers the effects of various factors on development, including temperature and moisture, nutrition and the effects of fish composition on infestation.

Development conditions

Temperature and moisture

A general account of the life cycle of *Dermestes* and *Necrobia* is given by Haines and Rees (1989) and a summary of developmental conditions by Blatchford (1962). *Dermestes maculatus* and *D. frischii* develop most rapidly at temperatures between 27°C and 35°C and at a relative humidity (r.h.) of 75–90% (Amos 1968; Azab *et al.* 1972b). Highest survival to the adult stage occurs at 25°C and 75% r.h., with lower temperatures and humidities increasing larval mortality (Amos 1968). Development continues to a maximum of 38–40°C and a minimum of 20°C and 30% r.h., though a combination of the minimum temperature and humidity would not support multiplication. Taylor (1964) showed that larval development would not take place below 55% r.h. at a temperature of 21°C. However, larvae can go through a different number of moults, from four at optimum conditions to as many as nine in more extreme conditions, though percentage emergence will not generally be adversely affected.

Longevity of beetles at 35°C is relatively short, and humidity and salt content (up to 60%) have little effect on life expectancy. At lower temperatures (< 30°C) beetles live longer at higher humidities and live longer on unsalted than salted fish (Amos & Morley 1971).

The oviposition period in *D. maculatus* is extended as the temperature decreases from 35°C to 21°C but the greatest number of eggs is laid at 27°C (Azab *et al.* 1972b). Changing r.h. from 55% to 75% has almost no impact on oviposition; maximum numbers are laid at 27°C and 75% r.h., though the oviposition period decreases with decreasing r.h. However, females with access to free water lay more eggs than those without, even under conditions of relatively high r.h. (Dick 1937; Taylor 1964). The absence of either drinking water or animal protein severely limits egg production (to less than 10%) and the period of oviposition.

Although free water reduces development time for *D. ater* this was not the case in *D. maculatus* (Jacob & Fleming 1985), but it did have a marked effect on the utilisation of nutrients during growth so that heavier adults resulted; free water access decreased larval period except for the last instar which was in any case longer. Humidity also had little influence on the development period of the egg but temperature did; below 25°C the duration of the egg stage increased from 2 to 4 days and thereafter continued to increase as temperature fell.

Dermestes maculatus can survive in conditions of low moisture (Fraenkel & Blewett 1944) (substituting metabolic water) but the trade-off is poorer development and access to free water is essential for regular oviposition (Osuji 1975c) and metabolism (Azab *et al.* 1972b). Osuji (1974a) showed that if other factors remain constant, low moisture content results in low infestation by dermestids.

Beetle infestation begins when fish has been dried down to 40% moisture content or less, the optimum for infestation being 15%. The extent of damage may be related to both moisture content (Osuji 1974a) and lipid content (Osuji 1974b). In India, *D. ater* thrived in dried salted fish with a moisture content of 14–20% and moisture content was a vital factor responsible for infestation (Blatchford 1962).

Dermestes maculatus was found to be the major pest of dried *Channa* sp, during the summer in Manipur, India (Barwal & Devi 1993); this local fish was sun-dried, smoked or charred but not salted and it was more heavily infested than imported salted fish. This beetle became active in March, increased its activity until August and then declined until it became inactive in December. Peak activity coincided with maximum temperature, humidity and rainfall during July and August.

At Ibadan, Nigeria, Osuji (1974c) showed that although *D. maculatus* was the predominant pest on a range of species of dried fish, *N. rufipes* was also rather abundant at particular times of the year. Both species were more prevalent during the warm dry seasons between October and March, when temperatures rose to above 30°C, than in the cool wet period between May and July when temperatures reached 26°C. All life stages of both beetles were found throughout the year; adults, pupae and late instar larvae of *D. maculatus* were found in the head capsule while early instar larvae and larvae of *N. rufipes* had burrowed into the musculature. There was no measurement of fish water activity nor an attempt to correlate r.h. with the development of infestation.

In Bangladesh the longevity of adult female *D. maculatus* reared on dried unsalted *Cybium guttatum* (mackerel) decreased as temperature increased from 20°C to

30°C (Begum *et al.* 1979) but development rate was shortest at 30°C. Egg incubation period also depended on the fish type. Paul *et al.* (1963) found that the duration of the egg stage was inversely proportional to the ambient temperature and that larval development was dependent on temperature, moisture content and food source.

Nutrition

Dermestes maculatus

This beetle appears to prefer fish with a high lipid content that would be unsuitable for other insects. Osuji (1974d) screened 30 samples of four fish genera, *Clarias*, *Citharinus*, *Heterotis* and *Synodontis*, which were collected dried from Ibadan market in Nigeria, for the presence of insect pests. He found that the numbers caught were directly correlated to the lipid content of the fish so that *Clarias*, with a mean of 82 individuals per 100 g (lipid content 16.4%), had most insects and *Citharinus*, with a mean of 9.9 individuals (lipid content 12.3%), had least. It remains to be shown whether the observation genuinely results from only differences in lipid content or whether other factors that could affect beetle development also had an impact.

However, Levinson *et al.* (1967) and Clark and Bloch (1959) have been able to rear *D. maculatus* on semi-synthetic diets almost completely free of lipids but containing cholesterol. Sterols are essential for insect growth, for hormones that regulate metamorphosis and cell membrane synthesis. Like other carnivorous insects, this species cannot survive on vegetable matter alone (Woodroffe & Coombs 1979) because unlike other insects it lacks the biosynthetic mechanism for transforming plant sterols to cholesterol (Kaplanis *et al.* 1965). Woodroffe (1965) and Bergmann and Levinson (1966) identified steroid requirements for *Dermestes*. Without cholesterol in the diet, *D. maculatus* shows a high level of mortality (Gay 1938). This explains why Osuji (1978) found the beetle to be unable to complete development on palm kernel meal and did poorly on wholemeal flour.

It has been suggested by Blatchford that dietary cholesterol analogues, which inhibit insect development, might prove useful in pest management (cited in Noland 1954). For example, Balogun and Ofuya (1986) found that B-sitosterol added to the fishmeal diet prolonged larval development period of *D. maculatus*. They postulated that this compound interfered with the metabolism of cholesterol by competing for metabolic sites.

These beetles also need a good supply of protein and minerals for optimal development. Osuji (1978) compared development of *D. maculatus* on ten different diets, five of which were dried fish (*Tilapia*, *Citharinus*, *Bagrus*, *Mormyrus* and *Clarias* spp), bloodmeal, wholemeal wheat flour, commercial fishmeal, bonemeal and palm kernel meal. Development (larval and total) was faster, the number of larval instars fewer and fecundity greater on the dried fish than on the other preparations except for the commercial fishmeal. The dried fish samples had substantially higher lipid content (10–20%) than bloodmeal (0.1%) and wholemeal flour (0.8%) but not significantly greater than palm kernel meal (10%) or bonemeal (11%). The crude protein content was 43–46% for dried fish, much higher for blood meal (79%) but lower for wholemeal flour (22%), palm kernel meal (14%) and only bonemeal (3%) was not protein-rich. All the media had moisture contents of more than 10% except bonemeal, which was very dry (3.4%). The author suggests the differences in life parameters can be attributed to the generally superior nutritional qualities of the dried fish. Similar observations were made by Azab *et al.* (1972b) who found different diets also affected development parameters. Dried rabbit skin, dried cheese and dried milk were not as satisfactory as dried meat and dried fish; on the former diets larval development was prolonged and mortality increased.

These carnivorous beetles also need an adequate supply of vitamins and development may be limited if vitamins are depleted or in the presence of vitamin antagonists (Levinson *et al.* 1967). Diets lacking nicotinic acid, pantothenic acid, biotin, folic acid or all the B vitamins hindered pupation and caused cannibalism as well as high larval mortality. Omission of thiamine or riboflavin prevented the development of some adults and greatly prolonged the larval period. There was no evidence of larval requirement for inositol. Only antagonists of folic acid (aminopterine) and biotin (avidine) produced effects resembling deficiency of the vitamin. However, addition of neopyrithiamine was more marked than the omission of thiamine; the antagonist completely prevented development to adult and significantly reduced the larval stage.

Necrobia rufipes

Adults feed on the surface of dried fish, and they lay their eggs, up to 2000 for each female, in crevices in the fish. Eggs hatch in 4–5 days and the larvae go through three to four moults, developing in 25–35 days under optimum conditions before pupation (Haines & Rees 1989). The complete life cycle takes 2–3 months (Mallamaire 1955).

Although larvae feed on dried fish, *N. rufipes* is a predator of other insect larvae and its development is much improved by access to prey (Ashman 1962). Adults have been observed feeding on moulds, especially *Penicillium* and *Aspergillus*, growing on the surface of mouldy products, but it is most likely that ingestion of fungal hyphae was incidental (Lepesme 1939).

Methods for control of beetle pests

Insect pests can be controlled by the application of synthetic insecticides, and 30 different contact insecticides have been used on fish and fishery products (Walker 1987). In some countries cheap, domestic and agricultural insecticides, formulated and packaged locally, are used by fish processors and sellers. However, these chemicals present a toxic hazard to both operators and consumers if used inappropriately.

At present, there are 15 insecticides for which maximum residue limits (MRLs) have been established for food commodities. These include: organophosphate insecticides such as fenitrothion and pirimiphos-methyl; pyrethroids including permethrin, deltamethrin and fenvalerate; naturally occurring pyrethrins and its synergist, piperonyl butoxide; and the insect growth regulator, methoprene. However, only three compounds have MRLs established for dry fish: pirimiphos-methyl at 10 mg kg^{-1}, pyrethrins at 3 mg kg^{-1} and piperonyl butoxide at 20 mg kg^{-1}.

Frequently, extremely hazardous insecticides are employed to treat cured fish. Among the 30 insecticides known to be used on fish and fish products, 19 are non-approved chemicals (Walker 1987), these include DDT and lindane (particularly in South Asia and Africa), which can accumulate in the body, and highly toxic trichlorfon (Dipterex) and dichlorvos (Nuvan) (Bangladesh, Thailand, Indonesia, Philippines) (e.g. Rattagool *et al.* 1990; Walker & Greeley 1991. Many insecticide packs provide little if any information regarding the active ingredient, and an unlabelled insecticide powder in The Gambia had, on residue analysis, contents that varied from time to time at the whim of the packager (Wood 1983). Of more than 230 samples of dried fish collected from all over Bangladesh all contained DDT, many having residues above 50 mg kg^{-1} (Walker & Greeley 1991).

Despite this misuse of insecticides, there are no records in the literature of accidental poisoning occurring during fish processing, handling of the cured products, or subsequent to consumption. However, this lack of information does not mean that accidental poisonings do not occur, rather they are not recognised as such; either the symptoms of acute poisoning are not known or chronic poisoning is too slow to be identified or associated with dried fish handling or consumption.

Pirimiphos-methyl (Actellic: Syngenta) is a chemical used around the world as a grain storage protectant. Its use for cured fish is to be welcomed. Pyrethrins synergised with piperonyl butoxide are restricted mainly to those countries in which the plant from which the active components are extracted, *Tanacetum cinerariifolium* (formerly *Chrysanthemum cinerariaefolium*), is cultivated, mostly in East Africa. Worldwide, there is a dearth of chemical products approved as fish protectants. There has been recent research to identify alternative synthetic compounds that would be safe to use on fish. Additional work has been undertaken to develop other methods, including insecticides derived from plants and the application of salt, heat or irradiation. The following paragraphs provide an account of the status of this work.

Contact insecticides

Experimental work

The development of synergised pyrethrins for protection of cured fish against beetle damage began in the 1960s. In Kenya (McClellan 1964) and South Arabia (Green 1967) dried fish were dipped in insecticidal emulsions to control beetle infestations. *Tilapia nilotica*, *Lates* sp and *Citharinus citharus* were dipped into 0.018% pyrethrins and 0.18% piperonyl butoxide while drying and were quite acceptable to consumers. The compounds gave good protection against beetles for 5 weeks (McClellan 1964). Similarly, emulsions of pyrethrins down to 0.02% were used successfully as dips to kill all adults and larvae of *D. frischii* infesting dried fish and prevented further insect development during the whole 6 weeks of the experiment (Green 1967). There was also a marked repellency effect as fish remained free from insect bodies. However, in the same study it was also found when dipping previously uninfested wet or dry fish, in pyrethrins (or malathion), the treatment was not successful as they both became heavily infested after 2–5 weeks by migration from other infested commodity. No clear explanation could be offered for this apparent difference in performance of the insecticides.

Brining was compared with dipping or dusting synergised pyrethrin or malathion as methods for protecting four species of fish in Zambia – *Clarias mossambicus* (barbel), *Tilapia* sp (bream), *Synodontis zambezensis* (squeaker) and *Hydrocynus vittatus* (tiger fish) (Proctor 1972). Fish were immersed completely for 1–2 s in pyrethrins (0.009% or 0.018%) or malathion (0.0625% or 0.125%) prepared in river water and stored after shaking in hessian sacks to remove excess liquid. Pyrethrum dust, at a maximum of 1 g per 900 cm^2, was applied to the sacks after they were filled with fish. Dip treatments were more effective than any of the dust treatments. Pyrethrin exerted a repellency effect as few insects, live or dead, were found on these treatments. Malathion was not repellent. Brining alone (up to 10%) did not provide protection, neither was there a repellent effect. Brined fish were, however, preferred by the taste panel to un-brined fish even though dipping in pyrethrins had no effect on taste.

The research described in the preceding paragraph included testing of malathion which is not an insecticide approved for use on dried fish. However, it does have an MRL for cereal grains of 8 mg kg^{-1}. In the test, malathion residues varied considerably, especially at the higher dosage, and were generally too high; the lower dose treated fish contained a minimum of 26 mg kg^{-1} even after 8 weeks compared with 51 mg kg^{-1} after 2 weeks. It was suggested that the fat content of the fish may affect uptake of insecticides, which might result in retention of high residues in oily fish (Proctor 1972). This illustrates the danger of assuming that insecticides approved for use on grain can be applied in the same way to fish. The penetration of malathion into fish flesh has been investigated and showed penetration 2.5 mm below the surface. In simulated dipping experiments in the laboratory with malathion, insecticide residues on shark, tuna and queenfish were generally well in excess of levels which would be regarded as acceptable (37–245 mg kg^{-1} after 8 weeks). However, samples treated in Aden and analysed 4 weeks later showed much lower residues but still generally in excess of what would be acceptable. It was concluded that, although malathion broke down rapidly on shark, relatively high residues may remain in other types of fish for much longer periods (Green 1967).

In examples where lower dosages have been used in order to maintain residues at acceptable levels, good control has not been achieved. In Malawi, the mixture of 0.018% pyrethrins and 0.036% piperonyl butoxide, which was used to control blowflies in *Lethrinops* sp, produced residues of up to 2.4 mg kg^{-1} pyrethrins and 4.6 mg kg^{-1} piperonyl butoxide after 4 days of sun drying. Although blowflies were controlled, subsequent

dermestid infestation was not prevented (personal communication cited by Proctor 1977).

In Kenya, where the source plants for pyrethrins (*T. cinerariifolium*) are cultivated, cost-effective treatments can be undertaken by dipping *Labeo horie* in emulsions containing different proportions of synergised pyrethrins for 4 s (Gjerstad 1989). Although these treatments gave good control of *D. maculatus,* the trials produced residues of up to 30 mg kg^{-1} for pyrethrins and up to 102 mg kg^{-1} for piperonyl butoxide, far in excess of MRLs.

Organophosphate insecticides (OPs), which are widely used as grain protectants, have been assessed for their potential use on fish. Under laboratory conditions, 25°C and 70% r.h., *D. maculatus* adults and full-grown larvae (and a range of other stored grain insect pests) were exposed to filter papers treated with a range of OPs (Tyler & Binns 1977). *Dermestes maculatus* larvae were not as susceptible as adults, or as some of the other beetles tested, and not many insecticides produced 100% kill after 5 days' exposure to 100 mg m^{-2}. Fenitrothion produced complete kill of larvae but pirimiphos-methyl did not. Furthermore, *D. maculatus* has the capacity to develop resistance to insecticides. Twelve out of fifteen strains tested from around the world have shown some degree of resistance to lindane, an organochlorine widely used to protect hides and skins against this pest, though there has been no evidence of resistance to OPs (Binns & Tyler 1978).

Other laboratory experiments, at 27°C and 75% r.h., demonstrated effective control of *D. maculatus* could be achieved for 5 weeks with pirimiphos-methyl emulsions containing 0.0063% and 0.0125% active ingredient (a.i.), which gave acceptable residues (Taylor & Evans 1982). In Francophone West Africa, there has been extensive testing with another OP, tetrachlorvinphos (Gardona: Dow-Elanco). Early work in Mali showed tetrachlorvinphos to be effective when applied at 0.0375% in protecting mixed batches of fish against mostly *D. maculatus* infestation (Guillon 1976). Fish were immersed in the emulsions for 1, 10 or 60 min and, although the authors claim the longest period of application was more effective, the data do not appear to indicate that the period of immersion was significant. By 11 days after treatment the highest residues measured were 7.6 mg kg^{-1} when fish were immersed in a 0.075% emulsion.

There have been similar experiments in Niger with tetrachlorvinphos (Bouare 1986). Mixed batches of fish were dipped in 0.0375% or 0.075% emulsions for 30 or 15 min, respectively, or in water. Other batches were sprayed with the same concentrations. Dipping the fish before smoking did not significantly reduce losses but spraying after smoking did. Results obtained by dipping were very variable and it was not possible to draw conclusions about this method; by 30 days 68% of controls were attacked, 55% of the low dosage and 25% of the high dosage, and by 90 days all fish were damaged whether treated or not (heat from smoking may have adversely affected dip treatment). Spray treatments were more effective after 30 days' storage; 90% of controls were attacked but only 20% and 5% for the low and high-dose treatments, respectively. As a result of this work, spraying with 0.5 g a.i. per litre of emulsion was recommended. Bouare (1986) obtained residues after 30 days of 2.5 mg kg^{-1} and 9 mg kg^{-1} after spray application. However, tetrachlorvinphos does not have CODEX approval for use on fish or on cereals and there is no MRL established. For these reasons it would not be appropriate to recommend the use of this chemical when there are alternatives which are equally if not more effective, and which do have international clearance. Moreover, in Mali 0.0375% emulsions of tetrachlorvinphos was found to be ineffective in protecting *Sarotherodon niloticus* against *D. maculatus* when stored for 120 days (Duguet *et al.* 1985).

Research to control blowfly invasion of fish during curing resulted in a MRL being established for pirimiphos-methyl of 10 mg kg^{-1} for dried fish (background information is given in Walker 1987). Thus the treatment of drying fish immediately after the catch has been landed to control blowflies must also take account of the subsequent residues remaining in the dried product which could possibly affect control of beetles. Walker (unpublished) compared the uptake of pirimiphos-methyl by fresh and dried, split rainbow trout when dipped for 30 s in emulsions of different concentration. The fresh fish were dried under an infra-red lamp for 36 h during which time their moisture content fell from 72% to 42%, finishing with about twice the moisture content of the dried product. This difference in dry weight mass (all samples began with the same weight) was reflected in the residues obtained; for 0.06% dip the mean residues were 7.8 mg kg^{-1} and 16 mg kg^{-1} for fresh and dry fish, respectively, and for 0.015% were 2.6 mg kg^{-1} and 6.6 mg kg^{-1}. Clearly, when applying insecticide for blowfly control it is important to be aware that the active ingredient will increase initially as the fish dries down and may exceed the recommended MRLs.

In Indonesia, Esser *et al.* (1990) assessed the residual efficacy of pirimiphos-methyl to control beetles after the insecticide had been applied as a dip to protect drying fish against blowfly. *Lutjanus sebae* (red snapper) and *Diagramma punctatum* (grouper) were split, gutted and brined overnight and then dipped for 15 s in a 0.03% emulsion or in water before being dried. They were then kept for 13 weeks in baskets in a store room. Treated snapper lost 5% by weight compared with 13% for the control and treated grouper lost only 0.5% compared with 4% for the control. All treated fish remained free from dermestid infestation, whereas adults and larvae were found on the controls. The treatments provided excellent protection against beetle damage but as chemical residues were not measured it is not known whether MRLs were within recommended tolerances.

The observation that pyrethrins repelled beetles led workers to assess related synthetic pyrethroids as well as alternative OP products as fish protectants. In the laboratory, at 27°C and 75% r.h., emulsions of the synthetic pyrethroid permethrin, containing 0.0063% and 0.0125% a.i., were not effective in controlling *D. maculatus* for a period of 5 weeks (Taylor & Evans 1982), and compared unfavourably to pirimiphos-methyl. However, in a trial at Lake Chad, northern Nigeria, permethrin did provide the same degree of protection as pirimiphos-methyl or chlorpyrifos-methyl when *Clarias*, *Heterotis*, *Gymnarchus* and *Tilapia* were dipped into 0.0125% emulsions, though the samples were only stored for 3 weeks (Taylor 1981). When applied to woven mats in which dried fish are normally bundled, at 0.5 g a.i. m^{-2}, permethrin gave better control than the other two insecticides over 7 days.

Experiments near Dakar, Senegal, confirmed that permethrin would not provide adequate control of *Dermestes* when the storage period was 3 months or longer. In this work, split dried *Plectorhinchus mediterraneus* were sprayed with or dipped into aqueous insecticide emulsions and stored for up to 24 weeks. Treatments containing another pyrethroid, deltamethrin, were the most effective in protecting against infestation from mixed populations of *Dermestes* and *N. rufipes* (Table 14.4) (Golob *et al.* 1995). Deltamethrin applied at 0.25 mg kg^{-1} gave good protection for up to 4 months' storage, but for longer periods it was concluded that the dosage should be doubled. There were very few insects on fish treated with deltamethrin due to repellency. Even at the highest dosage tested, 1 mg kg^{-1}, the residue found in the fish tissues immediately after treatment did not exceed the recommended MRL of 1 mg kg^{-1} for cereal grains (Table 14.5). Unlike the trials in Niger (Bouare 1986), dipping was found to be more consistent, and therefore predictable, than spray applications. Treat-

Table 14.4 Mean number of larvae of *Dermestes maculatus* and *Necrobia rufipes* developing on split dried fish (*Plectorhinchus mediterraneus*) dipped in various insecticide emulsions. Source: Golob *et al.* (1995).

Treatment	Dosage (mg kg⁻¹)	Duration of storage (weeks)				
		1	4	8	12	16
Deltamethrin	1.0	0	0	0	0	0
Deltamethrin + pirimiphos-methyl	0.5 + 5.0	0	0	0.3	0.5	0.3
Pirimiphos-methyl	10.0	0	0	2.3	0.3	2.0
Chlorpyrifos-methyl	10.0	0	1.3	4.3	0.8	0.8
Permethrin + pirimiphos-methyl	1.0 + 5.0	0	0	4.3	6.0	69.8
Permethrin	2.0	0	1.8	8.3	35.5	177
Control		1.5	119.5	19.5	27.5	93.7

Table 14.5 Mean insecticide residues (ppm) on split dried fish (*Plectorhinchus mediterraneus*)* dipped in various insecticide emulsions after 1 week or 12 weeks of storage. Source: Golob *et al.* (1995).

Treatment	Nominal dosage (mg kg⁻¹)	Duration of storage (weeks)			
		Dip treatments		Spray treatments	
		1	12	1	12
Deltamethrin	0.25	0.18 (72)†	0.55	0.14 (56)	0.18
	1.00	0.64 (64)	0.40	0.39 (39)	0.16
Permethrin	0.50	0.48 (96)	0.78	0.29 (58)	0.15
	2.00	1.74 (87)	1.75	0.72 (36)	0.72
Pirimiphos-methyl	2.50	0.54 (22)	0.40	0.79 (32)	0.10
	10.00	6.60 (66)	1.48	2.60 (26)	0.01
Chlorpyrifos-methyl	10.00	6.38 (64)	0.92	2.21 (22)	0.22

*Different fish were analysed at each occasion.
†Data in parentheses represent the residues as a percentage of the nominal dosage.

ments with pirimiphos-methyl or chlorpyrifos-methyl, which both have MRLs of 10 mg kg⁻¹ for use on raw cereal grains, also gave adequate protection of the dried fish for up to 6 months though the fish quality after this time was inferior to the deltamethrin-treated fish, as there was no repellency effect (Table 14.6).

Similar experiments were done on *Tilapia* sp in northern Kenya (Golob *et al.* 1987). The compounds tested included the organophosphates fenitrothion, iodophenphos, pirimiphos-methyl, synergised deltamethrin and pybuthrin (a commercial formulation of synergised pyrethrins). Different formulations of the active ingredients were used including emulsifiable concentrates, flowable concentrates and wettable powders. The fish were dipped for 4 s into aqueous emulsions of the insecticides which contained 0.01% or 0.02% a.i. except

for those containing deltamethrin which had 0.001% or 0.002% a.i. and pybuthrin, which was made up to give a 0.018% emulsion, a concentration recommended for commercial use. Fish were kept up to 6 months in a large dried-fish store and subjected to natural infestation from *D. maculatus*. All treatments gave good protection for 2 months when compared with untreated controls but only the two deltamethrin treatments, which exhibited significant repellency, gave adequate protection for 6 months. There was no difference in effect between the different formulations nor between the two concentrations used. Low dosages gave residues on the fish that were within the MRL for cereal grains and could be regarded as safe for human consumption.

Other trials in West Africa have also demonstrated the efficacy of deltamethrin. Duguet *et al.* (1985) found

Table 14.6 Mean percentage of the number of split dried fish (*Plectorhinchus mediterraneus*) dipped in various insecticide emulsions exhibiting insect damage. Source: Golob *et al.* (1995).

Treatment	Nominal dosage (mg kg^{-1})	Duration of storage (weeks)				
		1	4	8	12	16
Deltamethrin	0.25	0	0	3	15	26
	0.50	0	0	0	8	13
	1.00	0	4	3	8	7
Pirimiphos-methyl	2.50	0	15	25	50	51
	5.00	0	6	14	24	38
	10.00	0	4	19	35	24
Chlorpyrifos-methyl	10.00	0	4	26	38	31
Deltamethrin + pirimiphos-methyl	0.25 + 2.5	0	0	0	3	1
	0.5 + 5.0	0	0	0	1	10
Permethrin + pirimiphos-methyl	0.5 + 2.5	0	0	22	42	61
	1.0 + 5.0	0	6	31	57	72
Control		2	63	83	90	90

dip treatments to be effective protectants of *Sarotherodon niloticus* in Mali against *D. maculatus* infestation. Batches of 15 kg of wet fish were dipped for 10 min in emulsions containing between 0.001% and 0.01% deltamethrin alone or synergised with piperonyl butoxide and 0.01% to 0.1% pirimiphos-methyl. The lowest dose of deltamethrin provided protection for 120 days and 0.0025% for 180 days. Pirimiphos-methyl at 0.05% was also effective in preventing damage for 180 days but did not produce the repellency effect of the deltamethrin. No measurements of the chemical residues in the fish were made and, although the concentrations were similar to those used by Golob *et al.* (1987) in Kenya, the dipping period was much longer, 10 min rather than 4 s. The uptake of active ingredient by the fish was therefore likely to have been much higher than the permitted MRL.

In Thailand equally good results have been obtained. Brined *Arius* sp (catfish), *Megalaspis cordyla* (scad), *Scomberomorus commersoni* (Spanish mackerel) and *Chorinemus lysan* (queenfish) were dipped into emulsions of 0.03% or 0.06% pirimiphos-methyl for 15 s or into 0.003% deltamethrin emulsion twice for 3 s each time; there was no explanation for the difference in the treatment regimes (Rattagool *et al.* 1990). After sun drying for 2 days, samples were placed into paper bags and stored for 3 months. Other *Arius*, which had been smoked but not brined, were sprayed with the same treatments. During storage the fish was subjected to infestation mostly by *D. maculatus* but also *D. ater*, in the ratio of 4.5 : 1, and occasionally *N. rufipes*. In one trial

there was no infestation recorded either on the dipped *Arius* and *S. commersoni* or the controls that had been dipped into water. In a second trial with dipped *Arius*, controls lost about 5% in weight due to insect attack but the treatments provided complete protection. There was light *Dermestes* infestation on *Megalaspis cordyla* though complete protection was provided by the higher dosage of pirimiphos-methyl and the deltamethrin treatments. *Chorinemus lysan* showed evidence of heavy infestation of controls though the damage was light; all treatments imparted complete protection. Smoked *Arius* was the most heavily attacked with controls losing about 25% by weight.

The experiments described above, in Senegal, Kenya and Thailand, were all at either low r.h. or on fish samples with low moisture content. As Rattagool *et al.* (1990) conclude, effectiveness of the treatments was aided by the low moisture content (9.8%) of the fish in their trial. Whether the treatments would be as effective under more humid conditions, with fish having moisture contents approaching 20% or more, has still to be shown. Pilot trials at Mwanza, Tanzania, with fish of 15–22% moisture contents and at a time when the relative humidity was high during the early part of storage, indicated that control of beetles could be achieved. However, problems caused by fungal invasion became a major concern (P. Golob, unpublished).

Newer pyrethroids have been assessed. Dipping wet *Arius* into emulsions containing 0.003% a.i. alphacypermethrin (Fastac: Shell) provided complete protection against *Dermestes* sp for a 10-week storage period as did a 0.015% emulsion of pirimiphos-methyl (Esser *et al.* 1986), though levels of infestation in controls were low. Alphacypermethrin, similar in structure to the pyrethroid deltamethrin, is not approved for use on fish or cereals.

Cycloprothrin (a substituted pyrethroid of low mammalian toxicity) and deltamethrin have been tested for their ability to provide control of beetles when applied to packaging material rather than to the fish themselves. Each emulsifiable concentrate was applied in a starch solution by being painted on the outer surface of triple-walled paper bags that contained 700 g of anchovy (Madden *et al.* 1995). Application provided 0.05 g cm^{-2} and 0.5 g cm^{-2} of a.i. for deltamethrin and cycloprothrin, respectively. Painted bags were allowed to dry for 24 h before they were placed in containers with ten pairs of adults and 75 fourth instar larvae of *D. maculatus* or *D. carnivorus*. Deltamethrin inhibited penetration for 90 days and cycloprothrin for 60 days. It is not known whether there was any transfer of active ingredient from the paper to the fish.

Practical application of approved chemicals

In Zambia, it was demonstrated that an increase in net income of 60–80% could be derived by treating fish with insecticide (Proctor 1972). Four traders were taught to treat fish with pyrethrins emulsion and each given sufficient concentrate for 9 L of diluted emulsion. At the fisheries camp, each trader prepared the emulsion on the day he started to smoke-dry fish and used it to treat about half the fish processed each day for 6–8 weeks. Treated and untreated fish were then packed into separate bundles and taken to market. The effectiveness of the treatments were determined by taking 100 fish from a vertical section through each bundle and recording the number infested. The differences in net income provided a direct measure of the economic value of the treatment. Infestation in treated fish was in the range 0–14% compared with 18–100% for controls. A net financial gain from treating was obtained by all four traders. Differences in the infestation of treated fish appeared to be related to the weight of the fish treated and suggested that about 100 kg of smoked-dried fish was the maximum quantity which could be treated with a 9-L emulsion without incurring some risk of infestation. However, it should be noted that these treatments were undertaken primarily to protect fish against blowfly attack.

For treating drying fish against blowfly attack in Malawi, Meynell (1978) used a method of soaking, i.e. pouring insecticide emulsion (pyrethrum) over a washed 60 lb box of fish. The emulsion percolated down over the fish and flowed out of the bottom of the box. In commercial practice, the emulsion is collected and recirculated over many boxes using a pump or a series of watering cans. For each box, 5–10 L of emulsion treated 80% of fish. It took 30–45 s to treat each box, which is acceptable in practice. This method could be used for treating dried fish against beetles. Soaking avoids the double handling required by dipping although the latter may be more effective as *all* the fish are treated. There were indications that 30 L of 0.033% emulsion became exhausted by the time 60 boxes had been treated, but more work to determine such limitations is needed.

Fumigation

Only two fumigants are commonly used for the disinfestation of dried food commodities: methyl bromide and phosphine. However, sulphur dioxide has been used to fumigate storage rooms in which dried fish are stored in Russia (cited by Anon. 1966). Fumigants can only be used to disinfest commodities; they do not provide residual protection. Thus once the fumigation procedure has been completed the commodity can be reinfested immediately. For this reason, fumigants have not found common use for the treatment of fish or fish products. Unlike for contact insecticides, there are no MRLs for fumigants on dried fish but those for cereal grains have been used as guidelines. The use and status of fumigants as dried fish protectants has been reviewed by Friendship (1990).

Fumigant gases are highly toxic and very fast-acting. Great care is required when undertaking a fumigation and a high degree of training and skill is necessary in order to minimise risks and maintain efficacy. For these reasons, fumigation is really only appropriate for the treatment of large quantities of produce. The application of methyl bromide, in particular, requires specialised equipment and so its use has generally been restricted to commercial fish processing operations.

Methyl bromide has been tested and found to be particularly effective in the Sahelian area of West Africa. However, in recent years methyl bromide has been found to be a strong depleter of stratospheric ozone and will be phased out for all but a small number of essential uses by 2015 in developing countries (Vol. 1, p. 328, Taylor

2002). It is therefore very unlikely that this gas will have a role in the disinfestation of dried fish in the future.

Phosphine is much easier to apply and handle than methyl bromide. It is released from tablets of aluminium phosphide when these are exposed to water vapour in the air. In many parts of the world these tablets are readily available, can be purchased from retail traders and are cheap. They are commonly bought by individual consumers and farmers for protecting grain and other food supplies against insect pest damage during storage or as rat poison. There is no control on the supply of aluminium phosphide and no training is provided to help people use the gas correctly. Thus users have very little knowledge of the health risks involved or of how phosphine should be applied for maximum effect. The gas released from each tablet is potentially lethal to the applicator or persons living within the vicinity of the fumigation. Moreover, the fumigation of inappropriate enclosures such as gunny bags, woven baskets and mud-plastered dried fish stores will be ineffective as the gas will leak out of the structure. Such misuse will lead to the development of insect resistance. For health and safety reasons and to achieve effective control fumigation should only be undertaken by trained personnel.

Phosphine requires exposure periods of 5 days, sometimes more. Much of the work to identify optimum application rates was undertaken in Malawi and the Department of Fisheries now recommends fumigation of dried fish with phosphine under gas-proof sheets at 1.06 g m^{-3} for 5 days (Walker 1984), though commercial companies use dosages of 0.71–1.77 g m^{-3} over exposure periods of 3–5 days. However, Walker found that *D. maculatus* could survive a dosage of 1.06 g m^{-3} but that at 1.5 g m^{-3} for 3 days all insects died. A dosage of 1.3 g m^{-3} applied to dried fish in a fumigation chamber successfully controlled all insects in 6 days and did not leave any undesirable taste taints (Meynell 1977). In India, fumigated dried *Trichiurus lepturus* (ribbon fish) was fumigated with phosphine at a rate of 2 g phosphine for 5 days when a final concentration of the gas in air of 97.5 ppm was achieved (Friendship 1987). There are records of storage insect pests developing resistance to phosphine as a result of continuous exposure of insect populations to sub-lethal dosages. As a result, the UK Natural Resources Institute has recommended that a minimum concentration of 150 ppm should be maintained by the fifth day of treatment in order to kill all insect strains including those which might be resistant (Friendship 1990).

The fumigation of *Harpodon nehereus* (Bombay duck) with phosphine at 2.9 g m^{-3}, a dosage commonly employed for treating other food commodities, resulted in an initial metal phosphide residue of 0.11 mg kg^{-1} which gradually reduced to 0.01 mg kg^{-1} after 2–3 weeks of aeration (Harris & Halliday 1968). The results demonstrated that it was possible to fumigate dried fish with phosphine and that there would be no hazard from the accumulation of residues in the treated product.

Other methods of control

Inert dusts and other solid powders

Inert dusts are finely divided solid particles that exert their effect on target organisms by physical rather than by chemical means (Vol. 1 p. 270, Stathers 2002). Most inert dusts consist of aluminium silicate (in excess of 90%) and many are of organic origin, occurring as deposits known as diatomaceous earths, but some are synthetic.

Inert dusts are commercially available and are of similar cost to the OPs used for storage protection. They have been assessed for control of *D. frischii*, *D. ater* and *N. rufipes* (Kane 1967). Adults were exposed to filter papers treated with a wide variety of dusts. Some control was exerted particularly by those dusts having low bulk density. Several of the dusts showed activity when applied to dried fish of Arabian origin at a dosage of 0.03% but efficacy varied with the type of fish treated. As the oil content and moisture content increased the activity decreased. One of these dusts, Dri-Die® (SG-67), a silica aerogel containing 4.7% ammonium fluorosilicate, was used in trials in Aden to control damage by *D. frischii* (Green 1967). Two dosages were applied, 0.24% and 0.06%. The dust quickly absorbed oil and was ineffective in preventing infestation. Another unnamed dust (a silica aerogel containing 99.8% aluminium silicate) was tested in Zambia. The dust was shaken onto the surface of *Tilapia* to give 300 mg kg^{-1} (0.03%) w/w (Proctor 1972). The effect of this dust was compared with that of other treatments including pyrethrins. The inert dust was least effective and some treated fish were more severely damaged than controls.

Other trials in Indonesia by Madden *et al.* (1995) showed that Dryacide®, a commercially available dust used widely by the wheat industry in Australia for insect control, was ineffective in preventing beetle invasion when applied at 0.5 g per bag to the inner layer of all three paper surfaces of bags used to store dried fish. Aref *et al.* (1964) tried sulphur dipping with equally poor results.

Heat

Subjecting insects to excessive heat will kill them. Heating dried fish to 50°C (Nakayama *et al.* 1983) or 60°C for 30 min (Toyes 1970) killed all life stages of *D. maculatus*. In Mali, tests showed that exposing adults and larvae to 50–55°C for a few minutes would kill both *Dermestes* and *N. rufipes*, the latter being slightly more heat tolerant (Galichet 1960).

In the Kainji Lake area of Nigeria, dried fish are stored in chambers which are continually subjected to smoke, there is no sun drying and the practice of covering the fish results in almost no beetle infestation (Osuji 1974d). In areas where fish is moved directly from the drying kiln to the market and the storage period is short there is also minimal infestation. Use of drying kilns excludes insects because of the lethal temperatures generated and by dehydrating fish to levels detrimental to insect survival and development. Furthermore, periodic heat treatment in closed compartments has been shown to be effective in destroying all life stages of *D. maculatus* present in dried fish from Ibadan market (Toyes 1970).

When dried *Labeo horie* were spread out on concrete in the sun at Lake Turkana, Kenya, fish in one layer, 4 cm thick, increased in temperature between 11.00 and 17.00 hours from 43°C to 47°C at the bottom of the batch, and when the layer was 20 cm thick increased from 37°C to 42°C. At the higher temperatures all *D. maculatus* adults and larvae died but this was not so in the thicker layer of fish. However, the very high temperatures resulted in weight loss greater in one day than would have been experienced in 6 months' storage under ambient conditions due to water loss. Furthermore, the extreme surface temperatures obtained, 55–60°C, caused the exposed fish to become brittle which resulted in loss of weight as a result of fragmentation, the fish fetching a lower price than would normally have been expected. Thus, although solar drying was effective for controlling beetle pests it would need to be adapted to have less effect on the fish itself before it is commercially viable (Walker & Wood 1986).

Microwave exposure is a much quicker way of killing dermestids; 40 s of exposure killed all life stages except for eggs which took 150 s or longer. Infra-red exposure had a similar effect to microwaves (Nakayama *et al.* 1983).

Smoking during drying and the exposure of blowfly larvae to phenolic compounds which destroys eggs and larvae appear to have no effect on subsequent infestation by beetles (Walker 1988).

Ionising radiation

There are two methods by which ionising radiation can be utilised for the protection of dried fish. Firstly, adult beetles can be sterilised and released into the natural population to compete with normal adults and thus reduce the potential for the population to increase. Although this approach has been suggested (Walker & Greeley 1991) it has not been put into practice. The second method is to use irradiation as a direct means to kill insects by subjecting infested fish to a gamma-irradiation source. This method has been used experimentally but it requires the employment of expensive capital equipment, the use of which can only be countenanced for commercial operations, particularly those involving a high-value export product.

Several studies have assessed the effect of radiation on beetles. From Mali, samples of dried *Tilapia*, *Lates*, *Clarias* and *Labeo* were sent to France in polythene pouches and irradiated with dosages between 20 Gy and 500 Gy. They were then returned to Mopti, Mali, for assessment of insect mortality and organoleptic properties. There was no effect on taste but some survival of *D. maculatus* and *N. rufipes* adults and larvae on the lowest of the three dosages, but only three individuals survived 100 Gy; none of the survivors produced progeny (Boisot & Gauzit 1966). Other workers in France (Pointel & Phan van Sam 1969) suggested doses of 100–300 Gy would be effective for control.

Much of the work on irradiation has been done in Bangladesh where early trials were on dried fish (mackerel) stored in polythene pouches of different thicknesses. All insect life stages could be controlled with doses of 200–300 Gy but the fish became mouldy. This problem could be overcome if the fish were irradiated in jute sacks, the traditional method of storage, or by ensuring the fish was very well dried, to 13% or less, before being packed into polythene pouches (Ahmed *et al.* 1978).

However, irradiation has to be regarded as a similar type of treatment to fumigation in that it does not provide any residual protection. Therefore, the use of irradiation must be combined with other measures to prevent reinfestation; appropriate packaging is essential. In Bangladesh, Bhuiyan (1990) found that treatment of *Lepturocanthus savala* (ribbon fish), *Labeo gonius* (gonia) and *Cybium guttatum* (mackerel) with 1 kGy was effective in disinfesting samples of *D. maculatus*. Dried fish stored in polythene packages remained relatively free of insects up to 4 months, except where the packaging had been punctured, but unirradiated samples bore heavy insect infestations. Irradiated samples were also found to be of acceptable appearance and organoleptic quality. This experiment was repeated on a semi-commercial scale. Irradiated samples of 2 kg were placed in 0.5 mm polythene pouches and ten of these were put into cardboard boxes lined with the same thickness of polythene. The irradiated samples were stored successfully for 6 months (Shahjahan *et al.* 1996). Khatoon and Heather (1990) concluded that an average minimum dose of gamma-irradiation of 1 kGy would effectively disinfest hermetically sealed products of *D. maculatus*. Older life stages were less susceptible than younger stages. Larvae were more susceptible than other life stages apart from eggs that were less than 36 h old. As these treatments were applied in an atmosphere of nitrogen, the authors suggest that two to three times the dose would be required if the treatments were applied in normal atmospheric conditions.

In the Philippines, 225 Gy were sufficient to control all stages of *D. carnivorus* and did not affect the organoleptic properties of the dried fish *Parasarcophaga ruficornis* (striped mackerel) (Pablo 1978).

Aref *et al.* (1964) found 53–105 Gy to be lethal to adult dermestids and sterility but not death was achieved with 23 Gy. Gamma-irradiation affects the midgut epithelial cells of *D. maculatus*. Moderate cell disruption occurs at 50 Gy and at 200 Gy complete histolysis (Saidul & Rezaur 1996).

Pheromones

Pheromones, chemicals released into the environment by insects that affect the behaviour of other individuals, are used for pest management in agriculture. It may be possible to use them to protect dried fish by disrupting the mating behaviour of beetles or by creating alarm within the insect population.

During the last 20 years research has demonstrated pheromone production by *D. maculatus*. Males secrete a pheromone from an exocrine gland linked through a canal to the surface of the fourth abdominal sternite. The most active components of this secretion are isopropyl *Z*-9-dodecanoate, isopropyl *Z*-9-tetradecanoate and isopropyl *Z*-7-dodecanoate (Levinson *et al.* 1978, 1981). The pheromone attracts adults and promotes recognition of sexually mature males. Both sexes aggregate and feed on faecal material deposited by their conspecifics. Oleic and linoleic acids are effective feeding aggregants and were mainly responsible for the activity of the faecal pellets. These acids are perceived by hairs on the maxil-

lary and labial palps but not by the olfactory sensilla of the antennae.

Adults have a female-produced sex pheromone that excites males (Abdel-Kader & Barak 1979) but not other females. This male response was positively correlated with increasing age and females elicited the response maximally at 6–8 days of age. Males also responded to male-produced pheromones that can be extracted in hexane; females responded to a limited extent to this extract (Shaaya 1981).

Population density affects development of *D. maculatus* larvae; both low and high densities prolong the larval period. Experiments have demonstrated different development rates of larvae when in the presence of adults or by themselves. It appears that *D. maculatus* liberates in the faeces two compounds which influence growth and development (Rakowski & Cymborowski 1982). One compound, produced by the larva, accelerates larval growth; the other, produced by adults, inhibits larval development. Both these pheromones help to synchronise larval ecdysis. More recent work has demonstrated the effect of pheromone on locomotion (Rakowski *et al.* 1989) and how illumination affects responses (Rakowski 1988).

Insect growth regulators

Insect growth regulators (IGRs) have found use recently in the control of insect pests in crop production. IGRs are substances which either disrupt the development process of the insect by affecting moulting (these are usually analogues of insect juvenile hormone), or disrupt the formation of the exoskeleton by interfering with chitin synthesis. The effects of IGRs are chronic. They do not cause mortality of parent adults, rather they cause malformations during larval development so that the first generation offspring adults either do not emerge or die quickly.

Very little work has been done with IGRs in relation to control of dermestid beetles. However, diflubenzuron (Dimilin: Dow-Elanco) inhibits the action of chitin synthetase, preventing cuticle formation. *Dermestes maculatus* larvae exposed for 24 h to woven polythene treated with 100 mg m^{-2} diflubenzuron were completely controlled and this activity persisted for at least 12 weeks (Webley & Airey 1982). Early instar larvae exposed to fishmeal treated with 1% diflubenzuron dust were all killed within 3–10 days at a dosage rate of 1 mg kg^{-1}. All development was halted at a dosage of 3 mg kg^{-1} and there appeared to be some ovicidal activity at 5 mg kg⁻¹. These effects persisted for at least 8 weeks, the length of the experiment.

In experiments in northern Kenya to compare new conventional insecticides as protectants of dried *Tilapia*, diflubenzuron was found to provide good control of *D. maculatus* and *N. rufipes* during the 6-month storage period (Golob *et al.* 1987).

The only IGR that has an MRL is methoprene, a juvenile hormone analogue. Its MRL is 5 mg kg⁻¹ on raw cereals. Shahjahan *et al.* (1991) exposed kraft paper packaging impregnated with methoprene, with or without an additional thin polythene layer, to *D. maculatus* which was infesting dried fish. Morphological deformities and growth abnormalities were observed but the insects were able to puncture the packaging within 2–3 months.

Other IGRs do not have MRLs simply because they have not been applied for; the market for storage protectants, especially dried fish, is too small to warrant most chemical manufacturers applying for registration for this purpose. Nevertheless, these chemical have relatively low mammalian toxicities. Diflubenzuron and methoprene have acceptable daily intakes of 0.02 mg kg⁻¹ and 0.1 mg kg⁻¹ body weight, respectively, and acute oral LD_{50} for rats of > 4600 and > 34 000 mg kg⁻¹ (c.f. pirimiphos-methyl 2050 mg kg⁻¹). If cost effective, they could provide useful alternatives for protecting dried fish that has to be stored for a month or longer.

Plants and plant extracts

Information concerning the use of plants as protectants of fish during curing and once dried has been reviewed by Ward and Golob (1994). Many of the plants identified are used to control blowfly infestations and so only those used specifically for beetle control will be mentioned here. It should be noted that much of the information is anecdotal and caution must be taken when assessing the conclusions.

Neem (Azadirachta indica; Meliaceae)*:* In the Philippines, neem leaves are spread along the foot of walls and in corners of fish stores to prevent beetle damage to dried round scad, mackerel, nemipterids (threadfin bream) and other dried fishery products, particularly during the dry season. The method is said to be satisfactory but tedious. In Senegal, processors may place jute bags containing neem leaves underneath stored fish (Ward & Golob 1994).

Okorie *et al.* (1990) found that powder made from dried neem seeds applied to dried tilapia at 2–8% by weight prevented oviposition of *D. maculatus*. A treatment of 2% also proved to be larvicidal. Neem did not appear to have strong ovicidal properties, but did prolong the incubation period. Mathen *et al.* (1992) found that neem oil showed some repellent activity against beetles.

However, some adverse features of using neem have been observed. Neem seed powder produces a bitterness in taste which was removed by boiling (Okorie *et al.* 1990) and neem oil was observed by Mathen *et al.* (1992) to have a nauseating, objectionable odour that was picked up by both the packaging and fish.

Citrus peel: Don Pedro (1985) investigated the effectiveness of powders of the dried peel of orange (*Citrus sinensis*; Rutaceae) and grapefruit (*Citrus paradisi*) using chips of dried catfish (*Clarias* sp) and found that orange peel had greater insecticidal and repellency effects than grapefruit peel. Treatment of fish with 14.1% by weight of orange peel powder killed 50% of adult *D. maculatus* after 7 days: a 21.3% treatment killed 99% in the same period. At applications of 15% and 18% by weight orange peel powder reduced progeny development and slowed larval development. At 18% the number of emerging larvae was reduced by 60% compared with the untreated control. Of the larvae that did emerge only 32.7% and 37.1% of the 18% and 15% treatments, respectively, developed into F1 adults compared with 87.8% of the larvae from the untreated controls.

Subsequent work demonstrated topical toxicity of citrus peel oils is relative unimportant when compared with their activity in the vapour phase (Don Pedro 1996a, b). The volatile components possessed activity against all *D. maculatus* life stages, eggs being most susceptible and last instar larvae and pupae being least susceptible, though the differences were slight. In the presence of dried fish pieces the activity was greatly reduced as a result of sorption of the volatile components. Although citrus peel and its oil do have a significant vapour effect, their use in practice is limited by the difficulty in obtaining the large quantities required and by the fact that dried fish are normally stored in permeable sacks so that vapour activity would be dissipated rapidly.

Oils: Vegetable oils, particularly groundnut and palm oils, are known to be effective in controlling some pests of stored pulses (Golob & Webley 1980).

Nigerian fish merchants rub groundnut and other vegetable oils on dried fish for protective or cosmetic reasons (Don Pedro 1989, 1990). In Senegal, traders in Dakar retail markets coat dried fish with vegetable oil to protect it from insect infestation (Wood 1983). However, results from an experiment in the Lake Turkana region of Kenya, in which bottled cod-liver oil was applied to dried tilapia (*Oreochromis* sp) at a treatment level of 44 ml kg^{-1}, showed that after 45 days' storage there was little difference in insect infestation between treated and untreated samples (Walker & Wood 1986).

The insecticidal efficacy of oils such as groundnut, traditional coconut, industrial coconut, palm and shark liver was investigated by Don Pedro (1989, 1990). Oils were tested against *D. maculatus* under laboratory conditions, using strips of dried trout fillet (*Salmo gairdneri*). The oils were not found to confer any antifeedant, stomach or contact toxicity effects; they only deterred feeding temporarily. Progeny development was significantly retarded, as a result of the increase in egg mortality, but only at very high application rates of 112 ml of oil per kilogram of dried fish, nearly three times that used by Walker and Wood (1986). However, there was no mortality effect on other life stages and the number of larvae which successfully developed through to adults was not significantly different from the control. All the oils were similarly ineffective. It is worth bearing in mind that a level of 112 ml kg^{-1} means the fish would be thoroughly doused in oil.

Mathen *et al.* (1992) investigated the protection of dried silver belly (*Leiognathus* sp) from insects using cashew nut shell liquid and oils of coconut, neem, palm, gingelly, mustard, sunflower, safflower, castor seed, rice bran, groundnut and hydnocarpus. Oils were sprayed on to packing material, the gunny bags and on to the fish themselves. The authors stated that mustard oil was observed to be the best insect repellent, the treated sample remaining insect-free for 40 days, followed by hydnocarpus, sunflower and cashew. Unfortunately, few data were given, particularly regarding application rates and the species of insect used in the experiment, and therefore it is not possible to validate the observations.

Peppers: Peppers (*Piper* sp) are used traditionally in Africa as a means of repelling blowflies but there is no evidence to suggest a use on dried fish. Pepper is used domestically as a preservative for prawns in Kerala, India. When cured prawns were sprinkled with 1.5% by weight of pepper powder and stored in a mite-infested godown they remained insect-free for as long as the pepper smell persisted, about 7 weeks. The high price of pepper is said to have prevented further work (Pillai 1957).

Boscia senegalensis: This is an evergreen shrub or small tree usually growing to about 2 m, found in the arid zones of Sahelian Africa. In Senegal, the leaves are layered between dried fish to protect against beetle pests. Migrant Malian fishermen, Bozos and Somonos, in remote areas of the Niger Delta, used a coarse powder of *B. senegalensis* leaves mixed with fruit of *Capsicum frutescens* (Solanaceae) to protect dried and smoked fish from *Dermestes* and *N. rufipes*. This practice was reportedly occurring up to the early 1970s, before the introduction of the Mopti Fisheries Development Programme. It was then neglected because it was apparently inefficient although there is no evidence to suggest the method was thoroughly investigated. There is no indication of dose levels or the mix ratio of the two ingredients.

Dennettia tripetala (Annonaceae)*:* Seed powder of this plant has shown greater repellency to *D. maculatus* than pyrethrum. In small-scale filter paper tests, acetone and ethanol extracts also exhibited good repellency potential, though aqueous extracts did not (Egwunyenga *et al.* 1998). There has been no research into the activity of this plant for the protection of dried fish.

Salt

Salt affects the water and electrolyte balance of insects, disrupting activity across neurones and Malpighian tubules. Theoretically, high salt concentrations should adversely affect insect development and research has been undertaken for many years to determine whether increasing the salt content of cured fish can prevent infestation by beetles. Although the evidence is somewhat contradictory it appears that muscle tissue levels above 10% will eliminate problems due to *D. maculatus* but to control other species higher salt concentrations are required. The experimental evidence is described below.

Moderate salting hindered egg production of *D. maculatus* and *D. ater* in Indonesia and also increased larval mortality, increased development time by increasing the number of moults and shortened adult life. Strongly salted fish were almost uninfested but unsalted ones were seriously attacked (Kalshoven, cited by Blatchford 1962). However, beetles may become adapted to salt; in Aden it was thought that as sardines dried out they were susceptible to cross-infestation from other salted fish, which had become infested after some time (Blatchford 1962).

Salting of *Roccus chrysops*, a North American freshwater fish, to 13% or more prevented development and survival of *D. maculatus* larvae (Mushi & Chiang 1974). In Bangladesh, complete inhibition of development of this species occurred when the salt content was 20% (Rezaur *et al.* 1983).

Fillets of air-dried whiting, *Merlangius merlangus*, with a fat content of less than 0.5% and natural salt content of less than 2.2%, were dipped in brine and the effects of salt content on insect numbers and weight loss were recorded over a period of approximately 10 weeks. Infestation by *D. maculatus* and *D. frischii* under controlled laboratory conditions (27°C and 70% r.h.) resulted in weight losses in unsalted fish ranging from 15% to 41%. Increases in salt content up to 5% did not give any appreciable protection against infestation and there was an indication that brining to these levels may actually result in an increase in infestation. Salt content above 9% rarely protected fish from damage, but did reduce losses to less than 10% and significantly inhibited insect development (Wood *et al.* 1987). Salting was equally effective in preventing damage by either insect. However, the results are contrary to observations by Amos (1968) who found *D. frischii* were capable of development on fishmeal to which 25% salt had been added, and of Pablo (1978), who found *D. carnivorus* were able to tolerate salt contents of 12–17% in dried striped mackerel; these differences may have been due to different insect biotype responses. Feeding on unsalted anchovies in Indonesia, *D. maculatus* was found to be more fit than *D. carnivorus* but these differences were eliminated when the diet was changed to brined anchovy (Madden *et al.* 1995).

Earlier work, at a higher temperature (30°C) and relative humidity (75–80%) demonstrated similar effects (Osuji 1975c). Minced, dried *utaka* and *chisawasawa* from Lake Malawi, containing common salt concentrations of 9.2% and 10.2%, caused complete mortality of *D. maculatus* larvae. Salt concentrations at these levels, together with moisture content reduction to about 6%, effectively inhibited insect development, despite the relatively high lipid content (about 30%) of the fishmeal. Even at the lowest salt concentrations (just above 3%), larval development was prolonged, larval and adult body weight decreased and egg viability and fecundity were reduced.

Dermestes maculatus larval mortality increased with increasing salt concentration in dried hilsa (Ahmed *et al.* 1988) kept under ambient temperature (35°C) in Bangladesh. Complete mortality was achieved at 8.2% and 10.2% concentrations. At all salt levels mortality was greater than on untreated samples; at 1.3% salt content 47% of the larvae died within 3 days, though after 40 days mortality was less than 50%; in controls by 40 days mortality was less than 10%. However, dried fish containing less than 1.5% salt became rancid after 1 month but other samples with higher salt contents remained in good condition.

Although salt application will provide good control of beetles if the concentration is sufficient, heavily salted fish is not acceptable to some communities, especially in Africa away from coasts, and so salt content may have to be limited to 10%. At this level, some beetle infestation is inevitable (Proctor 1972).

Screening and barriers

If appropriate containers or packaging is used for the transportation and storage of cured fish then these alone would be sufficient to prevent insect infestation. Dried fish have been stored for up to one year in sealed polythene bags without serious loss of quality (Keshvani, cited by Proctor 1977) and other workers in southern Africa have recommended this method though the fish must be cooled after drying and the bags stored in a cool place to avoid sweating. Furthermore, sharp projections of dried fish can easily puncture polythene bags.

Impregnating packaging with insecticide could enhance protection. A synergised pyrethrins preparation, applied in a clay coating at approximately 50 mg m^{-2} and 500 mg m^{-2}, respectively protected flour for 9–12 months when applied to the outer layer of multi-walled paper sacks (Blatchford 1962).

In Indonesia it was found that screening anchovy in a market could reduce beetle infestation (Madden *et al.* 1995).

A traditional method of storing dried fish in Bangladesh, to prevent insect attack and maintain the consistency of the fish, is to keep the product in large earthenware pots into which fish oil was previously allowed to soak to prevent moisture uptake. The pot with fish is then buried underground up to its neck. Alternatively, the fish may be stored in trenches dug into the ground and lined with mats; the fish are covered with sand and earth (Anon. 1957).

Biological control

Classical biological control, the introduction of predators and parasites, is used extensively for the control of insect pests of growing crops. This is generally not the case for stored crops and neither is it practised for stored fish. There are very few specific predators of *Dermestes*, though parasites include the prostigmatan mite *Pyemotes herfsi* and the eugregarine amoebae *Pyxinia crystalligera* and *P. rubecula* on *D. maculatus* and *D. ater*. The beetle *Korynetes caeruleus* is a predator of *D. ater*, and the mite *Lardoglyphus konoi* is a competitor of this dermestid on dried fish (Haines 1991). The pyemotid mite belonging to the genus *Pigmophorus* attacks eggs and adults of *Dermestes* but whether it exerts any control is unknown (Tanyongana 1983).

Future developments

Post-harvest treatment of fish in developed countries is well advanced. However, most technologies are used in defined fishing situations and markets that are not easily transferable for use in many developing countries, particularly those in the tropics.

Poor infrastructure is a major constraint to the post-harvest quality of fish in developing countries. Thus facilities must be sufficient at all stages of post-harvest handling, during landing, chilling, storage, processing, distribution and sale to end-users. It is recognised in developing countries that there is a need for better quality products for local and regional markets and to increase exports to developed countries that have greater purchasing power and can pay in hard currency. However, improvements to this infrastructure do not necessarily demand large-scale investment. Appropriate developments would include providing better unloading and processing sites, construction of drying platforms on hardstanding, installation of latrines and running water, removal of waste material, access to micro-credit and training of producers. They can all make major contributions to reducing post-harvest losses, reducing the risks of food poisoning and improving incomes of small traders, particularly women.

Other initiatives could take advantage of materials and techniques for smoking that provide good-quality products with less use of firewood, or brining with lower levels of salt and faster fermentation of traditional fish sauces. The costs of preservation by heat treatment may be reduced by new flexible containers for 'canning' fish products. Research on other types of packing, the use of drying tents and solar furnaces should also be continued. Also necessary is evaluation of the effectiveness of indigenous technical knowledge and traditional methods used against infestation.

However, these processing improvements will need in-country research and development sustained by training programmes and good extension networks. If the size of the fish trade increases then it may become economically worthwhile to construct fumigation facilities at the principal sites.

The biology of the main beetle pests of stored fish is well known, but quantitative and qualitative information on losses is sparse and confined to situations where the storage period of the dried product is relatively short. Losses have not been assessed in fish stocks kept at schools, hospitals, prisons and other institutions where storage periods may be as long as 6 months. Nor are the effects of beetle attack on the nutritional composition of fish well understood. Opinions on value losses from producers and consumers are also lacking. Participatory rural appraisal methods could be valuable in teasing out attitudes to loss in post-harvest fish.

Maximum residue levels for pesticides on dried fish do not exist because manufacturers have not applied to establish limits. The dried fish market is so small and the cost of setting an MRL so expensive that the pest control companies do not believe the expense is warranted. This will probably remain their position. However, it is likely that compounds approved for agricultural products could be used safely for dried fish, as long as they were shown to be biologically active. Similarly, IGRs do not have MRLs simply because they have not been applied for. Nevertheless, these chemicals have relatively low mammalian toxicities. If cost effective, they could provide useful alternatives for protecting dried fish that has to be stored for a month or more.

Dipping fish in insecticide emulsions has shown some promise, but there is only limited evidence that such practices will protect fish against subsequent beetle attack. When determining the rate of utilisation of dipping emulsions, researchers must take account of the fish type, size and quantity to be treated, duration of dip and the concentration of insecticide emulsion. Apart from work by Meynell (1977, 1978) this information is lacking. Furthermore, the issue of treating dried fish away from the site of curing has not been addressed at all; apart from experiments to test efficacy of active ingredients there has been no attempt to develop practical methods of treating dried fish with contact insecticides.

Inert dusts may have potential for practical control of beetles but only for treatment of certain types of non-oily, very dry fish. Because inert dusts are so susceptible to changes in r.h. and moisture content of the commodity being treated, they lose effectiveness as moisture increases. More experimentation is required to test their efficacy. Pheromones may be useful in some situations. However, current knowledge, recently reviewed by Levinson and Levinson (1995), is not sufficient to enable pheromones to be used in pest control strategies for *Dermestes* at present. Nevertheless, this is a method that would be environmentally sustainable and should be pursued.

Thus there is much that could be done to raise fish quality standards for local consumption and for export. This may well mean that the future production of preserved fish in tropical countries will move away from being a low-technology, artisanal activity towards an industrial or semi-industrial enterprise undertaken in well-designed facilities with good management practices.

References

Abdel-Kader, M.M. & Barak, A.V. (1979) Evidence for a sex pheromone in the hide beetle, *D. maculatus*. *Journal of Chemical Ecology*, **5**, 805–810.

Ahmed, M., Bhuiya, A.D., Alam, A.M.S., *et al.* (1978) Radiation disinfestation studies on sun dried fish. 310–321. *Proceedings of the Indo-Pacific Fishery Commission.* 8–17 March 1978, Manila, Philippines. FAO Regional Office, Bangkok, Thailand.

Ahmed, M., Karim, A., Alam, Z., *et al.* (1988) Effect of salt concentration on larval mortality of *D. maculatus* in dried *Hilsa ilisha*. *Bangladesh Journal of Agriculture*, **13**, 123–126.

Amos, T.G. (1968) Some laboratory observations on the rate of development, mortality and oviposition of *Dermestes frischii* Kug. (Col., Dermestidae). *Journal of Stored Products Research*, **4**, 103–117.

Amos, T.G. & Morley, G.E. (1971) Longevity of *Dermestes frischii* (Kug.) (Coleoptera: Dermestidae). *Entomologist's Monthly Magazine*, **107**, 79–80.

Anon. (1957) A study of improved methods for the production and storage of dried fish, with particular reference to methods suitable for wet, humid seasons of the year. In: *Proceedings of the 6th Session.* 30 September–14 October 1957, Tokyo, Japan. Indo-Pacific Fisheries Council. FAO Regional Office, Bangkok, Thailand.

Anon. (1966) A practical guide to the control of pests of fish products. *Journal of Stored Products Research,* **2**, 174–176.

Anon. (1997) Pêches et Transports Maritimes: Au Sommet de la Vague. *Le Soleil*, Numéro spécial – Hors série, Juillet 1997.

Anon. (2000a) *Les Atlas de l'Afrique: Sénégal.* 5ème édition. Jeune Afrique, Paris, France.

Anon. (2000b) Résultats généraux de la pêche maritime sénégalaise. *Rapport annuel de la Direction de l'Océanographie et des Pêches Maritimes 1999.* Ministère de la Pêche et des Transports Maritimes, République du Sénégal.

Aref, M., Timberly, A. & Daget, J. (1964) Fish and fish processing in the Republic of Mali. 3. On the destruction of dried fish by dermestid insects. *Alexandria Journal of Agriculture*, **12**, 95–108.

Ashman, F. (1962) Factors affecting the abundance of the copra beetle, *Necrobia rufipes* (Deg.) (Col., Cleridae). *Bulletin of Entomological Research*, **53**, 671–680.

Awoyemi, M.D. (1991) Preliminary observations on the incidence of insect pests and related damage to stored dried fish in Kainji Lake area. *Insect Science and its Application*, **12**, 361–365.

Azab, A.K., Tawfik, M.F.S. & Abouzeid, N.A. (1972a) The biology of *Dermestes maculatus* DeGeer. *Bulletin de la Société Entomologique d'Egypt*, **56**, 1–14.

Azab, A.K., Tawfik, M.F.S. & Abouzeid, N.A. (1972b) Factors affecting development and adult longevity of *Dermestes maculatus* DeGeer (Coleoptera: Dermestidae). *Bulletin de la Société Entomologique d'Egypt*, **56**, 21–32.

Balogun, R.A. & Ofuya, G.E. (1986) Effects of three biologically active dietary substances on the development and reproduction of the leather beetle, *Dermestes maculatus* DeGeer. *Nigerian Journal of Entomology*, **7**, 28–41.

Barwal, R.N. & Devi, J. (1993) Pests of stored food commodities and their management with particular reference to hide and skin beetle, *Dermestes maculatus* DeGeer in Manipur. *Journal of Insect Science*, **6**, 189–194.

Begum, A., Seal, D.R. & Khan, A.T. (1979) Studies on the innate capacity of natural increase of *Dermestes maculatus* DeGeer at different temperatures. *Dacca University Studies*, **28**, 83–89.

Bergmann, E.D. & Levinson, H.Z. (1966) Utilisation of steroid derivatives by larvae of *Musca vicina* (Macq.) and *Dermestes maculatus* (Deg). *Journal of Insect Physiology*, **12**, 33–81.

Bhuiyan, A.D. (1990) *Packaging and storage studies of irradiated dried fish for commercial application.* Final Report of the Joint Research Project of the Bangladesh Atomic Energy Commission and Bangladesh Agricultural Research Council.

Binns, T.J. & Tyler, P.S. (1978) Lindane resistance in *Dermestes maculatus* Deg. (Coleoptera: Dermestidae). *Journal of Stored Products Research*, **14**, 19–23.

Blanc, A. (1955) La pêche artisanale: production,utilisation de la production, poisson frais, poissons conservés, qualités et défectuosités de la production. *Conférence Economique de la Pêche Maritime.* 12–14 Avril 1955, Paris. Centre National d'Information Economique, Paris, France.

Blatchford, S.M. (1962) Insect infestation problems with dried fish. *Tropical Stored Products Information*, **4**, 112–128.

Boisot, M.H. & Gauzit, M. (1966) Disinsectization of African dried and smoked fish by means of irradiation. *Application of Food Irradiation in Developing Countries. International Atomic Energy Agency Report*, **54**, 85–94.

Bouare, F. (1986) Prevention of losses of smoked fish (test on the use of tetrachlorvinphos) in Niger. 396–400. In: *Fish Processing in Africa.* Proceedings of the FAO Expert Consultation on Fish Technology in Africa. 21–25 January 1985, Lusaka, Zambia. FAO Fisheries Report **329**, Supplement. Food and Agriculture Organization of the United Nations, Rome, Italy.

Cachan, P. (1957) Les insectes du poisson seche et fume au Soudan. Report de Mission, Fevrier 1957. Unpublished Report. ORSTOM. Institute Français de Recherche Scientifique pour le Développement en Coopération, Paris, France.

Clark, A.J. & Bloch, K. (1959) Functions of sterols in *D. vulpinus. Journal of Biological Chemistry*, **234**, 2583–2588.

Daget, J. (1966) Insect infestation of African dried or smoked fish and the possibility of its control by irradiation. *Application of Food Irradiation in Developing Countries. International Atomic Energy Agency Report*, **54**, 78–83.

Dick, J. (1937) Oviposition in certain Coleoptera. *Annals of Applied Biology*, **24**, 762–796.

Diouf, N. (1987) Techniques artisanales de traitement et conservation du poisson au Sénégal, au Ghana, au Bénin et au Cameroun. COPACE/Tech/87/84. Comité des pêches pour l'Atlantique Centre Est. Food and Agriculture Organization of the United Nations, Rome, Italy.

Don Pedro, K.N. (1985) Toxicity of some citrus peels to *Dermestes maculatus* Degeer and *Callosobruchus maculatus* (F.). *Journal of Stored Products Research*, **21**, 31–34.

Don Pedro, K.N. (1989) Insecticidal activity of some vegetable oils against *Dermestes maculatus* Degeer (Coleoptera: Dermestidae) on dried fish. *Journal of Stored Products Research*, **25**, 81–86.

Don Pedro, K.N. (1990) The effect of mixed vegetable and animal oils on the feeding and development of *Dermestes maculatus* Degeer (Coleoptera: Dermestidae) larvae on dried fish. *Journal of African Zoology*, **104**, 23–28.

Don Pedro, K.N. (1996a) Fumigant toxicity is the major route of insecticidal activity of citrus peel essential oils. *Pesticide Science*, **46**, 71–78.

Don Pedro, K.N. (1996b) Fumigant toxicity of citrus peel oils against adult and immature stages of storage insect pests. *Pesticide Science*, **47**, 213–223.

Duguet, J.S., Brou Belou, C., Tamboura, R. *et al.* (1985) Evaluation of the efficacy of deltamethrin compared with pirimiphos-methyl and tetrachlorvinphos against *Dermestes maculatus* Deg. and *Necrobia rufipes* Deg. on *Sarotherodon niloticus* at Mopti (Mali). *International Pest Control*, **27**, 92–99.

Durand, M.H. (1981) Aspects sociaux-économiques de la transformation artisanale, du poisson de mer au Sénégal. Centre de Recherches Océanographiques de Dakar-Thiaroye Archive **103**, Dakar, Senegal.

Egwunyenga, O.A., Alo, E.B. & Nmorsi, O.P.G. (1998) Laboratory evaluation of the repellency of *Dennettia tripetala* Baker (Annonaceae) to *Dermestes maculates* (F.) (Coleoptera: Dermestidae). *Journal of Stored Products Research*, **34**, 195–199.

Esser, J.R., Hanson, S.W., Taylor, K.D.A., *et al.* (1986) Evaluation of insecticides to protect salted-dried marine catfish from insect infestation during processing and storage. Report **R3985**. Overseas Development Administration, London, UK.

Esser, J.R., Hanson, S.W., Wirayante, J., *et al.* (1990) Prevention of insect infestation and losses of salted-dried fish in Indonesia by treatment with an insecticide approved for use on fish. 168–179. *Seventh Session of the Indo-Pacific Fishery Commission Working Party on Fish Technology and Marketing.* 19–22 April 1988, Bangkok, Thailand. FAO Fisheries Report **401** Supplement. FAO Regional Office, Bangkok, Thailand.

FAO (1981) The prevention of losses in cured fish. FAO Fisheries Technical Paper **219**. Food and Agriculture Organization of the United Nations, Rome, Italy.

Fraenkel, G. & Blewett, M. (1944) Utilisation of metabolic water in insects. *Bulletin of Entomological Research,* **35**, 127–139.

Friendship, R. (1990) The fumigation of dried fish. *Tropical Science*, **30**, 185–193.

Galichet, P.F. (1960) La protection du poisson sec contre les dermeste, *Dermestes maculatus* DeGeer dans la bassin Tchadien. *Comptes Rendu Acadamie Agricole Francaise*, **46**, 404–410.

Gay, F.J. (1938) Nutritional study of the larva of *D. vulpinus*. *Journal of Experimental Zoology*, **79**, 93–107.

Gjerstad, D. (1989) Prevention of infestation by *Dermestes maculatus* Degeer in East African dried fish using pyrethrum and piperonyl butoxide. *Fishery Technology*, **26**, 25–29.

Golob, P. & Webley, D.J. (1980) The use of plants and minerals as traditional protectants of stored food products. Report **G138**. Tropical Products Institute, Slough, UK.

Golob, P., Cox, J.R. & Kilminster, K. (1987) Evaluation of insecticide dips as protectants of stored dried fish from dermestid beetle infestation. *Journal of Stored Products Research*, **23**, 47–56.

Golob, P., Guèye-NDiaye, A. & Johnson, S. (1995) Evaluation of some pyrethroid and organophosphate insecticides as protectants of stored dried fish. *Tropical Science*, **35**, 76–92.

Green, A.A. (1967) The protection of dried sea-fish in south Arabia from infestation by *Dermestes frischii* Kug. (Coleoptera, Dermestidae). *Journal of Stored Products Research*, **2**, 331–350.

Guèye-NDiaye, A. (1991) *Dermestes maculatus* (Deger) – Coleoptera-Dermestidae – principal déprédateur des produits halieutiques transformés au Sénégal. 71–78. In: F. Fleurat-Lessard and P. Ducom (eds) *Proceedings of the 5th International Working Conference on Stored Product Protection.* INRA, Bordeaux, France.

Guèye-NDiaye, A. & Fain, A. (1987) Note sur les Acariens des denrées alimentaires au Sénégal. *Revue de Zoologie Africaine,* **101**, 365–370.

Guèye-NDiaye, A. & Gningue, R. (1995) Le poisson transformé au Sénégal: techniques de production, rendement et état d'infestation par les insectes déprédateurs. *Bulletin de l'IFAN*, Cheikh Anta Diop, Dakar, sér. A, **48**, 107–115.

Guèye-NDiaye, A. & Marchand, B. (1989) *Lardoglyphus konoi* et *Suidasia pontifica*, déprédateurs des sardinelles braisées séchées au Sénégal – Etude en microscopie électronique à balayage. *Acarologia*, **30**, 2, 131–137.

Guèye-NDiaye, A., Golob, P. & Johnson, S. (1996) Utilisation de l'actellic contre les insectes du poisson séché en Afrique de l'Ouest. *Dossier Technico Economique Bonga* **023–96**.

Guillon, M. (1976) Controlling insect pests of dried fish, an experiment in Mali. *Span*, **19**, 127–128, 134, 137, 139.

Haines, C.P. (ed.) (1991) *Insects and Arachnids of Stored Products. Their Biology and Identification.* Natural Resources Institute, Chatham, UK.

Haines, C.P. & Rees, D.P. (1989) A Field Guide to the Types of Insects and Mites Infesting Cured Fish. Fisheries Technical Paper **303**. Food and Agriculture Organization of the United Nations, Rome, Italy.

Harris, A.H. & Halliday, D. (1968) Residues of fumigants in dried fish. Unpublished report. Tropical Products Institute, London, UK.

Hodges, R.J. (2002) Pests of durable crops – insects and arachnids. In: P. Golob, G. Farrell & J.E. Orchard (eds) *Crop Post-Harvest: Science and Technology. Volume 1 Principles and Practice.* Blackwell Publishing, Oxford, UK.

Indriati, N., Sudrajat, Y., Anggawati, A.M., *et al.* (1986) Insect infestation in dried-salted fish in Java. *Journal of Post-Harvest Fisheries Research,* **50,** 9–12.

Jacob, T.A. & Fleming, D.A. (1985) The effect of constant temperature and humidity on the egg period of *Dermestes maculatus* DeGeer and the influence of free water upon developmental period. *Entomologist's Monthly Magazine,* **121,** 19–24.

Johnson, C. & Esser, J. (2000) A review of insect infestation of traditionally cured fish in the tropics. Department for International Development, London, UK.

Kane, J. (1967) Silica-based dusts for the control of insects infesting dried sea-fish. *Journal of Stored Products Research,* **2,** 251–255.

Kaplanis, J.N., Robbins, W.E., Monroe, R.C., *et al.* (1965) The utilisation of B-sitosterol in larvae of the housefly, *Musca domestica* L. *Journal of Insect Physiology,* **11,** 251–258.

Kébé, M. (1994) Principales mutations de la pêche sénégalaise. In: M. Barry-Gérard, T. Diouf & A. Fonteneau (eds) *L'évaluation des Resources Exploitables par la Pêche Artisanale Sénégalaise.* 8–13 Février 1993, Dakar, Senegal. Colloques et Séminaires, Tome II. ORSTOM, Institute Français de Recherche Scientifique pour le Développement en Coopération, Paris, France.

Khatoon, N. & Heather, N.W. (1990) Susceptibility of *Dermestes maculatus* DeGeer (Coleoptera, Dermestidae) to gamma irradiation in a nitrogen atmosphere. *Journal of Stored Products Research,* **26,** 227–232.

Lepesme, P. (1939) The economic importance of some Corynetidae. *Revue Francaise d'Entomologie,* **6,** 17–20.

Levinson, A.R. & Levinson, H.Z. (1995) Reflections on structure and function of pheromone glands in storage insect species. *Anzeiger für Schädlingskunde Pflanzenschutz Umweltschutz,* **67,** 99–118.

Levinson, H.Z., Barelkovsky, J. & Bar Ilan, A.R. (1967) Nutritional effects of vitamin omission and antivitamin administration on development and longevity of the hide beetle *D. maculatus* Deg. *Journal of Stored Products Research,* **3,** 345–352.

Levinson, H.Z., Levinson, A.R., Jen, T-L., *et al.* (1978) Production site, partial composition and olfactory perception of a pheromone in the male hide beetle. *Naturwissernschaften,* **10,** 543–545.

Levinson, A.R., Levinson, H.Z. & Franke, D. (1981) Intraspecific attractants of the hide beetle, *Dermestes maculatus* Deg. *Mitteilungen der Deutschen Gesellschaft fur Allemeine,* **2,** 235–237.

Madden, J.L., Anggawati, A.M. & Indriati, N. (1995) Impact of insects on the quality and quantity of fish and fish products in Indonesia. 97–106. In: B.R. Champ & and E. Highley (eds) *Fish Drying in Indonesia.* Proceedings of an International Workshop. 9–10 February 1994, Jakarta, Indonesia.

Mallamaire, A. (1955) La desinsectisation du poisson seche en Afrique Occidentale Francaise. Unpublished report of the Bureaux de la Protection des Vegetaux, Dakar, Senegal.

Mallamaire, A. (1957) Les insectes nuisibles au poisson séché en Afrique. Moyens de les combattre. *Bulletin de la Protection des Végétaux,* A.O.F., **57,** 88–99.

Mathen, C., Unnikrishnan Nair, T.S. & Ravidranathan Nair, P. (1992) Effect of some vegetable oils on insect infestation during storage of dry cured fish. *Fishery Technology,* **29,** 48–52.

McClellan, R.H. (1964) A pyrethrum-dipping treatment to protect dried fish from beetle infestation. *Pyrethrum Post,* **7,** 30–40.

Meynell, P.J. (1977) Report on the preservation of dried fish. Unpublished report of the Ninth Meeting of the Stored Products Working Party Committee, Malawi.

Meynell, P.J. (1978) Reducing blowfly spoilage of sun drying fish in Malawi using pyrethrum. *Indo-Pacific Fishery Commission Symposium on Fish Utilization Technology and Marketing in the IPFC Region.* 8 March 1978, Manila, Philippines. FAO Regional Office, Bangkok, Thailand.

Mushi, A.M. & Chiang, H.C. (1974) Laboratory observations on the effect of common salt on *Dermestes maculatus* infesting dried freshwater fish, *Roccus chrysops. Journal of Stored Products Research,* **10,** 57–60.

Nakayama, T.O.M, Allen, J.M., Cummins, S., *et al.* (1983) Disinfestation of dried foods by focussed solar energy. *Journal of Food Processing and Preservation,* **1,** 1–8.

NDiaye, J.L. (1997) Une activité dynamique au sein d'un système complexe: rôle et place de la transformation artisanale dans le systeme pêche maritime au Sénégal. Etude de géographie économique. Thèse de Doctorat Université Paul Valery, Montpellier, France.

Noland, J.L. (1954) Sterol metabolism in insects. II. Inhibition of cholesterol utilisation by structural analogues. *Archives of Biochemistry,* **50,** 323–330.

Okorie, T.G., Siyanbola, O.O. & Ebochuo, V.O. (1990) Neem seed powder as a protectant for dried tilapia fish against *Dermestes maculatus* Degeer infestations. *Insect Science and its Application,* **11,** 153–157.

Osuji, F.N.C. (1974a) Moisture content of dried fish in relation to density of infestation by *D. maculatus* (Coleoptera, Dermestidae) and *N. rufipes* (Coleoptera, Cleridae). *West African Journal of Biological and Applied Chemistry,* **17,** 3–8.

Osuji, F.N.C. (1974b) Total lipid content of dried fish in relation to infestation by *D. maculatus* (Coleoptera, Dermestidae) and *N. rufipes* (Coleoptera, Cleridae). *Journal of the West African Science Association,* **19,** 131–136.

Osuji, F.N.C. (1974c) Comparative studies on the susceptibilities of different genera of dried fish to infestation by *Dermestes maculatus* and *Necrobia rufipes. Nigerian Journal of Entomology,* **1,** 63–68.

Osuji, F.N.C. (1974d) Beetle infestation in dried fish purchased from a Nigerian market, with special reference to *Dermestes maculatus* DeGeer. *Nigerian Journal of Entomology,* **1,** 69–79.

Osuji, F.N.C. (1975a) Recent studies on the infestation of dried fish in Nigeria by *Dermestes maculatus* and *Necrobia rufipes* with special reference to the Lake Chad district. *Tropical Stored Products Information,* **29,** 21–32.

Osuji, F.N.C. (1975b) Some aspects of the biology of *Dermestes maculatus* Deg. in dried fish. *Journal of Stored Products Research,* **24,** 167–179.

Osuji, F.N.C. (1975c) The effects of salt treatment of fish on the developmental biology of *D. maculatus* (Coleoptera, Dermestidae) and *Necrobia rufipes* (Cleridae). *Entomologia Experimentalis et Applicata,* **18,** 472–479.

Osuji, F.N.C. (1978) An assessment of the performance of *Dermestes maculatus* DeGeer (Coleoptera, Dermestidae) in some dietary media. *Entomologia Experimentalis et Applicata*, **24**, 185–192.

Pablo, I. (1978) Application of gamma radiation for disinfestation of sun dried striped mackerel (*Rastrelliger chrysozonus*). *International Congress of Food Science and Technology Abstracts*, **140**.

Paul, C.F., Shukla, G.N., Das, S.R., *et al*. (1963) A life history study of the hide beetle *Dermestes vulpinus* Fab. *Indian Journal of Entomology*, **24**, 167–179.

Pillai, P.R.P. (1957) Pests of stored fish and prawns. *Bulletin of the Central Research Institute, University of Kerala*, **V (III)**.

Pointel, J.G. & Phan van Sam (1969) Effets des radiations ionisantes sur le development de *Dermestes maculatus* DeGeer. *Journal of Stored Products Research*, **5**, 95–109.

Proctor, D.L. (1972) The protection of smoke-dried freshwater fish from insect damage during storage in Zambia. *Journal of Stored Products Research*, **8**, 139–149.

Proctor, D.L. (1977) The control of insect infestation of fish during processing and storage in the tropics. 301–311. In: *Proceedings of the Conference on the Handling, Processing and Marketing of Tropical Fish*. 5–9 July 1976, London. Tropical Products Institute, Slough, UK.

Rakowski, G. (1988) Effect of illumination intensity on the response of the hide beetle, *Dermestes maculatus*, to aggregation pheromone. *Journal of Insect Physiology*, **34**, 1101–1104.

Rakowski, G. & Cymborowski, B. (1982) Aggregation pheromone in *Dermestes maculatus*: effects on larval growth and developmental rhythms. *International Journal of Invertebrate Reproduction*, **4**, 249–254.

Rakowski, G., Sorenson, K.A. & Bell, W.J. (1989) Responses of dermestid beetles, *Dermestes maculatus,* to puffs of aggregation pheromone extract (Coleoptera: Dermestidae). *Entomologia Generalis*, **14**, 211–215.

Rattagool, P., Methatip, P., Esser, J.R., *et al*. (1990) Evaluation of insecticide to protect salted-dried, marine fish from insect infestation during processing and storage in Thailand. 189–204. *Seventh Session of the Indo-Pacific Fishery Commission Working Party on Fish Technology and Marketing*. 19–22 April 1988, Bangkok, Thailand. FAO Fisheries Report **401** Supplement. FAO Regional Office, Bangkok, Thailand.

Rezaur, R., Bhuiya, A.D., Islam, M., *et al*. (1983) Use of common salt as a controlling agent of dried fish pest in store houses. *Proceedings of the 8th Bangladesh Science Conference*. 5–9 February 1983, Dhaka, Bangladesh. Bangladesh Association for the Advancement of Science, Chittagong, Bangladesh.

Rollings, M.J. & Hayward, L.A.W. (1962) Aspects of the dried fish trade in Nigeria with particular reference to Lake Chad. *Report of the West African Stored Products Research Unit, 1961*, 115–120.

Saidul, I. & Rezaur, R. (1996) Gamma irradiation induced histopathological changes in larval midgut of the hide beetle *Dermestes maculates* (Coleoptera). *Polskie Pismo Entomologiczne*, **65**, 69–71.

Sembène, M. (1994) Effet de la teneur en eau et de la teneur en sel du poisson fermenté-séché sur son degré d'infestation par *Dermestes* spp.(Coleoptera-Dermestidae). Mémoire de Diplôme d'Etudes Approfondies de Biologie Animale, Université Cheikh Anta Diop, Dakar, Senegal.

Seydi, M. (1991) L'interprétation des résultats d'analyses microbiologiques et chimiques des produits marins transformés. In: *Amélioration des Techniques de Pêche Artisanale au Sénégal*. PRO-PECHE/ATEPAS. Amelioration des Techniques de Peche Artisanale au Sénégal. Food and Agriculture Organization of the United Nations, Rome, Italy.

Shaaya, E. (1981) Sex pheromone of *Dermestes maculatus* DeGeer (Coleoptera: Dermestidae). *Journal of Stored Products Research*, **17**, 13–16.

Shahjahan, N.R., Bhuiyuan, A.D. & Rahman, R. (1991) Effect of methoprene coated papers as an insect resistant packaging material on three stored product pests. *Nuclear Science and its Applications (Bangladesh)*, **3**, 23–26.

Shahjahan, R.M., Shaha, A.K. & Bhuiya, A.D. (1996) Radiation disinfestation and packaging studies of three dried fish. *Bangladesh Journal of Zoology*, **24**, 39–44.

Stathers, T.E. (2002) Pest management – Inert dusts. In: P. Golob, G. Farrell & J.E. Orchard (eds) *Crop Post-Harvest: Science and Technology. Volume 1 Principles and Practice*. Blackwell Publishing, Oxford, UK.

Tanyongana, R. (1983) Hide beetles. *Zimbabwe Agricultural Journal*, **80**, 35–36.

Taylor, R.W.D. (1981) Report on a visit to Nigeria to investigate insect infestation of dried fish from Lake Chad (October–December 1980). *Report of the Project: Improvement of Fish Processing and Transport on Lake Chad. NIR/74/001*. Food and Agriculture Organization of the United Nations, Rome, Italy.

Taylor, R.W.D. (2002) Fumigation. In: P. Golob, G. Farrell & J.E. Orchard (eds) *Crop Post-Harvest: Science and Technology. Volume 1 Principles and Practice*. Blackwell Publishing, Oxford, UK.

Taylor, R.W.D. & Evans, N.J. (1982) Laboratory evaluation of four insecticides for controlling *Dermestes maculatus* Degeer on smoke dried fish. *International Pest Control*, **24**, 42–45.

Taylor, T.A. (1964) Observations on the biology and habits of *Dermestes maculatus* DeGeer – a dried fish pest in Nigeria. *Nigerian Agricultural Journal*, **1**, 12–16.

Toyes, S.A. (1970) Studies on the humidity and temperature reaction of *Dermestes maculatus* DeGeer (Coleoptera: Dermestidae) with reference to infestation in dried fish in Nigeria. *Bulletin of Entomological Research*, **60**, 23–31.

Tyler, P.S. & Binns, T.J. (1977) The toxicity of seven organophosphorus insecticides and lindane to eighteen species of stored-product beetles. *Journal of Stored Products Research*, **13**, 39–43.

Walker, D.J. (1984) Report on two visits to Malawi to evaluate the use of the insecticide pirimiphos-methyl on fresh and cured fish. Unpublished report **R1221(A)**. Tropical Development and Research Institute, Slough, UK.

Walker, D.J. (1987) A review of the use of contact insecticides to control post-harvest insect infestation of fish and fish products. *FAO Fisheries Circular* **804**.

Walker, D.J. (1988) Control of insect infestation in fishery products in LDCs. 133–146. In: M.T. Morrissey (ed.) *Post-Harvest Fishery Losses*. Proceedings of an International

Workshop Held at the University of Rhode Island. 12–16 April 1987, Rhode Island, USA.

Walker, D.J. & Greeley, M. (1991) Cured fish in Bangladesh. Report on a visit to Bangladesh, November 1990, on behalf of ODA Post-Harvest Fisheries Project, Bay of Bengal Programme, Madras, India. Unpublished report **R1657**. Overseas Development Natural Resources Institute, Slough, UK.

Walker, D.J. & Wood, C.D. (1986) Non-insecticidal methods of reducing losses due to insect infestation of cured fish with beetle pests (Coleoptera). 380–389. In: *Fish Processing in Africa. Proceedings of the FAO Expert Consultation on Fish Technology in Africa,* 21–25 January 1985, Lusaka, Zambia. FAO Fisheries Report **329**, Supplement. Food and Agriculture Organization of the United Nations, Rome, Italy.

Ward, A. (1996) Quantitative data on post-harvest fish losses in Tanzania. The fisheries of Lake Victoria and Mafia Island. Unpublished report. Natural Resources Institute, Chatham, UK.

Ward, A. & Golob, P. (1994) The use of plant materials to control infestation of cured fish. *Tropical Science*, **34**, 401–408.

Webley, D.J. & Airey, W.E. (1982) A laboratory evaluation of the effectiveness of diflubenzuron against *Dermestes maculatus* DeGeer and other storage insect pests. *Pesticide Science*, **13**, 595–601.

Wood, C.D. (1983) Report on a visit to the Institut de Technologie Alimentaire (ITA) Senegal. Unpublished report **R1078(L)**. Tropical Products Institute, Slough, UK.

Wood, C.D., Evans, N.J. & Walker, D.J. (1987) The effect of salt on the susceptibility of dried whiting to attack by dermestid beetles. *Tropical Science*, **27**, 223–228.

Woodroffe, G.E. (1965) The sterol requirements of several species of *Dermestes*. In: *Proceedings of the 12th International Congress of Entomology,* 1964, London, UK.

Woodroffe, G.E. & Coombs, C.W. (1979) The development of several species of *Dermestes* (Coleoptera: Dermestidae) on various vegetable foodstuffs. *Journal of Stored Products Research*, **15**, 95–100.

Index